W9-CKT-494

Log Construction Manual

The Ultimate Guide to Building Handcrafted Log Homes

Robert W. Chambers

DEEP STREAM PRESS

Deep Stream Press
River Falls, Wisconsin

Copyright © 2006 by Robert W. Chambers

Cover and book design by Jason A. Wittwer.

Printed in the United States of America.
First printing: September, 2002
Second printing (revised): May, 2004
Third printing (revised, color section): February, 2006

About your safety: Home building is inherently dangerous. Builders risk serious injury, and even death, from accidents on the job. Throughout this book I try to promote safe working habits, but the author and the publisher accept no liability for your personal injury, property damage, or loss, from any actions that were inspired by this book. Recommendations in this book are not a substitute for the directives of professional contractors, builders, equipment or power tool manufacturers, or regulatory officials. *Please be careful.*

Library of Congress Cataloging-in-Publication Data
Chambers, Robert Wood, 1954-
  Log construction manual / Robert W. Chambers.
        p. cm.
Includes bibliographical references and index.
  ISBN 0-9715736-0-3 (pbk.)
1. Log cabins--Design and construction.   I. Title.
  TH4840 .C43 2002
  690'.837--dc21

                                        2002007696

Published by
Deep Stream Press
PO Box 283
River Falls, Wisconsin 54022
www.LogConstructionManual.com

Table of Contents

Preface

People have been building with logs for more than two thousand years, but in the last thirty years the craft of log building has changed dramatically. This is because there is a concentration of fine and innovative log builders in Canada and the United States, and because they belong to an Association that encourages all of us to share our ideas—the International Log Builders Association.

A short list of ideas that have changed handcrafted log building would include: the compression saddle notch, double-cut long-groove, underscribing, mitered joinery, log selection rules, sill-log layout, and accelerated log building—and all of these are less than 30 years old.

There are log builders who have not been quick to adopt the improvements—they are still building the same way that they first learned 30 years ago. And this is essentially the same way that log builders have been building for more than a thousand years. Except for the chainsaw and crane, you could bring a Viking log builder from 1100 AD into some modern log building yards and they'd fit into the crew with hardly a misstep.

For example, in many modern log building yards, only one person has authority to decide which log to use next. But when I teach hands-on workshops, students are selecting logs on their second day—it isn't rocket science. The skills and techniques of modern log construction can be learned, and now you are on your way to becoming a thoughtful and skilled log builder.

My work has taken me to the yards of many good log builders, and I have learned a great deal from them—*thanks guys*. Whenever I see another way of thinking about a problem, or making a cut, I try it out. Then, if it's better, I use it. I encourage you to learn the basics of log building, and then keep an open mind, too—you can help us continue to improve the ancient art of log construction.

I also want to thank my students for the influence they have had in my life and my craft. Their questions and enthusiasm have encouraged me to think about what we are doing, and then to find ways to understand it and to teach it. Like them, I am still a student of log construction.

Me in Boulder, Colorado, 2000. This is a full-scribe-fit *Accelerated* shell, not a *chinked* log home.

Me in 1959 (I'm on the left) and my sister Katy at a cabin in northern Minnesota, built by the Civilian Conservation Corps in the 1930's.

Building With Logs

Log homes have a bold and distinctive appearance. Because of this, some people are attracted to them, and some people aren't—log homes tend to polarize people, which is probably a natural reaction to something so bold. This makes my job easier, of course, because I don't have to convince you to build a log home—you are already interested.

Unlike brick, stone, timber frame, and conventional stick construction, handcrafted log homes use the only building material that is exactly the same on the inside as on the outside. (At least in climates that have a cold winter.) Logs are honest architecture. The whole structure—the skin, muscle, and bones—is exposed in a log home.

You can't see any straw in a straw bale house. The same goes for adobe, rammed earth, and cob—the real building blocks are always covered over. But not with logs.

I have found that many log home enthusiasts are detail-oriented people. They want all the information; and they feel perfectly able to judge, weigh, and balance for themselves. I promise to provide you with the details throughout this book —and also the reasons behind those details.

Two hundred years ago, there was probably a log builder nearby, or not far away, and learning was simple: you built the same way he did. Now, however, you have easy access to log building information from around the world. Keep in mind that while *opinions* are easily and freely shared; *knowledge* takes time and sometimes has a cost.

In the end, you'll seek help from many sources, and you'll have to try to reconcile the differences you find. Eventually your research must end—it will be time to pick up your tools—that's the only way to build your own log home. That time will come, perhaps soon. Right now, however, let's begin by asking: "why build a log home?"

CHAPTER TOPICS

Why build with logs
- Beauty & meaning
- Popular misconceptions
- R-value
- Air infiltration
- Thermal mass
- Embodied energy
- Log homes don't waste trees
- Renewable & sensible
- Life-cycle cost
- Natural materials

Types of log homes
- Handcrafted scribe-fit
- Handcrafted piece-en--piece
- Handcrafted chinked
- Manufactured log kits

Satisfying work

Why Build With Logs?

Figure 1: Log building the old way. Before the log can be hewn flat, chunks of wood called *juggles* are split off.

Figure 2: A log sauna in Latvia. Every wall is solid log, inside and out, through-and-through.

Photo courtesy: Karlis Apinis

Figure 3: Round log, round notch, Civilian Conservation Corps cabin built in the 1940's in Ely, Minnesota.

If you like log homes, then this question hardly needs to be asked. If you don't, then there may be no answer that can change your mind. But, either way, it is still a good question to ask. After all, log homes are expensive if you buy one, and require dedication and hard work if you build one for yourself *(Figure 1)*. Since log building can be expensive and difficult, it just makes sense to ask "why build with logs?"

Log homes are beautiful, durable, energy efficient, non-toxic, quiet, and strong. The organic shapes of natural logs have a rhythmic pattern that is attractive. Many people, when they talk about their dream log home, describe their *feeling*—a sense of calm, quiet, security, and serenity that they don't feel in other homes. The warm colors and smooth curves are soothing *(Figure 2)*. The words we use convey our emotional attachment to log homes: choosing logs as the building material is usually either an emotional decision, or an esthetic decision, or both.

I can spend hours looking at logs densely woven at corners, spanning distances for floor joists, admiring the geometry of trusses, and the roof with its layers of purlins and ridge. The interrelationship of all the parts, and the fact that they are naturally exposed and easy to see, makes log buildings endlessly fascinating. Frame walls built of 2x6's covered with drywall and paint, are boring in comparison.

Beauty and Meaning

Handcrafted homes use logs that still look a lot like the natural trees they came from. They have their original taper, knots, character, and personality, which is then tightly woven into the distinctive pattern of horizontal layers.

Log homes seem timeless, partly because so many cultures of the Northern Hemisphere have built with logs for thousands of years. Traditions, history, culture, and nature give log homes deep meanings that make them seem personally connected to us.

While you are on your log home quest, seek the technical information you are craving (starting with this book), but also spend time some soaking in the old traditions of log construction. Start with the log buildings of the Great Depression—there is a lot to admire in the park structures built in the 1930's. It was a time of smaller, cozy cabins with lots of interior log walls, short exterior walls, log gable ends, roofs of pole rafters, and low-slope ceilings *(Figure 3)*.

These structures have a scale, charm, and grace that is lacking in modern log homes that sometimes seem to have been designed mostly to be imposing: to impress us with their large size and great height *(Figure 4)*. It is difficult to make these log palaces with cavernous cathedral ceilings and jutting glass prows feel cozy or have a natural, human proportion.

Open your mind to log *cabins*. Learn from the Great Camps of the Adirondacks in New York—log structures from the turn of the century. Notice how a few pieces of bark and twigs add so much meaning when carefully selected and placed

(Figure 5). The complex and sheltering roof planes, dormers, gables, sheds, hips and valleys offer layers of interest from the outside, and plenty of niches and cubbyholes inside—they make you feel like a kid again *(Figure 6)*.

Finally, study folk houses built by cowboys and ranchers in America, and by farmers in Norway, Finland, and central Europe. Their scale is modest and human. Admire the rustic yet refined surface of old grayed logs, and the way each building intimately fits its site and has become part of it. Much of our attraction to log homes, and much of their beauty, comes directly from the long tradition of providing good shelter for families on the frontier.

Misconceptions

Log homes have been burdened with persistent misconceptions. Some of them might be in the back of your mind right now: "log homes are dark;" "settling is bad;" "log buildings won't last;" "log buildings use too many trees;" "log homes are difficult to heat;" "log homes are among the most energy efficient homes;" "log homes must be fire hazards—they're made of fuel;" "insects and decay are big problems;" "log homes are cheap and easy to build;" "log homes are inexpensive to buy." In this book I'll tackle these myths head-on. In fact, let's start now.

R-value

Wood is an excellent building material—one of the very best, in fact. It's resistance to transmitting heat or cold is good. Pine logs have an R-value of about 1.5 per inch (though some building codes require us to use R-1.2), which means wood is much better than brick (R-0.2), and half as good as fiberglass (R-3.1). Of course, the thicker the log wall, the better its R-value. A 6" thick milled log wall hovers around R-10, while a handcrafted home using 15" diameter wall logs may exceed R-22—maybe not superinsulated, but a very respectable number.

(For metric readers: a 25cm milled log wall has a metric R-value of 1.76 m^2KW^{-1}; and a 38cm handcrafted log wall would be valued at about 3.88 m^2KW^{-1}.)

In addition, wood is solid insulation, which means that water vapor has very little effect on its R-value. Space-filling insulation, like fiberglass and cellulose, is mostly air, and requires a plastic film barrier to keep water vapor out of it. When the humidity in fiberglass gets high, its R-value plummets down fast and far. Logs, because they are solid, not space-filling, do not need a vapor retarder.

Air Infiltration

Log homes must be tight to be energy efficient in cold climates. This is a big task when you consider that a modest log home can have several thousand feet of wood-to-wood joinery. To decrease air leakage and improve energy performance, use building gaskets inside the notches and long grooves. The best scribe-fitted joinery makes logs look like they grew together, but air, water vapor, and rain can penetrate joints that are visibly tight, and that's why gaskets are essential.

A strip of fiberglass batt inside the notches and grooves does not stop air, water vapor, frost, wind, insects, light, or rain. Unfortunately, some professional log home builders rely upon bare fiberglass. At the very least, all space-filling

Figure 4: A modern scribe-fit home at a ski resort town. The jutting gable roof line is sometimes called a butterfly. Certainly impressive and imposing, but is it cozy or inviting?

Figure 5: Adirondack-style footbridge made of scribe-fitted white cedar poles with the bark left on.

Figure 6: A classic with low pitch roofs, two eyebrow shed dormers; gable dormer, a tower with kicked hips; and a broken-back valley (on far left). The variety of roof planes and styles really makes this building interesting. It has charm and pleasing proportions—a human scale that is naturally welcoming, cozy, and appealing.

insulation must be protected from water vapor. If you use fiberglass or wool, then put it in polyethylene tubes or bags, or use gaskets along both edges of all the wood-to-wood joints.

Gaskets are effective at preventing air, vapor, and water leaks, and are easy to use. Foam tapes are manufactured especially for log homes *(see Resources Chapter) (Figures 7A &7B, and Color Figure C-26)*. There are also various backer-rod foam tubes and ropes that can be used. For ten years I used an EPDM rubber gasket that worked very well, though it is more expensive than foam tapes and rods. Now I have a 3-fin gasket manufactured for me. All gaskets are completely hidden from your view since they are installed along the interior and exterior edges on the *inside* of long grooves and corner notches.

Door and window openings require special attention. Before a window is installed, you'll see that the long grooves are open on their ends, even though sealed along their interior and exterior edges *(Figure 8)*. Fill the long groove copes with little pillows made from a handful of fiberglass in a plastic bag, which is then lightly sealed (so it won't pop when compressed). Fiberglass pillows can also help prevent air movement through corner notches and the settling spaces above doors and windows.

In four years, after settling has stopped, use a high-quality caulk to seal door and window trim around the log openings. Maintain these seals over time as needed.

Two more areas must be kept tightly sealed: the sill logs (between the floor or the foundation and sill logs) and the plate logs (the ceiling), and these areas must be handled with care. Homes with cathedral (vaulted) ceilings should caulk the ceiling vapor retarder (a polyethylene film) to the plate logs. Use a material called *acoustical sealant*—simply stapling the plastic sheeting to the plate logs is not good enough. Where vapor barrier sheets are joined in ceilings, the seams must be taped with the proper tape, not duct tape. Sill logs can be sealed to the floor decking with foam tape, rubber building gaskets, or with ripple-foam sill seal, which is commonly used in stick-frame homes.

Government-sponsored scientific studies of log homes in Minnesota, Maryland, Alberta, Idaho, and Alaska have found that with the details described here, log homes can be quite tight and energy efficient.

Thermal Mass

It is a scientific fact that some materials can store heat energy and radiate it later—stone and concrete are thermal heavyweights, while foam and fiberglass are featherweights. Logs fall in between these two extremes. I have studied the research on thermal mass effects of log homes and there is not much to benefit log home owners. There are definite, but very small, thermal-mass benefits to having interior log walls, but you really can't count on the thermal mass of exterior log walls to help keep a house warm. After all, a wall that is good at holding heat is equally good at holding cold!

Thermal-mass benefits are greatest in locations that have large differences between daytime and nighttime temperatures. When an interior log wall is warmed during the day, it can radiate the stored heat into the house at night,

Figure 7A: A rubber gasket being installed on the inside of the long groove to prevent air and water leakage.

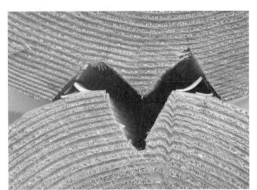

Figure 7B: Close-up of gasket in double-cut long groove. The rubber fins are an effective barrier to leakage. Also see Color Figure C-26.

when outside temperatures have dropped. But in climates that have cold weather that can last for days, and the temperature doesn't warm up much during the day, then the thermal mass of exterior log walls will be storing *cold* and radiating it into the house.

In homes where heat is delivered in batches (like wood stoves and passive solar) there are benefits to having thermally massive interior walls. But, homes that are heated with a continuous trickle of heat (like forced-air furnaces, baseboards, and heated slabs) do not benefit as much from the thermal mass of walls.

Despite the fact that thermal mass is not the holy grail that some log enthusiasts had hoped for, there's no reason to hang our heads. It is easy to build a log home that is warm in the winter, cool in the summer, does not have humidity or interior air-quality problems, and has good energy performance. Here's how: build the walls with large diameter logs, use high-quality gaskets to prevent air leaks, take care around door and window openings, and completely seal the ceiling vapor retarder to the plate logs, and the sill log to the floor.

To improve a log home's inherently good natural energy characteristics, super-insulate the roof or ceiling, buy high-performance doors and windows, and use efficient systems to heat and cool your home. Following these guidelines will help you build a log house that is energy efficient and comfortable.

Embodied Energy

Embodied energy is the amount of energy used to manufacture and transport building materials and then construct a house. Recycled timbers, adobe, and straw bales have low embodied energy, especially if they are gathered locally. Construction materials with large amounts of embodied energy include expanded polystyrene foam, steel, concrete, and aluminum (*Figure 9*).

Aluminum is the classic example of a material with a huge energy cost: mining heavy bauxite ore (in foreign countries), hauling it to a foundry blast furnace, and refining it into ingots, transporting it thousands of miles, only to be melted again to make into siding or windows. Luckily, not much aluminum is used in a home.

On the embodied energy spectrum, logs are much closer to the straw-bale end than they are to aluminum, but there has been no research on the embodied energy of log buildings. The embodied energy in homes built of sawn lumber is well known, and handcrafted log homes will come in far below that amount because they do not use kiln-dried wood, which is a big use of energy in sawn lumber. The embodied energy in rough-sawn lumber is about 850 Mj per cubic meter; but once it is kiln-dried and surfaced it climbs to 4700 Mj. Plywood, since it is processed even more, has 9400 Mj per cubic meter.

Further, many log homes use local trees, which reduces the energy used for transporting them. Finally, trees in North America are generally grown in the wild and without fertilizers, herbicides, or pesticides, all which are petrochemicals with lots of embodied energy.

Fuel is used to fell trees, and to transport them to your foundation. Logs are moved around your building site with a tractor or crane—which uses some fuel.

Figure 8: Fiberglass pillows can be stuffed gently into the long-groove coves before the buck and key (seen leaning on the wall to the right) is installed.

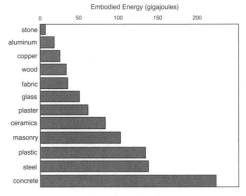

Figure 9: Table showing both the energy embodied in the various building materials and the quantity of each material used in a typical home. Aluminum has very high embodied energy, but not much is used; concrete has low embodied energy, but a large mass is used. Wood is used in large amounts, but still totals very little embodied energy. (1 gigajoule is approximately 1 million btu's.)

Of course, when you remove bark with a drawknife you'll feel like you have *personally* added an awful lot of embodied energy to the logs!

Not much oil and gas is used when building a 2000-square-foot log house, built with about 70 logs, 40-feet long and averaging 15" in diameter— about 8 gallons of gasoline to run chainsaws, and perhaps another 30 or 40 gallons for on-site equipment (a total of about 200 liters of fuels). Though the numbers are not precise, you can see that the energy embodied in many handcrafted logs home is extremely low.

Log Homes Don't Waste Trees

One of the most widespread and damaging myths is that log homes use extravagant amounts of wood. It does *appear* that "you could build a couple homes out of the logs that go into one log house," as I've heard people say. But, an average log home uses about the same volume of trees as a conventional, stick-framed house of the same size. As an example, let's compare a home built conventionally with framing lumber, to a house built of handcrafted logs.

An 1825-square-foot home, if built conventionally with 2x's, uses 13 thousand board feet (MBF) of sawn lumber. About 2,000 cubic feet of logs is needed to produce 13MBF of lumber (using the average softwood lumber recovery volume of 6.5 board feet per cubic foot of log). Assuming the logs are 40-foot long, tapered, and have a 15" midspan diameter, this amounts to 57 logs. (Please refer to the U.S. Forest Service *General Technical Report RM-199, An Analysis of the Timber Situation in the United States*, 1989-2040, by Richard W. Haynes).

The same home, if built of handcrafted logs, requires about 2200 lineal feet of logs—including a loft with log joists, logs in the roof system, and waste and trim for windows and doors. (An 1825-square-foot log home can be built with as little 1400 lineal feet of logs, and I have allowed an extra 800 lineal feet for an average log home built by an average builder.) Using the same log diameter and 40-foot lengths, this amounts to only 55 logs.

In summary, then, building an 1825-square-foot house would use 57 trees if it were conventionally framed versus 55 trees, or less, if it were a log house. The bottom line is that log homes do not waste trees, they are clearly average.

(For metric users: we are comparing two houses each with 170 square meters of living space. If built conventionally, this house would require 57 cubic meters of logs to produce the framing lumber; but, if built with logs, it would require only 55 cubic meters of logs.)

One cause of this long-standing myth is that the huge volume of waste produced by sawn and surfaced framing lumber is hidden from our view—it was left at the sawmill. Even the most modern sawmills are inefficient at converting trees into framing lumber: about 2.76 cubic feet of log are used to produce 1 cubic foot of framing lumber—which means that the other 1.76 cubic feet ends up as sawdust and shavings. Sawmills make more volume of *waste* than volume of *lumber*! Luckily, some of the waste produced at sawmills is put to good use making pulp and paper.

Renewable and Sensible

A second concern is how trees are grown and how they are harvested. Trees are a renewable resource, which puts them at the top of the list of sensible building materials. Trees can be planted, grown, harvested, and re-planted *(Figure 10)*. I have used trees as young as 35 years old, and they were excellent for building. Granted, that is still 35 times longer than is needed to grow straw for a straw-bale house, but considering that log homes have proven they can last hundreds of years, logs are a construction material that easily justifies a reasonable investment of both energy and resources.

One overlooked fact is that many log homes in North America are built using standing dead trees—that is, trees that died naturally from fire, bugs, or disease. It is certainly true that if dead trees were allowed to fall and rot, then their compost would improve the soil, but couldn't the same argument be made for leaving straw (or even the oats themselves!) in the farm field? But that's going overboard, of course. Living things can be grown and harvested by us to serve good purposes: food, clothing, and shelter. Our goal should be to use renewable resources sensibly, not to *never* use them at all.

Log home builders will need to be more careful about both the practical and the ecological implications of what we do. My own log home company now uses trees that come only from managed plantations, not wild forests. And it is only in the past two years that I have even been able to buy trees that are sustainably grown and harvested as certified by the independent Forest Stewardship Council. Sensitivity and sensible use must be the future of log home construction.

Life-Cycle Cost

Finally, there is longevity, or life-cycle cost. Log homes, when properly built and maintained, last for centuries, and can, if needed, be recycled into new homes. Stone may be the only building material that can surpass log and heavy timber in this regard. Log homes have proven themselves to be very difficult to knock over, burn down, blow away, wear out, or even become unfashionable *(Figure 11)*. Log homes last.

Natural Materials

It is good to use natural materials to build our homes. When it comes to clothing, we feel more comfortable in cotton than in rayon; and for a place to live, natural wood is healthier than the formaldehyde resins that off-gas from oriented strand board (OSB) and plywood.

And, in the end, all natural materials must come from nature. But that does not mean we shouldn't use them. The important thing is that we conscientiously do our best, and we are still learning just what that means, and how to do it.

Figure 10: Logs in a sorting yard with a clearcut hillside beyond. This is commercial tree farming on privately owned property—a managed plantation cut on a 38-year rotation. There is no erosion, and waste is minimal.

Hint   Forest Stewardship Council
www.fscoax.org
www.fscus.org

Figure 11: A 400-year-old log-cabin church on Lake Onega near Kizhi, Russia.

Types of Log Home

Figure 12: A modern handcrafted, full-scribe-fit home with full-compound dovetails. The hewn surface of the logs shows that a broadaxe was used.

Handcrafted Scribe-Fit

There are two types of log homes: *handcrafted* and *manufactured*, and each of these has a large number of varieties. Handcrafted homes are built by hand—a person fits each and every naturally-shaped log together.

Manufactured log homes use logs that are shaped by machines, and their fits come from milling every log to a uniform size and shape. There are various other products related to log homes, like log siding, but I think that a wall must be *solid* wood in order to be "log."

This book is about *scribe-fit* log construction, in which each wall log has a full-length long groove and corner notch, and fits to the naturally-shaped logs above and below without gaps. Also known as Scandinavian scribe, coped, chinkless, and full-scribe, this method is the highest form of the log builder's art.

There are a number of corner notch styles and long-groove shapes that are used, and the logs can be either naturally round or *hewn* (which means flattened on their vertical sides, *Figure 12*), but all handcrafted, full-scribe homes use tapered trees that substantially retain their natural shapes.

In handcrafted homes, logs never have visible end-to-end splices, or *butt joints*. And this style uses large logs—almost always more than 10" (25cm) in diameter, and frequently 2-feet (70cm), or more. Handcrafted home companies often build the log shell at their own yard, disassemble it, and transport it to the owner's foundation where they also reassemble the kit for the customer. Unlike some manufactured log homes, a handcrafted home is almost never assembled by the homeowners if it was crafted by a professional log building company.

Figure 13: Piece-en-piece buildings have no corner notches. They could be considered post and beam homes with in-fill panels made of scribe-fit logs. The panel on the left is being set by a crane, and will be tipped up until it is plumb.

Handcrafted Piece-en-Piece

Piece-en-piece handcrafted log homes have no corner notches—they are a post and beam structure with scribe-fit log panels that fill in between vertical log or timber corner posts *(Figure 13)*. The style is derived from historic French Canadian buildings, but its exact origin is unknown.

Handcrafted Chinked

A popular variety of handcrafted log homes has gaps between the lengths of the logs (where the long grooves would be in a full-scribe-fit home). The gaps are called *chinks*, which is from an Old English word (having a "chink in your armor" was not a good thing for a medieval knight). Groove areas are always chinked, and sometimes the corner notches are chinked, too.

The material used to fill the gaps is called *chinking*, and through the centuries various materials have been used: saplings cut into quarters and nailed into the gaps; horsehair, dry sphagnum moss, or oakum (made from hemp rope); clay; and cement. Most recently, synthetic materials that look like caulk mixed with sand have dominated the chinking market—and these are a big improvement over all the materials used earlier *(Figure 14)*.

Chinked handcrafted homes are not necessarily less expensive than full-scribe-fit log homes. There is less labor during construction because long grooves for chinked homes are not critical, but they have extra labor and material costs after the shell is erected. Because both the inside and outside of each log is chinked, a home with 3000 lineal feet of logs will have 6000 feet of chinking, and that could cost $20,000, if installed by a professional crew.

Manufactured Log Kits

There is a great variety of styles and types of manufactured log home kits. Some logs are small and squared, while others are larger and round. Ten-inch diameter is large for a manufactured log home, and many kits come with 8" (20cm), or even 6" (16cm) logs. No manufactured home uses logs that are as large as the logs found in most handcrafted homes.

Nearly every manufacturer uses its own log shape, or *profile*—double tongue and groove *(Figure 15)*, locking-V notch, D-log, hewn, and so on. In some cases, the profile is milled from a single tree, while in others smaller pieces of sawn lumber are glued together to form "logs." Very few manufacturers use tree-length logs—it is common for the logs to be shorter than 14-feet (4.3m), and as a result, the walls have visible butt joints where log pieces are joined end-to-end.

Manufactured log homes are often sold as do-it-yourself kits. That is, the logs have been pre-milled to their profile, and bundled together before being shipped to the customer. With some manufactured kits, every log is pre-cut to the exact length needed, and each piece is labeled for its order of assembly. In other kits, the homeowner or builder must cut, notch, and trim every log piece.

Milled kits have a wide range of costs. Some kits are quite inexpensive, while others actually cost more than some handcrafted, full-scribe shells. It is important to compare what is included, and what has been left out, of the kit—make sure you compare apples to apples.

Figure 14: A handcrafted building with both the notches and the grooves chinked. The groove gaps here are very small—some chinked buildings have 3" (75mm) of chinking between layers, or even more.

Figure 15: This manufactured log home kit has a double tongue-and-groove (T&G) between the logs. The logs came in random lengths, each with the T&G pre-cut, but the homeowner had to cut all the corner notches himself. These logs are solid (not glued up)—you can see that the heart of the tree is in the center of each log piece.

Satisfying Work

The focus of this book is building your own handcrafted, full-scribe-fit home, not buying a kit. Buy many of the methods described here can be applied to handcrafted chinked buildings, and if you decide to hire a handcrafted company to supply and erect your log home shell, you will be well prepared, as well.

Building your own log home is satisfying and challenging work. The basics skills of scribing and cutting can be learned in short order—perhaps just a few days. But handcrafted log construction is not easy—it requires practice, time, effort, and patience. Handcrafted construction is satisfying because the results of your work are there to see and enjoy *(Figure 16)*. And, at the end of the day, there are many logs that you will meet that you will never forget. No carpenter says that about 2x4's.

But beyond this, log building is a challenging craft and one that cannot be learned completely in a few years. After nearly twenty years spent working with (and thinking about) logs, I am still learning. There are deep rhythms and patterns behind weaving logs that we have only just begun to recognize and understand.

Figure 16: The joy and pride of building your own home.

Your Log Home Project

You want to live in a log home. You love the way they look, or the way you feel when you are in a log home. But where do you start? The budget? The design? Do you find a builder first? This chapter will help explain how to organize a log home project, step by step. You have to ask yourself some tough questions: How much can you afford? How many children will you have? Where will you live? Do you have what it takes to build it yourself? But if you ignore the tough questions, then your project will be very difficult for you. But if you answer the tough questions early, and then stay flexible throughout the project—adapting to conditions as they change, then you have a good chance to succeed.

CHAPTER TOPICS
Step by step to your log home

- Know your budget
- What things are attractive to you in a home?
- Know your land
- What do you *need* in a home? And what do you *want*?
- Check with local officials, and a lawyer
- Designing your home
- What do *you* want to do?
- So, how much will your home cost?
- Money to build your new home
- Before you start building
- A schedule for building your home

Step by Step to Your Log Home

Know Your Budget

How much can you afford to spend? This is where you must start. If you will have a mortgage on your new home, then a bank can provide you with a table that shows monthly payments required for mortgages of different sizes—this is helpful.

Do not start a building project unless you have a definite, practical plan to pay for all of it. Do not start a project when you can afford only to buy the logs. This seems simple, but many people get excited about buying trees and notching them into walls, and they forget that a well might cost $5,000, and a septic system $9,000. Trees are about the cheapest major part of a log home, so do not buy logs until your design, permits, and financing are completely arranged.

In general, and very rough, numbers a new 2x6 house built to your plans on land you already own might cost you $100 per square foot of living space. (That means a 2,000 square foot house would cost $200,000.) A handcrafted log home might be $140 per square foot, or more—in other words, 40% more than a custom frame house.

Yes, there are ways you can cut some costs, and I'll do my best to show you how to save money, but building your own log home will still cost you quite a bit more than building your own frame home.

What Things are Attractive to You in a Log Home?

Keep a scrapbook of photos clipped out of magazines or copied from books. Maybe it's a staircase you like, or a bay window, gable lighting at night, or the way a bed fits into the room. Instead of keeping 47 whole magazines each with two photos you like, cut the photos out and paste them into a scrapbook.

Figure 1: Make a scrap book.

I have met with families who bring out stacks and stacks of loved and ragged magazines—and they can never find the *one* photo they are looking for. I sit quietly as they curse the sliding pile. Cut out the stuff you like and collect it in one place. Add to it when you find more. Keep those magazines for your own use, just don't make the builder or architect use them.

Keeping a record of what you like will help you actually get some of these things in your home—whether you design your home or someone else does. If you hire a designer, your scrapbook will be their way of getting to know you, and your way of ensuring the house you get is the house you want.

Don't be concerned with organizing your scrapbook too much—go ahead and let the kitchens and baths mix. The important thing is to make a visual record of the things you like, and have it all in one place.

Know Your Land

Before you start designing a house, know your land. Do not just pick a house from a magazine or book and assume it will fit your property, or that it will fill your needs.

Is your land steep or is it flat? Are the best views to the south or to the west? Are neighbors really close on one side of your property? Where will your septic system be located? Does the building spot that has the best view never get sunlight until noon? How long will your driveway be if you build there? Does snow drift across that area? Does your electric company charge $10 per foot for trenching, and your house site is 1200 feet from the nearest power pole? Know your land in detail, and in all seasons, if possible.

In general, the rule is: do not build on your "best spot." After all, if you dig a huge 9-foot-deep-hole for your foundation at the best spot, then the best spot on your property has been dug up and has been hauled off in dump trucks—it is gone forever. Many experts recommend that you keep that best spot in your yard, not lose it in your basement.

Think about access—the driveway. Where will the well be? Where is the best morning and evening sunlight? What are your best views? Which area will give you privacy? Is there an rock cliff that would be expensive to stabilize? Are there wetlands that would give you a damp basement? Are there trees you would hate to see cut down?

Design the house to fit your land as well as your life-style and your budget. A "perfect" log house, cut from the magazine, that looks great in a mountain canyon will be completely wrong for the mountain top, or for the woodlot on the edge of a hay field.

What Do you *Need* in a Home? And What Do You *Want*?

Next, make a list of those things you absolutely need (as opposed to those things you want). Everyone's list will have: kitchen and bathroom. Now, how many bedrooms do you need (that is, how many people are always sleeping in your home, not counting visitors)? Do you need a formal, separate, dining room (is it required for entertaining? or do you just want one?) Consider every room: den, deck, storage, living room, utility room, laundry, loft. "Do we need that, or just really want it?"

Also consider those things you must not have in your home. For example, a two-story home could be a bad choice if you live with someone who needs a wheelchair.

Now, make a list of those things you really want, but do not need: stone fireplace, hot-tub, library, workshop, sewing room, extra bedrooms, wine cellar?

You get the idea: understand the differences between your needs and your wants. For example, if just two people are to live in a retirement log home, and adult children might visit once a year, then extra bedrooms may be wanted, but not really needed. Yet I have built 5-bedroom homes for retired couples and the spare rooms are used, maybe, two nights a year. It is difficult to look at your life objectively and with an eye to the future instead of the past, but you must try.

If you confuse needs and wants, or just refuse to consider the differences, then your house will definitely be expensive. Why? Because you will build more than you need. A new house often comes at a time in your life when things are changing for you personally: kids on the way; or grown kids leaving home; new job; in a new town; just retired (or planning for it). You need to consider what will be new with your new life, and then design and build a home that suits the changes that are coming.

Check with Local Officials, and a Lawyer

Get to know the zoning regulations controlling septic, wells, setbacks from property lines, fire truck access, heights of buildings, utilities, slope restrictions, etc. Either the town or the county will be able to tell you what restrictions apply on your land (*Figure 3*). Find this office and visit it now, don't wait until your floorplans are done and you want a building permit—that's too late to find out your house is too tall, or too small, or on too steep a slope, or too close to the alternate septic area, or the fire trucks cannot get up your driveway and you can't buy insurance, which means you can't get a mortgage. Ouch.

Figure 2: You probably don't *need* granite counter tops

While you are in these government offices, make a list of the permits you will need, how much they will cost, and what order you need to get them in. This will vary from place to place, but typical permits include: zoning ("land use"), septic (plumbing), and building. You may also need an electrical permit, road permit, well permit, energy audit, and structural engineer's stamp. Check with both the town and the county—sometimes they require separate permits.

Next, are there restrictions on the deed (the *title*) to your property? The zoning officials will not know about these—so you need to see the deed to your land and perhaps visit an attorney. There can be covenants and easements that affect your building project that no one knows about (*Figure 4*)! Developers sometimes restrict the height, appearance, and size of buildings, or the location on the lot. Electric power and gas companies have easements—the right to build—on millions of properties. Are they on your deed? Find out.

Designing Your Home

A professional can be a great help when designing your home—try and hire someone who is good at design, someone you can get along with well (someone who listens to you), and someone who has a successful track record with log homes. You can hire a draftsman, a designer, or an architect. An architect is a designer who has gone to architectural school and then worked for an architect for a certain amount of time.

Avoid a designer who has never worked on a log home before, or someone who really wants to design their first log home. Your project won't be helped if both you and the designer are beginners.

Please, don't hire a friend who is a civil engineer or a daughter who is an interior designer to design your home—they do not know about building homes, much less log homes—even if they are experts at roads or bridges, or office space. Stay friends by not hiring them, not by hiring them!

Design starts with a site plan (it's just a map) of your land, the list of items that you need in your home, and your scrapbook. Is your life-style formal or relaxed? How much privacy do you need inside your home? What areas will you use most often? Which rooms will be public and which are for family members only? If you cannot afford to have every wall be a log wall, then where do you want to put the log walls—in the private areas or public areas of the home?

One of the first questions the designer or architect should ask you is: what is your budget? Tell them honestly, because if you fib and tell them more than you can afford, then you are guaranteed to get a home that you cannot afford. If the designer does not ask, then you must tell them what your budget is—and hold them to it!

It is nearly certain that the design will cost more than you expected. In 20 years of building, I have not once seen a design come in that cost less than everyone expected. They always cost more. It is, first and foremost, the designer's or architect's responsibility to keep costs within your budget range. Do not accept an out-of-budget design and then assume you can find a builder to cut the costs. Cutting costs must *start* with the design.

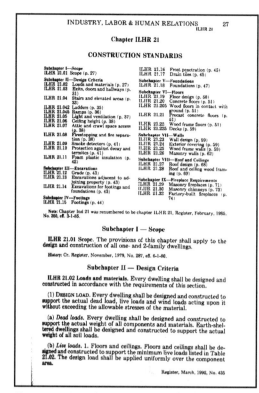

Figure 3: Page 1 of a Residential Building Code

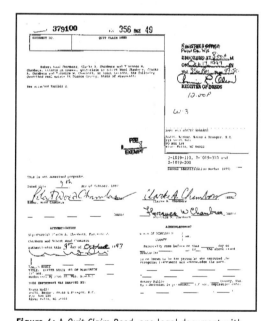

Figure 4: A *Quit Claim Deed*, one legal document with important information about your land. You need your property's Abstract for this kind of information.

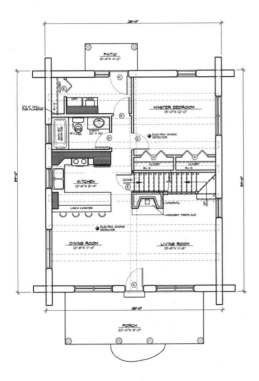

Figure 5: Main floor plan–one page of a set of blueprints. This plan would need a door and window schedule and their dimensions and locations, to be complete.

Whistling Crow home, courtesy: www.beavercreekloghome.com

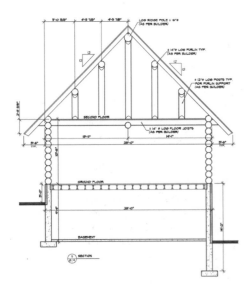

Figure 6: A section drawing, one of the pages in the set of blueprints.

Whistling Crow Home, courtesy: www.beavercreekloghome.com

Use the *Log Building Standards* (the "building code for handcrafted log homes," included in this book) to help you and your professional design your home—there is plenty of good information in there. Give your designer a copy. It is a very good idea to have your contractor and your log builder attend some of your design meetings—they will help you avoid problems and reduce costs, if you ask them for their advice.

If you can not or do not want to get professional design help, then look for a stock plan that closely matches your list of needs and will fit your site. I recommend the full-plan-sets produced by Robbin Obomsawin at Beaver Creek (www.beavercreekloghome.com), which can also be found in her numerous books. Consult with a log home builder about the plan you have chosen—this is especially important if you are using a manufactured log home plan to build a handcrafted home.

Many log home companies have in-house custom design services, or stock plans, or both—and you should expect to pay for them. But if you pay full-price for a construction set of log home plans, then it is reasonable that you can take them to a number of builders for price quotes.

It is possible to design a house and then not be able to buy the logs it needs—logs longer than 50-feet (15 meters) are difficult to buy and to ship, and in some places a 36-feet (11m) log is as long as you'll find. Design the walls to match the logs you have easily available locally, or the length your log builder can get.

Prepare blueprints that include at least the following information:

- *site plan that shows your lot lines, the driveway, location of well, septic, easements, and house*
- *foundation plan shows footings and foundation walls*
- *first floor plan that shows all the rooms and dimensions*
- *second floor (or loft) plan*
- *electrical/HVAC/plumbing plan or plans*
- *cross-section of typical wall (from footing to roof)*
- *construction details, as needed*
- *roof plan that shows the shape and pitch of roof, overhangs, valleys and hips, and any special construction details*
- *elevation of each outside wall (drawings of what the house looks like)*

You will need about 4 complete sets of the blueprints, plus one set for each company bidding on the project.

Design Fees

Draftsmen offer blueprint/design services for about $1.00 per square foot—so a full set of plans for a 2,000 square-foot (190 square-meter) house may cost about $2,000. (A list of architects, designers, draftsmen, and engineers is located in the Resources Chapter of this book.)

Architects sometimes negotiate their fees based upon a percentage of the total finished cost of the house, or by the hour, or by the square foot—it is negotiable, but almost always more expensive than a designer or draftsman. After all, they are professionals. And if they are good they are worth it.

Some, perhaps most, log homes also need professional engineering—usually for structural log beams like floor joists, roof purlins and ridge, and log or timber trusses. In earthquake country you will need engineering done on the log walls and foundation, too. Some larger architectural firms have their own engineers, or you can hire from the list I have provided.

You can also find your own engineer—make sure they have lots of experience with "wood as structural members." Use those words when hiring an engineer—many know concrete and steel, but only a few know wood. You want an engineer who knows and trusts wood or you will end up with exposed steel plates and bolts everywhere! And that's expensive and ugly, and I have seen it happen.

If the engineer tells you that you cannot span 18-feet (5.5m) with log floor joists, or that each log truss needs $1000 of custom steel, or your log builder refuses to cover a steel I-beam with log slabs, then it is time to find another engineer.

What Do *You* Want to Do?

What do you have the time to do? What skills do you have? What skills can you learn? What do you want to do? How much time and money can you afford to spend?

Considering all these questions will help you decide what your role in the project will be.

In general, there are three roles you could have:

   1) you build your own home (run a chainsaw, and pound nails), or

   2) you act as general contractor (you hire each subcontractor to do the work for you), or…

   3) you hire a general contractor to do it all.

There are combinations possible, too: for example, you could be the log builder and hire a general contractor.

There can be a cost savings if you do some of the work, but the savings are not as great as you might have heard. I have read that you can save 10%-20%, but in my experience this may not be true. You are not yet trained or experienced at log home building, or running a construction company, and that lack of experience will eat into some of the potential savings. A great contractor with 20 years of building experience works very hard to make a 10% margin—how do your skills compare?

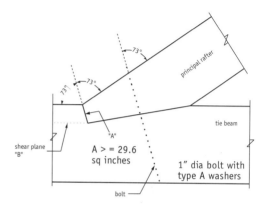

Figure 7: Rafter tail to tie beam joint in a log roof truss. An engineer that knows wood is needed for this kind of work.

Hiring a general contractor

Interview log home builders to find someone you like and can work with. See at least two of their homes that are 4 or more years old. Don't just ask for references, actually call those references.

Even if you hire a general contractor to do all the organizational work for you, you might still be the one to research and hire the log building company—this is usually (not always) a separate contract from your contract with the *general*. The general is usually a local who builds conventional (not log) homes, and the log builder often is not local—their business might be based 60 to 600 miles away, or even in another country.

Many log home builders will not act as general contractors—instead they supply just the *log shell*. A log shell means that all the logwork is done and assembled on your foundation and floor deck, but there are no doors and windows, and usually no roof.

A fully prepared log shell means that carpenters or plumbers will never need a chainsaw—the log builders have finished all that kind of work before they left. The door and window bucks are installed, the electrical holes and mortises for outlet and switch boxes are complete, the groove for second floor decking is cut into the perimeter log walls, purlins and ridge are cut flat on top and to final length, trim cuts in log walls for interior (frame) partition walls are complete, and the log ends of the notched corners are trimmed and sanded.

When interviewing log building companies be certain to ask what is included in the log shell package, because this varies widely from builder to builder. To accurately compare costs you need to know if the window and door splines and bucks will be installed, or not. Is there a kerf cut for foundation flashing? Do the purlins and ridge have grooves cut for sheathing to be set slightly into the log? Is trucking included?

The log shell is almost always built not on your foundation but in the log builder's construction yard. Once finished there, the shell is taken apart and the logs numbered and placed on semitrailers to be trucked to your land, where it is reassembled by the log builder's crew.

The International Log Builders Association (the new name for the CLBA) can give you a list of log home companies in your area, a good place to start, but not all the good companies are on the list, so you can also ask for the membership list (there might be a small charge), but the list is sorted by location to help you.

Being your own general contractor

Being a general contractor is hard work: you will spend many hours getting bids, checking references, answering questions, making early-morning and late-night phone calls, organizing material deliveries, negotiating prices, dealing with subcontractors who did not come when expected, or who did work wrong that must be corrected.

If you have a full-time job already, then it will be difficult for you to take on the additional job of being your own general contractor. Expect to work up to 20

Hint   Ask for references from log home builders—and then *call* those homeowners!

Hint   A fully prepared log shell means that carpenters & plumbers will never need a chainsaw. The log builders have finished all that kind of work.

Figure 8: This man learned how to scribe and cut, and he helped build his own home. He also acted as his own general contractor, which is a time-consuming and difficult, yet rewarding, job. He would say it was an experience of a lifetime.

hours per week for 8-10 months on your home. Then do the math: 15 hours per week for 10 months equals 600 hours. If you earn $20 per hour at your regular job, then you would have made $12,000 in this 600 hours. If you figure you are "saving" $5,000 by acting as your own general contractor, then you are actually losing $7,000 by acting as your own general contractor.

Check with your state and county about the laws concerning acting as a general contractor. You will be liable for paying Workers Compensation insurance on subcontractors who are not insured, and you may need a contracting license. You will definitely need business liability and property insurances. These are very different than homeowner-type policies.

Building your own log home shell

This is a great way to help build your own home. Could you read a book on frame carpentry and then build your own 2x4 house? Probably not. And log home building is much trickier. Attend a good hands-on workshop to learn and practice the skills you need to be your own builder.

Get training. See if you actually like running a chainsaw, and are any good at it. The International Log Builders Association keeps lists of workshops (250-547-8776), and see the list in this book.

If you decide to build you own log shell, then remember that you may not have income during the time you are building. If you have to take a leave of absence from your regular job in order to scribe and cut logs for 40 hours per week, then you won't be picking up paychecks during this time. Plan your budget accordingly.

Since you have little to no experience in building or working with subcontractors, you should expect to not build as efficiently as a professional contractor—this means it will take several extra months for you to finish your home. (And I know people who have been working for six years on their home and have not yet moved in.) Owner-built homes always take much longer than expected. And whenever a project takes more time than you expected then it will also cost more money than expected.

Figure 9: Hands-on workshops are very useful, and fun, too—just ask any of my students. Some have said it is one of the best things they have ever done. Most of my graduates build their own homes, and many have become professional log builders.

If you want to scribe and cut logs for your project, there is no substitute for a hands-on workshop—learn to work like a professional, not an amateur.

When choosing a workshop to attend, be certain to get and *follow up* on personal references—not every workshop teaches up-to-date methods, or enforces safety. A large Internet presence is no replacement for talking with people who have attended the school you're considering. Some workshops could be a waste of your time and money.

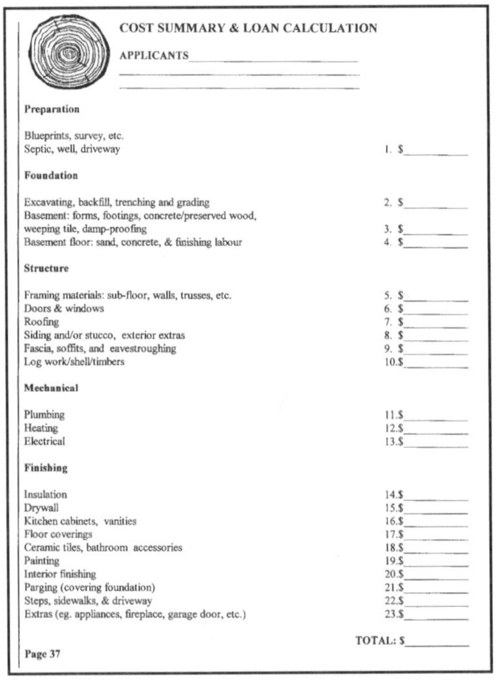

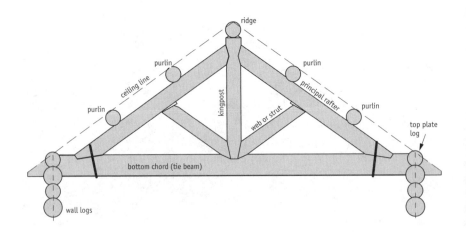

COST SUMMARY & LOAN CALCULATION

APPLICANTS _____

_____

Preparation

Blueprints, survey, etc.
Septic, well, driveway .... 1. $_____

Foundation

Excavating, backfill, trenching and grading .... 2. $_____
Basement: forms, footings, concrete/preserved wood,
weeping tile, damp-proofing .... 3. $_____
Basement floor: sand, concrete, & finishing labour .... 4. $_____

Structure

Framing materials: sub-floor, walls, trusses, etc. .... 5. $_____
Doors & windows .... 6. $_____
Roofing .... 7. $_____
Siding and/or stucco, exterior extras .... 8. $_____
Fascia, soffits, and eavestroughing .... 9. $_____
Log work/shell/timbers .... 10.$_____

Mechanical

Plumbing .... 11.$_____
Heating .... 12.$_____
Electrical .... 13.$_____

Finishing

Insulation .... 14.$_____
Drywall .... 15.$_____
Kitchen cabinets, vanities .... 16.$_____
Floor coverings .... 17.$_____
Ceramic tiles, bathroom accessories .... 18.$_____
Painting .... 19.$_____
Interior finishing .... 20.$_____
Parging (covering foundation) .... 21.$_____
Steps, sidewalks, & driveway .... 22.$_____
Extras (eg. appliances, fireplace, garage door, etc.) .... 23.$_____

TOTAL: $_____

Page 37

Figure 10: This is from the booklet *From Land to Lockup*, for sale only from the International Log Builders Association (250-547-8776).

Figure 11: Log trusses look great, but they cost much more than supporting the same roof with posts and log purlins.

How Much Will Your Home Cost?

Once you have a basic floor plan and you have decided what your role will be, then you can get rough estimates of costs. These will likely not be a firm *bid* yet— that is, not a price that the builders will guarantee. Some builders will figure costs on a per-square-foot basis, while most will want more information—a sketch of the floor plan, or more. For a useful spreadsheet to use for tracking quoted prices, get a copy of the booklet *Log Homes—From Land to Lockup, a Guide to Handcrafted Log Home Construction*, published by the International Log Builders Association (*Figure 10*).

In my own experience, architects have not been an accurate source of construction costs—9 times out of 10 their estimate is too low. You need to talk with builders and contractors to get a better idea of costs.

You will not pay for these rough estimates, but just because they are free, you should not take advantage of builders, either. Preparing a bid takes time, and a good builder, a quality builder, will not slap together a bid. If you get bids from 10 builders it is clear you are shopping price only, and don't care much about the quality of construction and reputation of the log builder—you'll hire whoever is cheapest.

You do not need to send your floor plan to more than 2 builders at this time. It takes 20 hours, or more, to prepare a price quote—a smart builder won't invest this time unless they have a chance of actually getting the job. You might want 3 firm bids before you hire a builder.

If cost is too high

If these early estimates are within your budget, then you are doing okay. If your total budget is $100,000, try and keep log shell costs less than $30,000—about 25% to 30% of the total you intend to spend.

If the cost estimates are high, then you will need to go back to the drawing board. Here's what to look for. Check your lists of "needs" and "wants" again. Is a 30 x 40 workshop in your "needs" list, but it is for your jewelry-making hobby, not your job? Will two people live in the house, but you have 5 bedrooms listed as "needed" (because all your adult children might visit one Christmas)? Do you eat in the kitchen area every night?—Then don't build that formal dining room. Unless you are wealthy, do not build rooms that you will use just once or twice a year.

Inexperienced builders and designers will probably suggest that you cut back on square footage, and this may be necessary, but it is not the only way to save money.

Is that cathedral ceiling with log trusses really needed for your great room? Post and purlin is a cheaper way to support the same roof with logs. Perhaps the garage could be 2 x 4's and not log (that's a good place to save money!).

Reducing the number of notched corners is a very good way to reduce costs (*Figure 12*). My own home has 4 notched log corners, but I have gotten blueprints to bid on that had 4 log corners in the laundry room alone. Eliminate unnecessary roof dormers—they add cost even if they have no logs at all.

There are plenty of ways to save, and experienced builders and designers can help you find those ways to make the most of your money. Consider delaying some costs until later: don't finish the basement yet; add the roof dormers in the future; build the deck two years from now; buy the elk-antler chandelier and the 6-burner restaurant-style range when you win the lottery. Wait to build the $20,000 fieldstone fireplace.

Specification sheets

Once you have a basic plan that is within your budget, start thinking about the details. Before you start construction, you should specify every detail of the building, especially if you will be hiring a general contractor. Every floor covering, every window model and manufacturer, every bathroom fixture and kitchen appliance, every tile brand, every light fixture, every set of door hardware should be specified by someone before they end up in your home. The *spec sheets* are usually attached to the contract you sign and are part of the legal document.

If there are items you have not yet specified, then include a dollar allowance in the spec sheet instead: for example, "kitchen countertop allowance $2,600." Then if the actual price is less than or more than this amount you will still be credited for $2,600 in the contract.

Finishing costs vary widely—toilets sell for as little as $45 and for as much as $2,500. Every little detail makes a big difference in the final cost of your home: tile versus linoleum; hardwood floor versus carpet; custom-made light fixtures versus local lumber yard.

Saving money during construction:

Remember that the cost of your home will tend to climb over time, as you are building it. I have never heard of a home that came in for less than it was budgeted, but I know of many that cost more.

Stay flexible during construction. If you have enough money, then being flexible might mean allowing changes that boost price. If money is tight, and the project

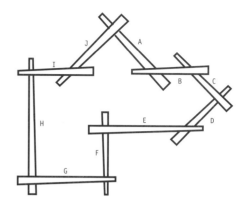

Figure 12: This is a complex, and expensive, way to enclose 1600 square feet: 10 notched corners, including two that are "prows." You could enclose the same area with just 6 notched corners—you try.

Figure 13: Stay on schedule—it will save you money.

is costing you more than you thought, then being flexible means eliminating things you had thought you would be able to have.

You must keep track of how much of the job is complete so that you can use flexibility to help you. If you don't know how much the house cost you until after you move in, that is WAY too late to save any money! You need to know your position every month during construction, or even more often than that.

Limit the changes you make. *Change orders* are always more expensive to add-on than they are to design-in to begin with. Moving a window can cost an arm and leg. And for any change order that is necessary, get a price in writing. Know your costs, do not guess at them.

A good design, that came from a design process in which you were involved, is a great way to avoid changes. If you don't have much experience with blueprints, then make a cardboard model of your home, and carry it out to the land—it will help you see potential problems, and find great opportunities.

Stay on schedule. Delays in time mean higher costs. If you are acting as your own builder or your own contractor, then you will be responsible for scheduling the work of others and keeping it on schedule—a big challenge even for experienced contractors. There is construction and job scheduling software available to help—and get a wall-sized calendar. Owner-builder projects almost always take more time than estimated, especially if the owner is doing the logwork, too.

Doing things twice (because they were done wrong the first time) always costs more than doing them right once. Even if your brother-in-law is an MIT electrical engineer, this does not mean that he will be an inexpensive, or even an adequate, electrician. If the electrical inspector makes you replace all the switch boxes (because they have too small a volume for the electric code), then your hoped-for savings will be lost and you will be behind schedule. This means you lose money twice. Save money by being smart, not by being cheap.

Pay Your Bills

Pay the bills on time, if you have no questions about them. If there are questions, then get them answered promptly to your satisfaction, and then pay on time. Paying late, or refusing to pay (if only because you don't like to part with your hard-earned money) is a poor way to treat people you made promises to, and will often cause other problems on the work site.

The subcontractor's respect for you, and maybe for the work they do for you, can be affected. Worse, the builder can file a legal lien on your property which binds your mortgage—not good. Avoid these problems, and keep the project moving smoothly by treating people fairly. Business relationships during a 10 month construction project can become tense. Keep your cool, and a level head, and expect the same from everyone who works for you.

It is also customary for you to ask for a lien waiver each time you make a payment to a contractor, builder, or subcontractor. This protects you from being "liened" for work you, in fact, paid for.

Deal with serious disputes by involving a neutral third party: you can hire a nonbinding consultant experienced in construction dispute resolution, and you should have a clause for mediation and for formal arbitration in your contracts—they are cheaper than a lawsuit, and can help bring happier, and quicker, results.

Money to Build Your New Home.

Construction loans cost more than mortgages, but until your home is completed and you have a certificate of occupancy from the local building inspector, and the bank has seen your home, you probably cannot get a mortgage.

Construction loans are riskier for the banks, and that's why they charge you more. In fact, construction loans to owner-builders are so risky that some banks do not offer them at all. The bank, after all, is betting a hundred thousand dollars, or more, on you successfully completing your home on time and on budget—and you have no experience. Think of it this way: would you lend money to someone to do something that they have never done before? Well, that's what your asking a bank to do.

The bank may also require you to provide 25% or more of the cash (or other collateral) needed to build the home. They do this to reduce the amount of money they have at risk.

What the bank or lender needs

To act as your own general contractor and get a construction loan, you will need to assure the bank that you are knowledgeable, organized, and capable, not just eager. Get to know a couple local bankers and lenders. Find out what they require from you in the way of blueprints, specification sheets, subcontractor bids, insurance and bonds, personal financial statements, tax returns, permits, surveys, values of comparable log homes nearby (*comps*), clear title to the land, etc.

If you can provide them with everything they want, then you have a better chance of getting their money. Do not guess what they want, ask them what they want from you—they'll be glad to tell you.

Banks will not give you all the money you need up front. Work in partnership with the banker/lender to prepare the draws you will need. As you prepare to dig the foundation they will provide you with some money—the first "draw."

The amount of money you can borrow varies from place to place, bank to bank, and your credit history. Basically, the bank wants to know that you can afford to pay them back. Some banks lend a maximum of, for example, 28% of your annual adjusted gross income ("agi"). Some figure out what monthly payments you can make (probably including future property taxes and insurance premiums). An amortization schedule can give you a rough idea of how much you can borrow given a maximum monthly payment.

After the foundation is in and capped, the banker may come out for a look and then okay your second draw: perhaps to pay for the log walls and roof. When you show that this part is done, you can get the next draw, and so on. Carefully consider how much money you will need and when you will need it so that you don't run out before your next draw.

MONTHLY PAYMENT
Necessary to amortize a loan   7%

TERM AMOUNT	24 YEARS	25 YEARS	28 YEARS	29 YEARS	30 YEARS	35 YEARS	40 YEARS
$50	.36	.36	.34	.34	.34	.32	.32
100	.72	.71	.68	.68	.67	.64	.63
200	1.44	1.42	1.36	1.35	1.34	1.28	1.25
300	2.16	2.13	2.04	2.02	2.00	1.92	1.87
400	2.88	2.83	2.72	2.69	2.67	2.56	2.49
500	3.59	3.54	3.40	3.37	3.33	3.20	3.11
1000	7.18	7.07	6.80	6.73	6.66	6.39	6.22
2000	14.36	14.14	13.60	13.45	13.31	12.78	12.43
3000	21.54	21.21	20.39	20.17	19.96	19.17	18.65
4000	28.72	28.28	27.19	26.89	26.62	25.56	24.86
5000	35.89	35.34	33.99	33.61	33.27	31.95	31.08
10000	71.78	70.68	67.97	67.22	66.54	63.89	62.15
15000	107.67	106.02	101.95	100.82	99.80	95.83	93.22
20000	143.56	141.36	135.93	134.43	133.07	127.78	124.29
25000	179.44	176.70	169.91	168.04	166.33	159.72	155.36
30000	215.33	212.04	203.89	201.64	199.60	191.66	186.43
35000	251.22	247.38	237.87	235.25	232.86	223.60	217.51
40000	287.11	282.72	271.85	268.86	266.13	255.55	248.58
45000	323.00	318.06	305.83	302.46	299.39	287.49	279.65
50000	358.88	353.39	339.81	336.07	332.66	319.43	310.72
55000	394.77	388.73	373.79	369.68	365.92	351.38	341.79
60000	430.66	424.07	407.77	403.28	399.19	383.32	372.86
61000	437.84	431.14	414.57	410.00	405.84	389.71	379.08
62000	445.02	438.21	421.36	416.73	412.49	396.10	385.29
63000	452.19	445.28	428.16	423.45	419.15	402.48	391.51
64000	459.37	452.34	434.95	430.17	425.80	408.87	397.72
65000	466.55	459.41	441.75	436.89	432.45	415.26	403.94
66000	473.73	466.48	448.55	443.61	439.10	421.65	410.15
67000	480.90	473.55	455.34	450.33	445.76	428.04	416.36
68000	488.08	480.61	462.14	457.05	452.41	434.43	422.58
69000	495.26	487.68	468.93	463.77	459.06	440.82	428.79
70000	502.44	494.75	475.73	470.50	465.72	447.20	435.01
71000	509.61	501.82	482.53	477.22	472.37	453.59	441.22
72000	516.79	508.89	489.32	483.94	479.02	459.98	447.44
73000	523.97	515.95	496.12	490.66	485.68	466.37	453.65
74000	531.15	523.02	502.92	497.38	492.33	472.76	459.86
75000	538.32	530.09	509.71	504.10	498.98	479.15	466.08
76000	545.50	537.16	516.51	510.82	505.63	485.54	472.29
77000	552.68	544.22	523.30	517.55	512.29	491.92	478.51
78000	559.86	551.29	530.10	524.27	518.94	498.31	484.72
79000	567.04	558.36	536.90	530.99	525.59	504.70	490.94
80000	574.21	565.43	543.69	537.71	532.25	511.09	497.15
81000	581.39	572.50	550.49	544.43	538.90	517.48	503.36
82000	588.57	579.56	557.28	551.15	545.55	523.87	509.58
83000	595.75	586.63	564.08	557.87	552.21	530.26	515.79
84000	602.92	593.70	570.88	564.59	558.86	536.64	522.01
85000	610.10	600.77	577.67	571.32	565.51	543.03	528.22
90000	645.99	636.11	611.65	604.92	598.78	574.98	559.29
95000	681.88	671.45	645.63	638.53	632.04	606.92	590.36
100000	717.76	706.78	679.61	672.14	665.31	638.86	621.44
105000	753.65	742.12	713.59	705.74	698.57	670.80	652.51
110000	789.54	777.46	747.57	739.35	731.84	702.75	683.58
115000	825.43	812.80	781.55	772.95	765.10	734.69	714.65
125000	897.20	883.48	849.52	840.17	831.63	798.58	776.79
150000	1076.64	1060.17	1019.42	1008.20	997.96	958.29	932.15
175000	1256.08	1236.87	1189.32	1176.23	1164.28	1118.00	1087.51
200000	1435.52	1413.56	1359.22	1344.27	1330.61	1277.72	1242.87

Figure 14: An amortization schedule for a 7% mortgage. $80,000 for 30 years will cost you $532.25 each month.

Figure 15: Your house is under construction. Keep the worksite as clean as possible—this will improve everyone's efficiency.

You may have to hire a general contractor in order to get a local construction loan. It all depends upon your local bank's policies and the way you appear to them as an owner-builder. Are you prepared—or confused? Are you organized and capable—or do you whine about all the paperwork they are making you do? Do you know the difference between a zoning permit and a building permit? Do you have a realistic budget and time schedule? Have you prepared your finances because you won't have your usual income while you build?

Before You Start Building:

Here are some of the things you need before you start building. Many log home shell contractors will require this of you before they deliver the log shell.

Temporary electrical service—you can build entirely with gas-powered generators to supply electricity for tools, but if you are going to have electric service to your house anyway, then call the power company and have it installed early in the project—you will be very glad you did.

Telephone—it is difficult, though possible, to run a building project without a phone on site. But expect delays to be caused by this: concrete trucks get lost and can't find your building site and they can't call you for directions. It is a good idea to install the telephone network 'block' in a weathertight cabinet that can be locked at night. Cellular service can substitute.

Toilet—you may need to rent a portable toilet. Neighbors will get tired of you and your workers after a while. And remember, you are going to live next to these people for a long time, and this is their first impression of you!

Tool and material storage—building-site burglaries are unfortunately common. Find a way to store and protect valuable items. Insure them, too. Even building materials are attractive to thieves—plywood is expensive and relatively easy to steal.

Driveway—building projects are hard on driveways. Have a good road-base installed and expect to have to fix the road again before you move in. Remember that cranes, and semi's delivering logs can't make tight corners, cross weak culverts, fit under overhanging trees, or negotiate soft spots. Cut trees back well away from the edge of the road. Your driveway must be good enough to easily accept a 70' long truck that weighs 80,000 pounds.

A loaded semi weighs 40 tons—a driveway that feels rock solid under a pickup truck can turn to jelly under a concrete truck.

Prepare a staging area to handle materials—you will need a large, relatively flat, stable space within about 50-feet (15m) of the foundation to store and stage materials and to allow trucks to enter, remove materials, turn around, and leave.

Figure 16: Enjoy your new home.
Logwork by Moose Mountain.

A Schedule for Building Your Home

A basic timeline for a building project:

- *buy land*

- *get hands-on training in scribing and cutting logs (if you are considering building it yourself—do not wait until the logs arrive!)*

- *design home—make blueprints*

- *get bids for initial costs from all subcontractors*

- *re-work plan or design to adjust costs*

- *buy logs, if building it yourself, or*

- *hire log builder and start log walls at his site*

- *get permits, insurances, arrange financing*

- *clear building site and staging area*

- *stake nearby property lines*

- *clear driveway, install culverts and good road base*

- *excavate for foundation and septic tank*

- *pour footings and build foundation walls*

- *get your foundation inspection*

- *bring electric service to your foundation or somewhere nearby*

- *cap foundation with first floor framing and subfloor*

- *waterproof and backfill foundation*

- *log walls erected by you or your log builder*

- *roof deck, roofing, and gutters installed*

- *rough-in plumbing*

- *bring well supply pipe into basement*

- *pour basement and garage concrete slabs*

- *second floor framing and subfloor*

- *temporary stairs to basement and 2nd floor*

- *fireplace and chimney*

- *exterior frame walls and gable ends (if any)*

- *exterior sheathing and finished siding (if any)*

- *structural inspection*

- *insulate exterior frame walls (if needed)*

- *exterior doors, garage doors, and windows installed*

- *exterior door and window trim*

- *porches, walkways, and decks*

- *interior frame walls*

- *plumbing, HVAC, and electrical rough-ins and inspection*

- *insulate roof*

- *sheetrock interior walls and ceilings*

- *start landscaping*

- *install cabinets, fixtures, appliances, & interior doors*

- *install plumbing fixtures like sinks and tubs*

- *install tile and floor coverings (except carpet)*

- *install toilets (must be installed after bathroom tile floor)*

- *finished stairs installed*

- *paint, wallpaper, and varnish walls*

- *install and finish interior trim*

- *install finished electrical fixtures (lights and switch plates, etc.)*

- *install carpet*

- *final inspection and certificate of occupancy*

- *move furniture*

- *now...kick back and enjoy your new home!*

How to Find, Buy & Handle Trees

In this chapter you will learn what to look for when you go shopping for building logs, where to look, and how to buy logs and get them delivered. My advice is: buy the best logs that you can afford. Building a log home is easier and faster, and the results are superior, if you use great logs.

Log buildings start and end with trees. The best trees are large, straight, have few knots, not much taper, have straight grain, are free of sapstain and decay, and are long enough to span from corner to corner in one piece. While few trees are good enough for building log walls, fewer still are suitable for floors and roofs made of logs. Good building logs are rare.

It is still possible to buy good trees, though they are not cheap, and you must allow enough time to find them. We will start by learning what makes a tree good for using in a log home (and what is undesirable, as well), and then explore how you can find and buy good trees.

CHAPTER TOPICS

Getting to know trees
- Tree diameters
- Tree length
- Straight & bowed
- Spiral grain
- Dead or green trees
- Taper
- Knots & burls
- Decay & sapstain
- Log grading
- Age of trees

Selecting, buying & preparing your trees
- When to cut trees
- What species of trees to use
- Where to find trees
- How many trees to order
- Log measures
- Removing the bark

Getting to Know Trees

Figure 1:
Left: Medium-size logs—about 14" average diameter.
Right: Large logs—about 23" average diameter.

Butt diameter plus tip diameter, divided by 2, equals the average diameter.

Tree Diameter

A good log has an average diameter of more than 14" (35cm). Average diameter is the diameter half way along the length of a log—at its midpoint. It is possible to build using smaller trees, but the *Log Building Standards* (a copy is in the back of this book) specifies that trees with *tip* (small-end) diameters of less than 10" (25cm) are usually not suitable. I have used small logs for an historical replica project, but generally I recommend larger logs.

Small logs have notching problems that we will explore in detail later —but briefly, small tips don't have enough wood to cover large butt ends of logs. For my own building projects I usually use logs that are 12" (30cm) in diameter at the tip, and sometimes a little bit bigger.

If you cannot safely lift a log with your equipment, then it is too big. I have seen log walls with trees that averaged over 26" in diameter (just four logs to make a wall taller than 8-feet, 2.5m)—and they have a distinctive appearance that might not appeal to everyone, but even they were not too big to lift, move, or notch.

Large logs can save time and money. Much of the labor that goes into building log walls is spent peeling, scribing, and cutting logs. If you can build walls that are 9-feet tall with eight courses (*rounds*) of logs instead of twelve courses, then you will have 50% fewer notches and 50% less lineal feet of long-grooves to cut and logs to peel—so, nearly a 50% savings in labor and materials (*Figure 1*).

Small logs also do not provide much insulation in the walls. Dry pine is generally accepted to have a value of R-1.5 per inch, and thicker walls provide more insulation than walls built of small logs.

Tree Length

Before you start a project, find out the maximum length log that you can get. Do not assume that you can easily get 60-feet (18 meter) trees, for example, or you could end up designing a home that you cannot build. The maximum length of logs you can get will depend upon where you live and what species of tree you use (*Figure 2*). But you should also call a local trucker and ask about the rules and costs for hauling logs that are over 50-feet. Home design must be based, in part, upon knowing the length of logs that you can buy, haul, and lift.

If some walls will be longer than the longest logs you can get, then jog the walls in or out to provide corners so you can use shorter logs. If you must end-splice, then follow the rules found in *Log Building Standards*, and limit the number and the position of splices to ensure that you have strong walls and corners.

All end-splices must be covered by notches, and it can be useful to have two midway stub walls instead of just one, so that end-splices can be staggered on each round of logs. If every single log is spliced at one intersecting wall or stub wall, then it could be possible for the wall to hinge there, weakening the wall.

Figure 2: These trees are long, straight, don't taper much, and have few knots—Western larch in Idaho, about 50' to the first branches.

Most of the logs used in North American homes are between 36' and 50' long, and 40' is perhaps the most common length. Logs longer than 50-feet (15m) that are straight enough and large enough to use in one piece are rare—and expensive.

Straight & Bowed

Straight logs are easier to use than bent logs. Crooked logs may make an interesting design statement as a post, beam, or in a roof system, but for log walls, straighter really is better. It takes experience to accurately judge how straight a tree is while it is still alive and standing—*on the stump*, as it's called. But once a tree is cut down, rolling it over on the ground will give you a very good idea of just how straight or bowed it is. Straight logs roll smoothly; bowed logs flop around.

Very few trees are truly straight, and so a little bow, also called *sweep*, is okay. Here's a quick way to judge the amount of bow. Stretch a string or long tape measure so it is straight from the middle of one end of the log (half the butt diameter) to the middle of the other (half the tip diameter). The string should not come outside the log at any point. Ridgepole logs should be even straighter than this, if possible.

Avoid trees that bow in two directions, or planes—this is sometimes called *wobble* (*Figure 3*). If a log noticeably bows left and right *and* up and down (and many logs do), then don't buy it unless you are going to cut it up into shorter pieces. A log with a single bow will lie flat and stable on the ground; but a log with two bows will never be stable as you roll it over. To help you visualize this concept, bananas bow in just one plane, not two, so they would be okay for log walls (plus they are very easy to peel!).

The best trees are round in cross-section. Some trees are oval, or egg-shaped—which can be caused by growing on a steep slope. Oval trees are often bowed, too, and are more difficult to use than round logs (*Figure 8*).

Spiral Grain

Under a microscope, wood is made of small cells—*fibers*—that look like drinking straws: the cells are long, narrow, and hollow. The wood fibers of straight-grained trees line up parallel to the length of the tree. But fibers can also spiral around the trunk like a barber pole. The spiral may go either to the left or to the right, and it is very important to know the difference (*Figure 4*).

Left-hand spiral-grain trees change shape dramatically—they twist more tightly as they dry. The more severe the angle of the spiral, the worse the twisting. Twisting makes it difficult to keep left-hand spiral-grain logs tightly fitted: the long grooves get large gaps (*Figure 5*). Lefties are also much weaker in bending than either right-hand or straight-grain trees—they aren't as strong or as stiff.

It is easy to detect grain-slope in logs that have started to dry because the checks, or surface cracks, clearly indicate the slope of the grain. For logs that are completely green and un-checked, look for the small depressions above and below a knot—if they line up with the trunk, then the grain is probably straight, if not, the dimples indicate whether the tree is left- or right-grained.

Figure 3: Coconut palms would never be used for log building, of course, but these dramatically illustrate bow in more than one direction—an undesirable characteristic for wall logs.

Right-hand spiral grain   **Left-hand spiral grain**

Figure 4: This standing-dead spruce has spiral grain— the checks (cracks) in the tree follow the fiber slope like the stripes on a barber-pole.

Put your right hand over the tree. If the grain follows your fingers, it's right-hand; if it follows your thumb, it's a lefty.

Figure 5: Logs with left-hand spiral grain wind themselves tighter as they dry, ruining the fit of the notch and long-groove. Here is a classic lefty: the near side of the long groove is tight, but the far side has a wide gap—the log has twisted. The logs above and below the lefty are straight-grained and they fit tightly.

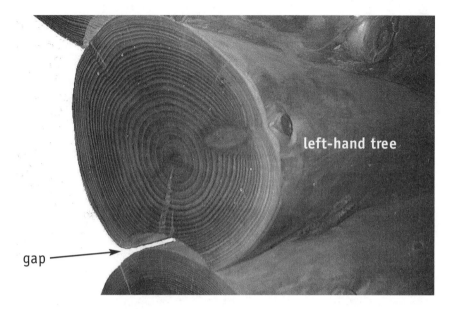

left-hand tree

gap

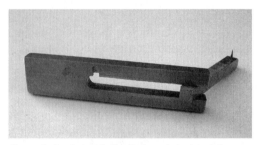

Figure 6: A spiral grain "scribe" can help detect the slope of wood fibers in green trees. (See Tools Chapter for information on where to purchase this item.)

You can also use a spiral-grain *scribe*: a small needle attached to a freely swinging arm. Drag it along a peeled tree, and the needle follows the wood fibers *(Figure 6)*, leaving a slight scratch.

You can tell right-hand from left-hand by placing your right hand on the log so that your forearm points towards either end of the log. (It does not matter if you stand at the butt or at the tip.) If the grain follows your fingers, then the tree has right-hand grain, if the checks, or the grain scribe, follow your thumb, then the tree has left-hand grain *(Figure 4)*.

Spiral grain is genetic—it is a natural, inborn condition of certain trees. Trees that have spiral grain produce seeds that grow into spiral-grain trees. The reason that nearly all the trees on a north slope are, say, right-spiral is because trees that grow near each other are very likely to be close relatives.

When right-hand trees dry they don't twist as much as left-hand trees. Here's why: almost every tree that has spiral grain started its life as a lefty. If a tree is left-hand on the surface then it is probably left-hand through-and-through, and so it twists severely when it dries. If a tree is right-hand on its surface, however, then it *switched* grain direction from left to right—and the opposing layers of grain balance the shrinkage and twisting stresses and it doesn't twist much.

Log Building Standards (Section 2.A.4) provides a full description of how much spiral grain is allowed, and where spiral grain trees can be used in a building. For example, left-hand spiral grain trees should not be used for floor joists or for structural roof beams because they are unreliably weak in bending—they can have less than half the bending strength of right-hand or straight-grain trees.

Dead Trees or Green Trees?

Both green and dead trees are good for log home construction, so you can safely buy the best logs you can find and afford—whether they are dead or green. Living trees are full of moisture—both inside the wood cells and between the cells. After the tree dies, or is cut down, it loses its moisture slowly over time.

Most standing-dead trees were killed by insects, fire, or disease. They can remain sound and standing for years before they fall over—which is why they are called "standing dead trees." At very high elevations, dead trees can stand for decades before they fall over.

Dead trees weigh less than green logs of the same size, so shipping costs can be lower if you buy dead trees. But dead trees already have checks, and this can limit how you use a log: rotating it so that the bow is in a good position might put a check facing *up* on the outside of the wall, where it can catch rainwater, promoting decay. Or, sometimes the scribe line falls into a check and you cannot make a mark or provide a tight fit.

Dry wood is harder than green wood, and more difficult to cut—it tends to splinter more easily. Dead logs usually have sand and dirt in their checks, meaning you'll spend more time sharpening your chainsaw. Standing dead trees are always sapstained, and are more likely to have some decay and bugs—so inspect them carefully. I know builders who order up to 40% extra when using dead trees so they can cut off decayed portions.

Dead trees are not *dry* trees (though they certainly are drier than green trees). I have heard loggers brag that their trees have been dead for 40 years and are "as dry as a bone." They may feel light in weight, and look dry and brittle and checked, but that does *NOT* mean that they are finished shrinking or the walls won't settle. Standing dead in the rain and snow means they still have moisture to lose, and sometimes a lot of it. Standing dead trees will shrink and compress—in other words, they settle.

Green logs are excellent for building—in fact, I often prefer them to dead trees because they are generally not sapstained, have no checks, and do not splinter as easily when cut. Green logs can be kerfed to promote checking where it cannot be seen and where it will help keep the fits tight over time—this is a substantial advantage of using green logs.

If you want a natural honey-color, then you should use green logs. Winter-cut trees stain more slowly than summer-cut trees because sapstain fungi grow best when temperatures are above 50° F (10°C). Summer-killed green trees can sapstain quickly in hot and humid conditions, and you may have to treat them with a chemical that prevents sapstain.

Sapstain does not hurt the logs, but some people don't like the appearance of sapstained wood—streaks and blotches of green, blue, gray, and brown. It is impossible to remove sapstain colors from the wood, though you can disguise them somewhat by using a tinted finish. The colors do fade somewhat as the logs dry. Don't try to drawknife sapstain off either, because it infects the entire depth of the sapwood—in fact, only the tree's heartwood is immune to sapstain.

What about buying green logs and drying them before you start building? It is technically difficult to dry whole logs in a kiln. I know of just one kiln in North America that is commercially drying whole logs for scribe-fit log home

Figure 7: Green trees are almost always delivered with the bark on. These are 'tree-length' meaning mixed lengths, and all with about the same tip diameter (these are 11"—28 cm).

Hint Standing-dead trees absolutely WILL shrink and settle

Wood can lose almost half its weight before it even starts to shrink. Moisture content must be below 30% for shrinkage to begin.

Figure 8: Some trees grow oval or egg-shaped, instead of round. While oval trees can be used, it is easier if the logs are nearly round in cross-section.

construction. (This company exports many homes to Europe where customs agencies were requiring either dry logs or fumigation to eliminate insects, and using toxic chemicals was undesirable for their clients.) So while it might be nice to use kiln-dried whole logs, it is impractical.

Whole logs can be air-dried in certain climates, but I don't see any advantage to trying this. Even in great conditions (a climate that is desert-dry, warm, and high elevation) it can take more than 6 years to dry a 12" diameter log. Bigger logs will take even longer. As the logs air dry you have to prevent them from turning gray, and they will *check* (crack). Really, the best place for green logs to dry is scribe-fitted in a wall, well above the ground on a foundation, and covered by a tight roof. In other words, build the walls with green logs and let them dry in place.

Despite widespread misunderstanding (fueled by the marketing of some log home companies), both green and dead logs shrink, and a home built of either will settle. The amount of settling can be less with dead logs, but settling is not eliminated. If you ignore the fact that dead logs settle, then you risk serious problems with the house.

Settling and shrinkage end only when the moisture content of the log is at equilibrium with its location *(see the chapter on Settling)*. More to the point, you do not need to avoid settling in order to build a good quality log home, you just need to know how to allow for settling—a topic we will cover in great detail later in this book.

Taper

Trees become more slender the farther they get from the ground. The big end is called the *butt*, the small end is the *tip* (or *top*). Taper is natural, and gives log homes much of their distinctive handcrafted appearance. A few log builders specialize in using highly tapered logs, and their buildings are impressive accomplishments. Most builders, however, prefer logs with less taper because they are easier to use.

Taper varies with species and location: Western and Southeastern logs tend to have less taper than most Midwestern and Northeastern logs. Excellent logs taper 1" every 10' (a 4" difference in diameter from tip to butt in a 40' log). In Wisconsin, where I live, trees that taper 1½" to 2" per 10' (15mm per meter) are more common (6" to 8" difference from tip to butt in a 40' log).

Logs that taper more than 3" for every 10' (25mm per meter) of length can be a challenge for inexperienced builders, though it is certainly possible to use highly-tapered logs *(Figure 9)*.

Trees that grow in dense, tightly-packed stands tend to have little taper. Those that grow in the open are usually very tapered and have lots of branches. And infertile soil often grows trees that are slender and less tapered.

Some trees have large butt-swells near the ground, but are less tapered once you get past the bottom 6 to 10 feet (2-3 meters). It is usually best to remove the swollen portion of the trunk, and use only the less-tapered part of the tree. Some loggers, though not all, can be convinced to keep the first *butt log* and sell it to a

Hint For more information on wood drying and shrinkage, I recommend the book *Understanding Wood* by Bruce Hoadley

Figure 9: The log corner on the right has some logs that taper dramatically: a few 5" tips and 25" butts. Logs with less taper are easier to use. This is an Eastern white cedar home in Minnesota.

sawmill, leaving you with the rest of the tree for your log building. If you must buy the whole tree, then some butt logs can be used for posts and short wall logs, or can be sawn into timbers.

Knots & Burls

Trees have limbs that hold leaves or needles. The limbs tend to grow in groups around the circumference of a tree, called *branch whorls*. For some softwoods, pines, spruces, and firs, each branch whorl is about one year of growth.

When limbs are cut off, the parts left on the trunk are called *knots*. All trees have knots, but some have more than others, and some have large knots while others have small knots. Western larch and Western red cedar tend to have small knots, while ponderosa pine usually has large knots. Trees that grow in the open tend to have large knots, while those that grow in dense forests often have fewer and smaller knots.

You cannot order trees from a logger "with very few knots"—ordering logs is not like ordering pizza (half with double cheese and half without pepperoni, please!). The choice to have large or small knots in wall logs is primarily esthetic, and one of convenience. Large and numerous knots make trees tougher to peel, and more difficult to scribe and cut. But some builders love the natural, gnarly shapes of knots and branches and actually prefer bumpy trees.

Large knots and numerous knots do weaken logs that are used as beams: floor joists, purlins, ridgepole, and the like. You should consult an engineer if you have specific questions, but in general, a few sound knots of 2" (50mm) diameter are not typically a problem. Knots that are 3" or more in diameter, knot whorls spaced closer than one foot , and dead knots could be a concern for log beams.

Burls are lumps caused by a virus that infects a living tree. Lodgepole pine in the Northern Rocky Mountains seem to be especially prone to burls *(Figure 10)*. If you like burls, ask the logger to keep his eyes open for them.

I love the appearance of natural logs, but it is definitely easier and faster to peel, scribe, and cut logs that are smooth, not bumpy. Other builders feel that the extra time to carefully scribe all the ins and outs, jigs and jags, is well spent and shows off their skill and craft *(Figures 11 & 12)*. You can make up your own mind on this!

Figure 10: Some people love burls, and some don't.

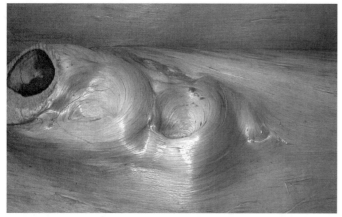

Figure 11: Knots are a challenge to scribe and fit.

Figure 12: Even gnarly logs can be scribed to fit tightly to each other.

Figure 13: Sapwood has much higher moisture content than heartwood, as can be seen in this frozen Eastern white pine. The frozen sap makes the sapwood look dark here. Younger trees, and faster-growing trees, usually have a higher percentage of sapwood and less heartwood. Here a tenon is laid out and cut on the end of a post.

Figure 14: Sapwood is the light-colored layer surrounding the darker heartwood.

Decay & Sapstain

The sapwood of all trees is susceptible to stain and decay by mildews and fungi. The heartwood of some species, Western red cedar and larch, for example, resists decay. But natural decay resistance of heartwood is not really important for log homes—after all, if the sapwood rots away who cares if the heartwood is sound?

To make log cabins last a long time they should be designed and built so that the logs stay dry. Wide roof overhangs and gutters are the best place to start; and get the sill logs at least 20" (50cm) above soil. Do not plant shrubs too close to logs—they keep the logs from drying out after rainstorms by slowing air circulation. Never use untreated logs to support decks and balconies—a mistake that even professional log builders make. And do not let the purlins or ridgepoles extend out beyond the roof. The secret to longevity is keeping the logs dry.

Remember that the sapwood of no tree species is decay resistant—even naturally-resistant logs like cedar and larch are wrapped with a layer of sapwood that can easily rot (*Figure 14*).

In some regions and climates, where decay or wood-destroying insects like termites are severe threats, you may need chemicals to help preserve exterior wood. There are some anti-mildew and anti-fungal chemicals available now that are safe and effective when applied to the surface of logs. The days of pressure-treating whole logs with "penta" are behind us, thank goodness.

Log Grading

A growing number of building code officials and mortgage lenders want logs to be *grade stamped* for use in log homes. The Log Home Council and Timber Products Inspection are two organizations that conduct this service in accordance with a standard known as ASTM D3957-90(1996)e1 (www.astm.org) (contact information is in the Resources Chapter). Grading provides some assurance to the builder and homeowner that the logs that end up in their home will be structurally strong, or free of certain defects. I know very few professional handcrafted builders who get logs graded unless required to do so by the local building inspector, but its is likely that log grading will become more common in the future.

Age of Trees

How old are good log building trees? Trees grow fast when conditions are good: soil, climate, genetics, and nutrients. I have used trees as young as 35 years old, and as old as 300 (knocked to the ground by a windstorm, and then salvaged). Don't worry too much about the age of your trees.

Fast-growing trees may not be as strong, and may shrink and check more than slow-growing trees. But, in a pinch, old or young will do. I prefer good second-growth and slow-growing plantation trees with about 8 growth rings per inch, or more. I think virgin, first-growth trees are precious and should not be wasted.

Selecting, Buying & Preparing Your Trees

Figure 15: Broad-leaf trees, and all trees that have grown in the open, are usually not suited for log building: they branch too much and don't have just one good trunk, or *bole*.

When to Cut Trees

Living trees that are cut down in the winter are best for log building: they shrink less, check less, and sapstain less—though they are more difficult to peel. Summer-killed trees are full of sap and swollen with water, and though they peel very easily, they also sapstain quickly. It does not matter as much when dead trees are harvested, though skidding on winter snow helps to keep them clean and free from the grit that can get packed into the checks if they are harvested in the summer or spring.

In North America, most log builders have little control over when trees are cut. In some countries, loggers and builders are more sensitive to the season and even the time of month for felling. In Latvia, for example, it is easy to buy winter-cut trees, and it is also possible to get trees that were felled during the new moon—the two weeks before a full moon. Latvian builders feel that trees cut during the new moon are less likely to sapstain and check *(Figure 16)*, and I have seen impressive differences between trees cut in one part of the lunar month rather than another.

Green trees cut down during their growing season are susceptible to sapstain, and chemical preventatives may be necessary to keep logs a natural honey-color.

In Scandinavian countries, the tradition was to *girdle* standing trees (remove a belt of bark) and let them die and dry over a number of years. There are log buildings that are more than 500 years old in these countries that are still in great condition. There may be advantages to girdling and new-moon cutting, and we would be smart to consider these old ideas with open minds—it's just that we are in a rush to build, at least when compared to 16th century Norwegian, Finnish, and Russian craftsmen.

Figure 16: These pines were cut during the new moon, and the bark was removed. Over the summer, the surface turned olive green from mold, but when peeled (right) they were perfect and unstained.

Figure 17: Needle-leaf trees, especially when they grow in dense stands like this, produce good building logs. These are European larch in a New Zealand plantation—they are only forty years old.

Figure 18: Douglas fir grown in a plantation—nearly every tree is a fine building log. In a natural forest, however, sometimes only 10% are good enough to be used for log building. Also see Color-Figure C-16.

What Species of Tree to Use

Most log homes are made of needleleaf trees like pines, cedars, spruces, larches, and firs. Most broadleaf trees are not used because they branch severely at a relatively short height, and lose their single, central trunk, or *bole*. Needleleaf trees, in contrast, have boles that stay large and whole for the full length of the tree—a desirable feature for building log walls (*Figure 17*).

In 18th- and 19th-century America, oak, walnut, and elm were often used for log homes, but walls were shorter then, and broadleaf trees from virgin forests had longer boles, and were less tapered. At the present, aspen (also known as poplar or *popple*) and tulip poplar (tuliptree or yellow poplar) are the only broadleaf trees that are used much for handcrafted log homes in North America, and they can both be good building logs.

Tree species vary widely in their strength and stiffness, so for beams and long spans, some species are better than others. Check the bibliography in this book for detailed information, or consult with an engineer who has experience with wood as a structural material. In general, cedar and true firs are at the weak end of the needleleaf trees, pines and spruces are in the middle, and Douglas fir and Western larch are among the strongest and stiffest North American conifers (*Figure 18*).

There are many good trees that can be used for log homes—do not ignore good, local trees in favor of expensive, distant logs. Start by looking locally for good trees no matter what the species. People often ask me: what is the best tree? I do not think cedar is better than Douglas fir, or spruce is superior to lodgepole—I look for quality logs, no matter what the species, and I have not been disappointed with any species I have used (*Color Figure C-16*).

My advice is: focus on quality, not species. Any tree that is long, large, straight, smooth, has little taper, is slow growing, and has straight grain is a good building log.

Professional builders sometimes prefer one species over all others, and this is okay. After all, a cabinetmaker may specialize in oak, but that does not mean that maple, walnut, and mahogany are no good for furniture. For a professional log builder, it can be an advantage to know one species very well—but that is not a reason for *you* to limit your choice.

Where to Find Trees

There are a number of ways to find trees—and you may have to use them all because good building logs are simply not common. Even some loggers do not know a good log building log from a poor one. This means that *you* need to be the expert.

If you are not in the logging business already, building log homes is not a good reason to start. Felling trees is dangerous, difficult, and can be expensive. I am going to assume that someone else will be cutting, skidding, and trucking your trees.

Provide a clear description of what you want: specify the lengths of the trees you need, the minimum tip diameter you will accept, the amount of taper and bow allowed. Make it clear that logs that bow in two directions, logs with left-hand spiral, and logs with decay are unacceptable. Do not let loggers or log suppliers talk you into unreasonably lowering your standards, but keep in mind that trees do not have to be perfect to be used.

Sorting for quality is done *at the landing*, which means: near where they were cut down. You do not want to pay expensive trucking costs for logs that you cannot use because they are too small, too bent, or too lumpy. Try to ensure that you can actually use every log you receive. I meet each tree face-to-face before I buy it: you can't tell what a tree looks like over the telephone! So, plan on traveling when you shop for logs.

How do you find logs? First, find out where good trees grow near you, and look there. Federal, state, and county forests have many trees, and so start with the government offices that are local: your county forester, state or provincial department of natural resources, local National Forest office, or state forestry office. Some government offices are very helpful: Oregon and British Columbia have programs to help sell building logs.

A local district or regional office will be of more help than the headquarters in the capitol because you want to talk to people who get out in the woods and know where the trees are. You need to talk with people who wear dirty boots to work each day. Ask these people who is cutting the sort of logs you want, and who has a sale coming up for harvest.

Where to Look for Building Logs:

- Local district government offices
- Log home builders
- Log home builders associations
  - www.bclogandtimberbuilders.com
  - www.logassociation.org
- Sawmills
- Loggers
- Truckers that you see hauling good logs
- Log brokers
- Newsletters and magazines
- Private consulting foresters
- Private tree-grower associations
- Native Indian nations

Figure 19: Some sawmills have plenty of good building logs, but it is can be difficult to get them to sell their best logs to you. These ponderosa and spruce logs are at a mill on an Arizona Indian reservation. Few of the trees you see here are building-log quality.

Figure 20: You'd be fortunate to find trees this beautiful. Luckily, you can use trees that are not this big or straight for your log building. But using good trees does make building much easier. These are Douglas fir growing in a managed plantation.

Figure 21: Thousands & thousands of building logs—long, straight, not much taper—at a sawmill, waiting to be 2x4's, in the mountain west.

Another source for logs is a local log home builder. Get the names of builders from the log builders associations listed later in this book. If the builder will not tell you where you can buy logs (they guard their sources carefully for good reason: good logs are hard to find), perhaps they would be willing to sell logs to you—some log builders also act as log brokers because to get the logs that they need they must buy more than their business can use. So they sell some extra logs.

I have only rarely been able to buy good logs from sawmills. Increasingly, small sawmills have trouble buying enough good logs to stay open, and large mills that have plenty of logs may not want to deal with you—you are just too small a customer *(Figure 21)*. Sawmills are unlikely to sell their best logs, but it happens now and then.

Private land is becoming one of the best places to shop. Private consulting foresters are listed in the Yellow Pages, and they often know about good trees on privately-owned land. Your state or province may have an association of private forest owners: call their office, and ask for help—they're in the business of growing good trees and finding a market for them, so they should be eager to help you. You might be able to place a classified advertisement in their newsletter.

Classified ads in newsletters like *Log Building News* (250-547-8776) also produce results. Or try any newsletter or magazine that is read by loggers or log home builders. There are Internet web sites for forest owners, so do a search for them.

If necessary, go outside your local area to where good trees live: the Rocky Mountains, Pacific Northwest, Great Lakes, or the south Atlantic coast. When you see logging trucks on the highways that are carrying the sort of logs you want, write down the names and telephone numbers from the doors. These trucks are driven by people who work all week hauling what you want, and they know who is cutting and who is selling.

Look in the Yellow Pages for "log brokers," and "loggers." You may not find them listed in your home town, but in logging country, the phone book is thick with them. In general, you want to talk with the smallest logging companies: 'ma and pa' outfits—the ones that have a skidder, some chainsaws, and a logging truck *(Figure 22)*. Most of the loggers I buy from don't have a computer, and one doesn't even have a telephone. Big logging companies don't have time to deal with small-volume customers, though it's tempting because they have lots of wonderful logs.

Some of the Indian nations in the mountain west from Arizona through to British Columbia have logs and logging companies—call the tribal offices for details.

There are laws that control the export of whole trees from Canada and importing trees to the United States, but when the exchange rate is favorable, as it is now, it's definitely worth considering.

When you find logs and are arranging to buy them, ask for references, and follow them up. Log home builders, loggers, sawmills, and log brokers, like all fields of business, have their share of unreliable people. Know your supplier's reputation, arrange to make a partial payment, and then pay the balance after you receive the

Figure 22: A small, self-loading truck with some great logs—get the name and phone number from the door!

logs. A common arrangement is to give the truck driver a check for the full balance due once he has delivered and unloaded—then he gives your check to the logger when he gets home.

Logs are moved from the skidder *landing* to your building site by trucks—either flatbed semi's or logging trucks. Railroads are rarely used. Ask the logger for advice on over-the-road trucking—they don't usually send their pole-trucks down the highway more than 100 miles. There are firms that specialize in hauling power poles for local utility companies—check the Yellow Pages. In a pinch, you can move them on a semi flatbed trailer that has stakes and straps, or steel bunks.

Some trucks are self-loading, meaning they have a crane-like device *(a knuckle-boom or clamshell, see Figure 22)* to pick up logs. But if your trucks cannot load themselves, the logger should be willing to load them for you at no additional cost. Remember that you will have to unload the trucks when they arrive at your site—if it takes you more than 1 or 2 hours to unload you should expect to pay the trucker an extra fee. Most truckers are paid by the mile for delivering logs, and costs vary widely, so shop around.

Thinking about cutting trees from your own land? First, cruise your woods with an expert to see if you have enough good trees. After twenty years of building, teaching, and consulting, I have met just one landowner with enough good logs. In some stands, only 5% of the trees are good enough for building. And keep in mind that you will be living in a clear-cut if you harvest all your trees!

How Many Trees to Order

Start with your floorplan to make a list of logs that you need. Make separate lists for wall logs, floor joists, roof parts, posts, and specials. In North America, loggers often buck trees in two-foot increments, so round all lengths up to the nearest even foot. You may not be able to buy a 37' log, for example, so order 38'.

To figure how many logs to buy for each wall you need to know how many rounds (or courses) of logs it will take to reach the total wall height you want *(Figure 23)*. To do this, take the average wall log diameter (tip diameter plus butt diameter,

Figure 23: Count the number of logs in each wall. This cabin needed 9 logs in order to make walls that were 9'-8" (2.9m) tall, or slightly taller. Using logs this diameter, 8 rounds would have made walls that were too short, so I ordered 9 logs per wall. This cabin was entirely built by students in one of my 4-week workshops.

Figure 24: An actual list of logs for a project. Every log length from 6' to 50' has a row, by two-foot increments. The number (the column marked "#") of logs of each length that are needed is listed by purpose: wall, roof, floor joists, and posts. For example, the floor will need 10 logs 22' long, for a sub-total of 220 lineal feet. And we also need one 22' log for the walls—a total of 11 logs 22' long.

Add short logs onto medium-length logs to consolidate your order. The thirteen 6' logs could be added to the six 32' and seven 34' logs.

LENGTH	WALLS #	Feet	ROOF #	Feet	FLOOR #	Feet	POSTS #	Feet	TOTAL #Logs
50			3	150					3
48			1	48					1
46	1	46							1
44	9	396							9
42	13	546	3	126					16
40	2	80							2
38									
36	3	108							3
34	7	238							7
32	6	192							6
30									
28	1	28							1
26	8	208							8
24	7	168							7
22	1	22			10	220			11
20			1	20	8	160			9
18									
16	17	272	3	48					20
14			1	14					1
12			2	24					2
10							5	50	5
8							8	64	8
6							13	78	13

Totals=   2304 Walls   430 Roof   380 Floor   192 Posts
Grand Total=3306 lineal feet
List does not yet include any extra logs, about 5% to be added

Figure 25: Trees at the "landing," which means: in the woods, very near where they were cut down. The road to the landing is often a muddy one-lane track with little room to turn around. Don't expect every trucker to take his rig to your landing! Some truckers stay on the interstate.

divided by 2—it is a guess at this stage since the logs are not yet at your building site), and subtract one inch (the log diameter you lose when you cut a long groove in each log). Divide the desired wall height by this number and then round up to the nearest whole number to get the number of rounds needed.

For example, if you want the walls to be 9'-8" tall, and the average log diameter is 15", then divide 116" (9'-8") by 14" (average diameter 15" minus 1" loss-to-scribe) and you get 8.28. Then round 8.28 up to the nearest whole number, and you need 9 courses of logs to make the walls at least 9'-8" tall.

To figure the length of wall logs, start with the floorplan—it should show the dimensions from center to center of the log walls or the size of the subfloor. To this length, add the average diameter of one log, plus twice the length of the *flyway* (the log outside the notch is called the flyway), and then add at least one more foot for trimming to final length.

If a wall has doors and windows, then you can order shorter logs for this portion of the wall, and then cut and expand the log to fit as you build. But the sill logs should be full length in one piece, and the logs above the doors and windows will be one piece, too *(see Figure 23).*

Short logs should be consolidated and then cut from longer logs as needed. For example, do not order 13 pieces of 6' logs—even if your list says you need them *(Figure 24)*. Instead you can add 6' to the seven 34' and six 32' wall logs that you need, and order 40 ' and 38' logs instead. It is difficult to truck a load of logs that are of widely mixed lengths, so find ways to order longer logs.

Continue your list of logs with floor joists, posts, and the roof structure, and add a few extra logs for mistakes, or to allow for logs you don't use because of flaws: sweep, diameter, or taper. I usually order an extra 5% to 8% of lineal feet for an average-size home—and I get the extra logs mostly in the longer lengths.

Some loggers do not buck logs to the length you want—they sell logs *tree length*, take it or leave it. If you buy logs tree-length, then make certain you have enough total lineal feet of logs and enough long logs to build the home.

As you can see, you need accurate plans for the wall and roof design before you buy logs. Do not order logs until you have finished the house plans. If the length of a wall is changed from 10' to 12', then you need to change your log list.

Log Measures

There are a number of ways that loggers measure the quantity of logs they are selling: lineal foot, thousand board foot (MBF), ton, metric ton (tonne), cubic meter, cunit, truckload, cord, and so on. Each logger will have one favorite method that he uses. It is confusing to have all these units, but there is little that you can do to change the way someone sells logs. You must adapt to the way logs are being sold by converting any unit to lineal feet.

Lineal feet is easy to understand. A log that is 42' long from butt to tip has 42 lineal feet. Lineal feet is also very useful for log builders because your log list is in lineal feet. I don't know anyone who can say that a roof system with two trusses, two purlins, one valley, three posts and a ridge will require 8 cords of logs, or 2.4 tons, or 5.1 cunits, or 4.1 MBF. But it is easy to figure how many lineal feet will be used: just study the blueprints and add the length of each piece that you need.

A professional log builder may feel comfortable buying by the truckload, or ton, or cord. But what if you order by the ton and end up 3 logs short? The local lumberyard will not have what you need, and the truck coming from Montana will cost you the same whether it is full or has just three logs. To avoid this problem, convert any unit being used by the seller into lineal feet of log.

For example, if the seller uses tons as his unit, then measure how many lineal feet are in a truckload of logs that are the diameter you want, and ask how many tons are on that truck. Simple division will tell you how many lineal feet are in a ton. Average this for two or three truckloads, and your conversion factor will be more accurate.

Is $7,000 per truckload for green lodgepole pine a good deal? Find out how many lineal feet are on a 'truckload.' If a truckload averages 1700 feet of logs—that would be $4.11 per lineal foot, which may not be a bad price, if they're great logs. Use this technique to figure how many truckloads, or tons, or cubic meters you need to buy, and what it will cost.

Convert everything to lineal feet, for your purposes. But when you talk with the person selling the logs, speak only their language—no matter what units they use.

Figure 26: Not a self-loading truck—more of an 'un-load it yourself' truck, at a workshop in St. Petersburg, Russia.

Hint Convert everything to lineal feet, for your purposes. But when you talk with the person selling the logs, speak only *their* language – no matter what units they use.

Removing the Bark

You need to remove all the bark and cambium (the thin stringy layer between the bark and the wood). As we all know, there's more than one way to skin a cat. The same goes for logs.

Any way you do it, peeling is hard work. The bark of living trees that are cut in the summer is not firmly attached to the wood—the cambium layer is swollen

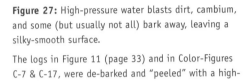

Figure 27: High-pressure water blasts dirt, cambium, and some (but usually not all) bark away, leaving a silky-smooth surface.

The logs in Figure 11 (page 33) and in Color-Figures C-7 & C-17, were de-barked and "peeled" with a high-pressure washer.

with water and separates easily from the wood. You can use a shovel or peeling spud, to knock the bark off *(Figure 29)*. With some species you can pick up a strip of bark and peel it off just by walking along the log! But you'll probably need to use a sharp drawknife—it's a time-proven method, and it makes a finish that looks good and is easy to scribe.

Some builders like the smooth appearance left by high-pressure water, but this is not a very widespread method *(Figure 27 and Color-Figure C-17)*. You'll need a gas-powered unit with at least 3000 psi (and probably more) and a rotary-point nozzle *(turbo,* one brand is called). One problem with water-peeling is that every tiny imperfection shows up—every dent, bruise, and scratch is accentuated.

Some log home companies use automated peeling machines, usually with curved steel blades rotating like a power planer driven by a large electric or gas motor *(Figure 28)*. No owner-builder would go to this expense for one log home.

Figure 28: An electric peeling machine with curved steel planer blades that rotate at high speed.

A sharp drawknife makes an attractive finish, and is effective and cheap, even if it is slow and hard work. First, buy a drawknife that is made for peeling logs—the best ones were designed by Ed Shure and forged by Barr Quarton *(see Tools chapter)*. These knives are big, wide, heavy, and razor-sharp. If you try peeling logs with the small, old, dull drawknife in your garage you risk never getting to the point of scribing and notching a log.

If you prefer to sit on the log while you peel it, then put the log in V-blocks near the ground. If you want to stand next to the log while peeling, then put it on sawhorses or skids so it is about waist-high. Drawknife peeling is a smooth rocking motion that uses the shoulders, back, and even legs (if you're standing). If you use just your arms to pull the knife you'll tire quickly.

Slice the bark off—don't drag or pull the knife as if you were shoveling snow. Here are some useful hints. Instead of holding the knife's edge at 90° to the length of the log, trying cocking it either to the left or to the right—this skew angle lowers the "apparent bevel" of the blade, and this actually makes the knife seem sharper *(Figure 29)*.

Once you have the feel of the rocking motion with a skew blade, then add in another action: start each pull with the end of the sharp edge that is closest to you, and slice through to the other end of the edge. In this way you use every inch of the edge as you draw it towards yourself to *slice* the bark in long strokes.

Think of it this way: you can't push a razor-sharp knife through a tomato. You need to smoothly slice. Don't saw the edge back and forth through the bark—take just one controlled swipe that uses every inch of the drawknife's edge.

As your arms tire you'll change over to short, choppy strokes that make the log rough looking, and ugly. Stop and rest; sharpen the knife. When you get back to work, take slices that are 12" to 24" (30cm to 60cm) long—this makes a finish that looks great, and it forces you to use your shoulders, back, and legs because you can't make a long slice with just your arms *(Color-Figures C-18, C-19, C-21, C-25, & others)*.

Develop the rhythm, and in a few days you'll get stronger and faster. Or, you'll be ready to teach a couple young guys what you've learned about peeling like a pro.

Figure 29:

Left: You can knock the bark off summer-killed trees with a spud or a shovel.

Right: Slicing off bark with a razor-sharp drawknife. Take long slices—the results look a lot better.

Also See Color Figures C-18 & C-19

Log Wall Basics & Log Selection

The essence of handcrafted log building is that logs are scribed to fit each other. The task of fitting two fifty-foot, thousand-pound, hand-peeled logs together without any gaps seems impossible. In fact, as a student, I can still remember thinking: this just can not be done. And yet logs can fit so tightly that they look as if they grew together in the forest.

Hand-peeled logs have bumps, knots, flat spots, bow and twist, and they taper: big at the butt end, smaller at the tip. In this chapter we will cover the basics of making scribe-fit log walls—but not how to use scribers and chainsaw (we'll get to that soon). In a nutshell, you need to learn how to cope with log taper when selecting which log to use next. Dealing with taper is one of the most important skills of log building.

This book is about log building, not carpentry, and so I am assuming that you already know basic carpentry: the difference between level, plumb, and square; which is the gable end of a house, and how to frame a floor deck on top of a concrete foundation. If you need to review carpentry basics like these, then please go to the Resources Chapter for my recommendations of good books.

CHAPTER TOPICS

Log walls
- Log diameters
- The log list
- Big & small
- Log wall basics

Selecting the next log
- Corner order
- The selection rules work for all sets of logs
- Using the rules to select which log to use next
- How close is close enough
- Other things can affect which log you select to use next
- Automatically corrects itself
- Wall logs for interior walls

Log selection Q&A

Log Wall Basics

Log Diameters – Know Your Logs

Before you start building walls you need to know just how big your logs are, on average. You also need to decide which logs will be set aside for the roof and floor *(Figure 1)*.

Go through and mark important logs like the ridges and purlins so that you don't use them in the walls. A great ridge log is hard to find—and I can promise that you'll be tempted, at some time, to use it in the walls! Also set aside floor joists, top plate logs, truss parts, and other important logs that should not be used in the walls.

Next, measure the diameter of every wall log and make a list. The logs do not have to be trimmed, or *bucked*, to final wall length in order to be measured for diameter. Here's what I mean: if you have a 40-foot log that you might use by cutting it into a 14-foot wall log and a 20-foot wall log, and the last 6-foot piece will be part of a stub-wall, then you can measure just the log's tip diameter (small end) and its butt diameter (large end) and put those two numbers on your list. You do not need to measure the log diameter at the tip, 20' from the tip, 36' from the tip, and at the butt. Wondering why this is true? It's because we use average diameters, and the average of the whole log is exactly the same as the average of all the shorter lengths you might cut the log into.

If your logs are not yet peeled, then just measure the wood inside the bark at each end. Logs that are obviously oval, not round, in cross-section, can be measured the long and short directions, and then write down the number that is about halfway in-between these two *(Figure 2)*.

If a log is seriously *flared* (the bell-shaped swelling near the base of the tree), then measure its diameter in a couple feet from the end—two framing squares make a handy log caliper. Or trim those parts off the logs before you measure them, if you are going to trim them anyway.

Write the tip and butt diameters of each log on the log ends to make finding logs easier later on—you would be amazed how many times you will measure the same log, if you don't do this! A big bold "21-12" on both log ends will quickly tell you this log has a 21" butt and a 12" tip— ½" is close enough for these measurements. Also write the log length on the log ends *(Figure 3)*.

The Log List

Write the diameters on a sheet of paper as you measure the logs. When you have measured all the wall logs, find the average tip diameter of your set of logs by adding up all the tip diameters and dividing by the number of logs *(Figure 4)*. Then add all the butt diameters and divide by the number of logs to get the average butt diameter. Finally, add the tip average and the butt average and divide by 2 to get the average log diameter. You will use these three averages frequently as you build.

I run a log building yard, and so I take another step to improve the list. I mark

Figure 1: The logs have arrived. First, set aside important roof logs, and do not use them in the walls. Then get to know your wall logs.

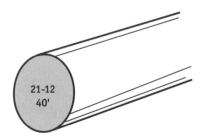

Figure 2: Measure the diameter of oval logs in both directions, and write down the average on the list.

Figure 3: Write the tip and butt diameters, and length, on the ends of every wall log.

each log with a serial number that is keyed to a computer spreadsheet that is a complete inventory of all the logs. Spreadsheets are an easy way to find averages, and keep an updated list of the logs remaining as you build the walls. In fact, as I eliminate logs from the list the spreadsheet can automatically re-figure the averages based on the logs that are actually remaining. Once I know what size log I want to use next in a particular place, I look at the spreadsheet—the logs are sorted by tip diameter first, and then by length.

Big & Small

It is easier to build log walls when each log is fairly close to the average log size. Sometimes people write to tell me that some of their wall logs are large and some are small—and they can tell the difference just by looking at the logs (it's so obvious that there is no need for a tape measure).

If this sounds familiar, then I suggest you make two lists: small logs on one, and large logs on another list. Figure the averages separately. Use the larger logs first (lower in the walls), and then, when you run out of large logs, change over to the smaller logs as the walls get taller.

Log Wall Basics

Before getting to the rules for selecting logs to deal with taper, let's quickly consider the basics building of log walls.

1 – Butts and Tips Alternate

First, butts and tips alternate each time we add another log to keep each wall approximately level *(Figure 5)*. Putting all the butts at one end would make a wall really slope!

2 – Log Diameter is Lost to Scribing

If we built with logs that were all machine-lathed to be perfectly straight and exactly the same diameter (say 14") we could stack these logs one on top of another and gain exactly 14" of wall height for every log we added—five logs would make a wall 70" tall (5 x 14 = 70); six logs would be 84" tall, and so on *(Figure 6)*.

If we scribe-fitted this same set of lathed logs so that we gained 13½" with each round (that is, we lose ½" per log to the scribe), it would be still be easy to precisely control corner shoulder heights because there is no taper.

In the real world of handcrafted log building, however, enough wood must be removed to close the widest gap between the two naturally-shaped logs, and the widest gap is created by the completely unique shapes of two logs *(Figure 6)*. Crooked, bumpy, and bowed logs lose more to the scribe than straight ones because more wood must be removed to close the widest gap.

A 14" naturally-shaped log will result in only about 13" of wall height when scribe-fitted—this is a *loss to scribing* of 1". Scribe-fitting a 14" log onto a wall will produce about 13" of height gain. The second fundamental of scribe-fit logwork is: you lose some log diameter when you scribe and cut the long groove in the bottom of each log—about 1", which is 25mm.

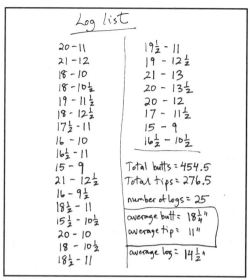

Figure 4: Make a list of the butt and tip diameters of all the logs you will use to build your walls. Add all the butt diameters together and divide by the number of logs—this equals the average butt diameter. Do the same for the tips. Now also find the average log diameter (some call it the midspan diameter). Measure all the wall logs to get the averages, not just a few logs, as in this list.

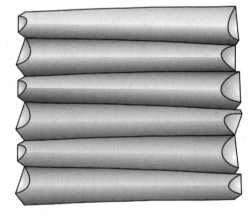

Figure 5: On each wall, you alternate the direction that tips and butts point every time you add another log. This helps keep walls level.

This does not mean that the grooves are 1" wide—it means that they average 1" deep (*Figure 7*). Figure 26 shows that straight, smooth logs have more uniform grooves, and slightly less loss to scribe. Bumpy, crooked logs have wild grooves that are deeper, and have slightly more loss to scribe. But both smooth and bumpy logs can be made to fit beautifully. A loss-to-scribe of 1" works for all logs.

Figure 6: Each of these log walls is made of 5 logs that average 14" diameter—but the walls they make are not the same height. Why? On the left, perfectly lathed logs are resting on top of each other—they just touch. In the middle, lathed logs are scribed to lose ¹/₂" per round—they have identical, small, long grooves. On the right, naturally-shaped logs need scribe-fit long grooves that vary in width and depth. Natural logs lose about 1" (25mm) per log to cutting the groove. Enlarged view on the right shows the loss to scribe of one groove.

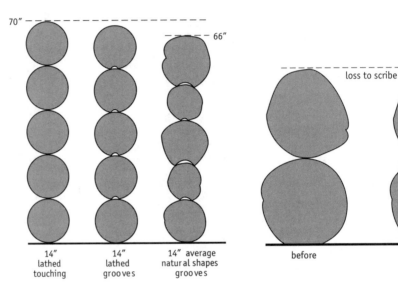

Figure 7: The long groove (also called lateral groove) varies in width, and follows contours of this log and the one it has been scribed-to-fit that is below it in the wall. These logs are straight and smooth, so the groove has quite a uniform width and depth.

3 — Rotate the log straight-side up

Here's a helpful tip to keep from losing too much log diameter to the loss to scribe: when positioning each log for scribing, rotate it so that its straightest surfaces are on top and on the bottom. The amount of wood lost to scribing is minimized by having the straightest sides face each other—put the bumps to the inside or outside of the building where they won't create wide gaps. Rotate each log so the bow, or sweep, bulges out on the wall, rather than in, up, or down.

Some log builders like to position the bow up-and-out, instead of flat out—and this can work well, too. One advantage is that the long grooves fall more in a vertical line (plumb above each other) throughout the length of the walls, and this can make installing doors and windows a bit easier, especially if a 2x4 spline, or key, is used. When the bow is flat out, then the grooves vary more from plumb. But, if an angle-iron spline is used, it is so much narrower that installing it plumb is usually easy.

4 — Rough-notch to even the gap

Another way to gain the most wall height possible with each log is to rough notch before you final scribe. This moves the widest gap towards the center of the log's length—midway between its end-notches.

Once you get the widest gap about centered on the wall, then you will gain the maximum wall height possible with each log (*Figure 8*). When the widest gap is

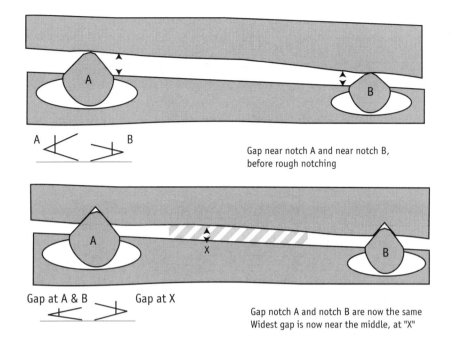

Gap near notch A and near notch B,
before rough notching

Gap at A & B   Gap at X

Gap notch A and notch B are now the same
Widest gap is now near the middle, at "X"

Figure 8: Before and after rough notching. At first, the widest gap is near Notch A; after rough notching, the gap between the two logs is more uniform, and the widest gap is about in the middle of the span between the end notches. The best location for the widest gap after rough notching is anywhere in the gray area.

near an end notch, then the long groove gets very wide and deep at the one end (builders call them dugout canoes!), and the loss to scribe is large—in other words, you waste wood.

Here's how to do it. Before rough notching, the widest gap is usually near one of the end notches. Rough notching lowers the end with the big gap more than it lowers the end with the smaller gap. Use two different rough-scribe settings to make the shape of the space (the gap) between the two logs more uniform, and the widest gap more nearly centered along the length of the wall. There will be much, much more about rough-notching later.

Summary

1) *Decide which logs will be used in the walls, and which in the roof, measure the wall logs, and calculate the average tip and average butt diameters.*

2) *On each wall, tips and butts alternate in every round to help keep walls fairly level.*

3) *Hand-peeled logs have unique shapes: bow, bumps, twists. When you scribe-fit two logs together along their lengths, you lose some of a log's diameter to cutting the long groove—about 1" (25mm) of height is lost.*

4) *Each log is rotated on the wall to put its straightest faces up and down (instead of in and out), and then it is rough-notched (pre-notched) to even the gap between the two logs that are about to be scribed.*

Figure 9: The rough notch has been cut and the log is rolled back right-side up on the wall.

Selecting The Next Log

Having covered the basics, we are now ready to consider log selection—you have a large pile of logs with varying shapes, tapers, and diameters. How do you decide which one of them to use next? To answer this question, I invented the Log Selection Rules that are now used by most log builders.

You will eventually use all your logs somewhere in the walls—but where is the best place to use the logs that are bigger than average? Where can you use logs that are smaller than average? Does it matter?

Yes, it definitely matters. Without my Selection Rules, log builders can have trouble getting a tip to cross (be notched over) a butt. Sometimes it is not a problem, but sometimes all work stops while someone figures it out—and the solutions (like reversing a log on the wall, butt for tip, or finding "the biggest tip we got!") cause future problems higher on the wall. After all, eventually you run out of the "biggest tip!"

Butts are big and tips are small, and if a butt leaves too tall a shoulder, then it can be impossible to find a tip big enough to safely cross over it *(Figure 10)*. I have seen it happen: sometimes a butt leaves a shoulder 13" tall, and the biggest tip of any log in the yard is 12½"—not even enough wood to get over the shoulder. There is no reason for this to happen. Let me show you how to avoid these problems.

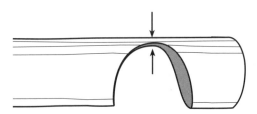

Figure 10: The most common log-selection problem: a tip that is notched over a large butt. This severely weakens the tip—the shoulder is too small, and there's not enough wood left to keep the end (flyway) from breaking off. Keep the T1 shoulder tall enough—that's the goal. Some builders call shoulder height the *head* or the amount *left up*.

Corner Order

Each wall alternates butt, tip, butt, tip (BTBT). When two walls meet (at a notched corner), the corner alternates:
butt, butt, tip, tip — butt, butt, tip, tip—BBTT—BBTT—BBTT *(Figure 11)*.

Corner order is always a repeating pattern of four different elements: first butt, second butt, first tip, and second tip, and then it repeats through this cycle again. We abbreviate these: B1, B2, T1, and T2.

A butt is not just a butt—we must know whether it is a *first* butt or a *second* butt.

Figure 11: Each wall alternates butt, tip, butt, tip (BTBT), and this makes the corner order BBTT as you count up from the bottom and the logs overlap each other, one after another. Butt 1 has different needs than butt 2, and so on. The arrows show the shoulder left by B2, the second butt, in the BBTT series. At the heart of log selection is keeping the B2 shoulder low enough that it will be easy to get a tip (T1) over it.

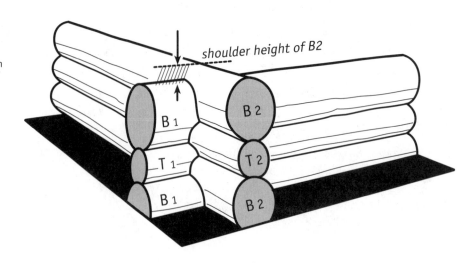

The first butt has different requirements than the second butt; the first tip is different than the second tip. When selecting the next log we must know where we are in the BBTT pattern.

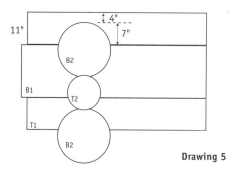

Drawing 5

Because the first tip is the most common and serious log selection problem *(Figure 10)*, my log selection rules start here. If the shoulder left by the second butt is too tall, then we won't be able to get a tip to go over it.

I will introduce log selection to you by using a "typical" set of logs, and then later we'll switch to formulas that can used for any set of logs—including all sets of logs that you will use. (For metric users, ignore the "inches" and just follow the arithmetic—adding and subtracting the numbers.)

Let's say you measured your logs and found that the average tip is 12" and the average butt diameter is 17". You want the second butt (BBTT) to leave a shoulder that will be easy for any first tip (BBTT) to cross. It would be best if first tips, once notched, still had a shoulder about 4 inches tall, so let's say that the rule for the first tip is: leave a shoulder about 4" tall *(Drawing 1, Figure 12)*. (As an exercise, please mark the un-labeled logs in these drawings as we go—B1, B2, T1, and T2.)

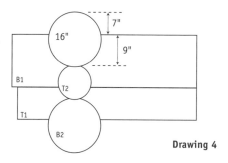

Drawing 4

Remember that we lose about 1" of diameter when we cut the long groove. Even though the tips average 12" in the log deck, they add only 11" of wall height (their *average gain*); and the butts that average 17" only add about 16" of wall height. All these drawings reflect the loss-to-scribe of 1" per log.

So, continuing on up in the corner, the first tip has left a shoulder of 4", and then you notch another log (it has to be a T2) over it that, on average, adds 11" of wall height (12" average tip minus 1" loss to scribe = 11"). You are covering a 4" shoulder with 11" of wood, so the shoulder that remains for T2 must be 7" (11" minus 4" = 7")—Drawing 2. Please study the drawings on this page to get a hands-on feel for what we are doing.

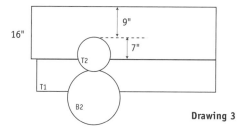

Drawing 3

Now you cross the 7" shoulder left by T2 with a butt (it has to be B1), Drawing 3. The average butt is 17", but the wall height you gain when you scribe a butt on a corner is 16", so you are crossing a 7" shoulder with 16" of wood. That means that B1 leaves a shoulder of 9".

Now you cross the 9" shoulder left by B1 with another butt (B2), Drawing 4. Once again, when we add a 17" butt to the wall we get only about 16" of wall height, so B2 leaves a shoulder of about 7" (16 - 9 = 7).

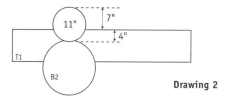

Drawing 2

B2 left a shoulder of 7" and you will cross it with a tip (T1) that on average adds 11" of height to the wall. This means that T1 will leave a shoulder of about 4" (11 - 7 = 4). And you are back where you started—you wanted T1 to usually leave a shoulder of about 4", and we started there, and now we are back to a T1 and it has a shoulder of 4"—Drawing 5. Great!

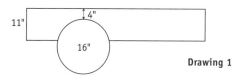

Drawing 1

If we had started the corner not planning ahead for the tip that must cross over a butt, then things would have turned out bad. We avoided log selection problems by starting with the shoulder that is left by the second butt (B2)—and keeping that shoulder low enough that a tip can cross it easily.

Figure 12: Example - average tip = 12"
average butt = 17"
average loss to scribe = 1"

Figure 13: Look at one wall:
it goes tip-butt-tip-butt-tip...TBTBTB

Look at the order of the overlap in the corner:
it goes tip-tip-butt-butt-tip-tip-butt-butt...
TTBB...TTBB

Desired Shoulder Heights

1st tip   Leave a shoulder about equal to one-third the average tip ($^1/_3$ T)

2nd tip   Leave a shoulder about equal to two-thirds the average tip minus the average loss to scribing ($^2/_3$ T - 1")

1st butt   Leave a shoulder about equal to the average butt minus two-thirds the average tip (B - $^2/_3$ T).

2nd butt   Leave a shoulder about equal to two-thirds the average tip minus the average loss to scribing ($^2/_3$ T - 1")

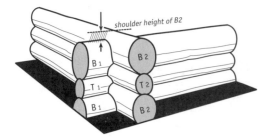

The Selection Rules Work for All Sets of Logs

You have just seen a log selection example that used 12" tips and 17" butts, but your set of logs will be different—how do you figure what shoulder heights you should use? Your logs might have 10" tips and 14" butts, or 9" tips and 19" butts. Now what do you do?

You always start with the first tip. In our example with tips averaging 12" we said: have T1 leave a shoulder of about 4". This is the same as saying: for T1, leave a shoulder that is about one-third the average tip. One third of 12" is 4".

If your tips average 9", then your shoulder height rule for T1 would be: T1 leaves a shoulder that is about 3" tall. If your tips average 15", then your T1 should leave a shoulder of 5" (one third of 15 = 5). If your tips average 13½", then your T1 should leave a shoulder of 4½".

The selection rule for FIRST TIPS is: T1 should leave a shoulder of about one-third the average tip diameter.

We will abbreviate these formulas to save space: "T" means "average tip diameter of your logs;" "B" means "average butt diameter of your logs;" and "L" means "average loss to scribe," which is 1" (25mm).

The log that gets notched over a T1 is T2. When we add T2, we are covering a shoulder that is ⅓T tall with T - 1" of wood, because every time we add a log to the wall, we do not gain it's full diameter as wall height, we lose a bit, so when we put a tip on the wall we get T - 1" of wall height (that is, it's actual diameter minus one inch).

This means that the shoulder left by a T2 is equal to ⅔T - 1". (This is algebra—which is just adding and subtracting things that stand for numbers. Take T - 1" and subtract ⅓T and you end up with ⅔T - 1"). This is much easier to see if you just plug in your average butt and tip diameters of your wall logs, and do not use the formulas—then you'll just be working with numbers, not letters.

So, the rule for SECOND TIPS is: T2 should have a shoulder about ⅔T - 1" tall. Try out this formula using our example with 12" tips: ⅔ of 12" is 8", now subtract 1", because our loss to scribe is 1", and you get T2 leaving a shoulder of 7". Compare this to Drawing 2 in Figure 12. It works!

Keep on going up the corner. The next log in the corner is a B1 (B1 always follows a T2). When we add a butt to the wall we will gain a B - 1" of wall height. So we have B - 1" of wood crossing over a shoulder left by T2 that is ⅔T - 1" tall. So the shoulder left by B1 equals butt minus loss minus ⅔ tip minus loss, or in shorthand: B - 1" - (⅔T - 1"). This equals B - ⅔T, and this is the rule for the shoulder left by B1.

Following B1 is B2. We have B - 1" of wood crossing over a shoulder that is B - ⅔T tall. The shoulder that this will leave is, on average, B - 1" - (B - ⅔T), which equals ⅔T - 1". The rule for B2 is: leave a shoulder about equal to two-thirds the average tip diameter minus the loss to scribe.

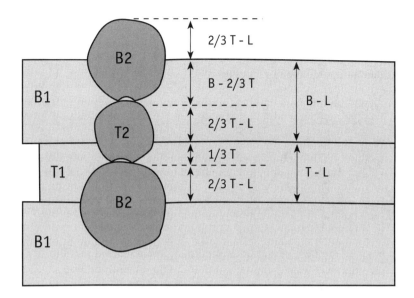

Figure 14: We want T1 to leave a shoulder of about $1/3$T. To do this, B2 must leave a shoulder of $2/3$T - 1". Now let's work our way up this corner.

The rules for T1, T2, and B1 follow directly from our choice for B2 because every adjoining pair of shoulders, when added together, must equal an average log diameter (either a B or a T) minus the average loss to scribing (L).

Can't quite follow this? This is the same drawing as the set in Figure 12, but this drawing works for *all* size logs, instead of just for set of logs with 12" tips and 17" butts.

You plug your average butt (B) and your average tip (T) diameters into these simple formulas and work out the desired shoulder heights for B1, B2, T1, and T2 for your set of logs.

Now we have the desired shoulder heights for all four types of logs: T1, T2, B1, and B2. Once you have measured your set of logs and know the average tip diameter and the average butt diameter you can just plug those numbers into the formulas and you will be ready to keep notched corners under control.

Out in the log building yard you don't use the formulas—just write the four numbers (like T1 = 4", T2 = 7", B1 = 9", and B2 = 7") on a scrap of plywood in an obvious place.

Note that the Selection Rules, though they based on the average diameters of your logs, are not guesses. They are real. If T1 is $\frac{1}{3}$T, then the shoulder height left by the average B2 really is $\frac{2}{3}$T - 1". Once you decide that T1 is $\frac{1}{3}$T, then all the other rules are determined—so, don't change the Rules willy-nilly.

Hint You can decide the shoulder height of T1, but then *ALL* the other shoulders must be what they will be.

My selection rules tell you what height shoulder to leave at each position in the BBTT series, but they do not tell you what diameter log you should use next. Let's start that right now. And it simpler than the last few paragraphs!

Using the Rules to Select Which Log to Use Next

Here is how to use the Rules to select which log to use next on your building. First, know where you are in the corner order: to do this, look at the last two logs notched—say they are a tip and then a butt (TB). Since the corner order is always TTBB, this means a butt has to be next, and it will be the second butt (TB<u>B</u>). So, you are selecting a B2 log.

From the Rules, a second butt should leave a shoulder of about two-thirds an average tip minus the scribing loss. If an average tip is 12", then two-thirds of that is 8". Now subtract the average loss to scribing (8" - 1" = 7"). The desired shoulder height for a second butt to leave is 7".

So, what diameter butt do we need to use in order to leave a 7" tall B2 shoulder? The answer is on the next page.

Steps For Selecting Logs

1) Know the average tip (T) and butt (B) diameters of your own set of logs.

2) Your average loss to scribing (L) is 1" (25mm).

3) What are you choosing? B1, B2, T1, or T2?

4) Use the Rules to find the desired shoulder height to leave with your next log.

5) Measure the existing shoulder that must be crossed.

6) Desired log diameter for the next log equals desired shoulder height (from the Rules) plus existing shoulder height (measured on the building) plus average loss to scribing (L). Add these 3 numbers together.

7) Repeat steps 3 to 6 for the other end of this log.

8) Go to the log pile knowing your desired tip and butt diameters for the next log. Also consider: the shape or profile of the log below, the height of parallel walls, and keeping walls level on top, when required.

Desired log diameter equals :
- *desired shoulder (from the Rules)*
- *plus existing shoulder (measured on the building)*
- *plus the loss to scribe (1", or 25mm)*

Add these three numbers together each time you are choosing which log to put on your wall next. Back to the example, add 7" (the desired shoulder for B2, from the Rules) to the existing shoulder (measured on the actual building), say it is 9", and therefore we need 16" of wood. Don't pick a log with a 16" butt, however, or you will end up short. Remember that about 1" of the new log's diameter will be lost to scribing and cutting the groove. So we actually need a butt that is 17" in order to get a 7" shoulder. Add three numbers: 7" + 9" + 1" = 17" to find our desired butt diameter.

The other end of this log is T2, and the Rules say the desired shoulder for a T2 is two-thirds of an average tip (8") minus the loss to scribe (1"), or 7". We measure the existing shoulder and it is 4". Add together the desired shoulder plus the existing shoulder plus the average scribe loss (4" + 7" + 1" = 12"). Now we can go to the log pile knowing that the log we want has about a 12" tip and a 17" butt, and, as always, has a shape approximately matching the shape of the log below, for easy scribing.

How Close is Close Enough?

Let's say you go to your log deck looking for a "12-17" log and you find an 10-15 and a 13-21, and nothing closer. Which should you use: the bigger one (13-21) or the smaller one (10-15)? It depends in part on whether the log you are choosing has a B1 or a B2.

If the log you are choosing has a B2, then don't go much bigger than the Rules recommend: remember that a T1 is next, and if the shoulder left by this B2 is 4" too tall (because it is 21" instead of 17" as the Rules suggested), then it could be difficult to get a tip to cross it next. You had better use the smaller of the two closest choices: use the 10-15 this time *(Figure 15)*.

If the log you are choosing has a B1, not a B2, however, then go ahead and use the bigger of the closest choices from your supply of wall logs—use the 13-21, in this example.

Figure 15: You are picking the next log and the Rules say a 12-17 is preferred. There is no 12-17 in the log deck, and the two logs that are closest are 10-15 (smaller than Rules say) and 13-21 (larger).

On the left is what would happen if you used the 13-21 log. On the right is what would happen if you used the 10-15 log. The 13-21 would leave a butt shoulder that is 6" taller than the 10-15. Which log should you use?

The answer depends in part upon where you are in the BBTT pattern. Is this butt you are choosing a B1 or a B2? If it's a B2 and a tip is next that extra 6" of shoulder is a big problem. If this butt is a B1, and so a butt is coming next, maybe you *could* use the larger log.

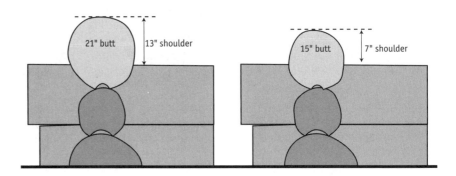

Don't be a slave to the numbers, just use them to help you stay out of trouble. In fact, think of your results from log selection as telling you which logs to *not* use next.

For example, I never re-shape a log to make it smaller so that it more closely matches the results of log selection. I almost never rough notch one end of a log down (*sink it in*, some builders call it, or *make a canoe out of it*) because it is bigger at that end than log selection suggested.

So what good are the log selection numbers? Simple: without them you wouldn't know whether to use the bigger log or the smaller log when you cannot find the perfect log. With log selection you do know which log is a better choice.

Other Things Can Affect Which Log You Select to Use Next

Selecting logs by their diameter is a great way to control shoulder heights, keep notches strong, and avoid selection problems like having to reverse a log on the wall—putting a butt where there should be a tip (we used to do this when we got into trouble). But, the diameters that the Rules suggest are not the only things to consider.

1- Put Big Logs Lower on Wall

Logs that are bigger than average usually look better lower on the wall, and logs that are smaller than average often look best a bit higher on the wall. Putting your biggest logs at the top of the wall, with small logs under them, looks tippy and unbalanced. (Though sometimes the plate logs are larger than average.)

2- Keep Parallel Walls the Same Height

Try to keep parallel walls about the same height as you build (within a few inches, or so)—it is not good to get up to your last round of logs and then find that one wall is 4" (100mm) shorter than the other. It's hard to put a roof on walls that are different heights! *(Figure 16)*

Hint   Use the Log Selection results to tell you which logs **not** to use this time.

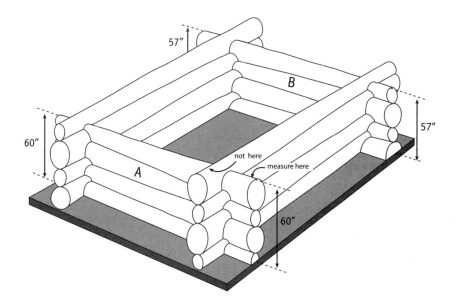

Figure 16: Walls A & B both have 4 logs, but they are not the same height, and they should be closer than this. By selecting the logs for both parallel walls (A & B) at the same time you can keep them close to the same height by adjusting the diameters recommended by the Rules. This is also how you can change the slope ("level-ness") of the top of the next round (the top of the logs you are selecting now).

Beginners often make the mistake of measuring the wrong log—you don't measure the *tallest* logs because you are not adding the next log to those walls. You are selecting logs for Walls A & B, and so it is the height of A & B that you need.

Also note that Walls A & B in this drawing have 4 rounds, not "3¹/₂" rounds.

Figure 17: A map to keep parallel walls about level. We are selecting logs X and Y—they are not yet on the building. Measure the current height of the walls you are adding onto (Wall X is now 69 and 72); to this add the log diameter the Rules say to use (Log X is 12-17); the answer (in squares) is how tall the wall would be if you used the log suggested by the Rules (for Wall X this is 84-86).

Now compare the parallel wall. Wall Y would be 82-83 if it got a 12-16 log next , like the Rules suggest. But then Wall Y would be shorter than Wall X by a few inches. Let's fix this. Slightly modify Log Y to be a bit larger in diameter than the Rules say in order to get Walls X and Y about the same height. Instead of a 12-16 for Y, try a 14-19.

In this example we are selecting logs that will be B1-T1 logs, so we decided to make log Y bigger than the Rules said. If we were choosing B2-T2 logs then we probably would have made Log X a few inches smaller than the Rules said (see How Close...? in this chapter). And sometimes you change *both* logs a bit in size, instead of just one log.

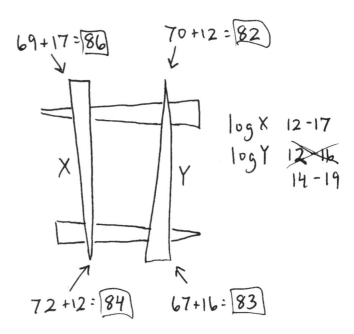

Figure 18: Making a map of how tall the walls are, and how tall they would be if you followed the results of your Log Selection Rules. This is the best way to keep parallel walls at about the same height, and level when you want them to be level. I draw a map like this for every other round of logs.

You could use a smaller-than-average log on a wall that is taller right now, and a bigger-than-average log on a parallel wall that is a bit low.

This is why I usually select more than one log at a time. In fact, I try to select *all* the logs within one course, or round, of logs on all parallel walls (this can mean two, or more, logs selected at the same time, depending upon your floorplan).

Measure the actual height of the ends of each wall, and then add the log diameters that Selection Rules say you should be using. Sketch a map that shows what the resulting heights would be *(Figures 17 & 18)*. In this way you can easily see how tall the walls will be if you used those logs, and then adjust them away from the Selection Rules, if necessary.

Remember that you need to sketch out this map, and adjust log diameters if necessary, for the north-south walls, and then after these have been cut and fitted, you need to do it again immediately for the east-west walls. A round of logs means adding one layer to all the walls, not to half of them.

3- Keep Particular Wall Logs Level
Using diameters that are different than what log selection recommends also lets you have walls level when you want them to be. It can be very useful to have window and door header logs, floor joists, and top-plate logs level, instead of sloped, on top. You can do this by using a log that has either more taper or less taper than average.

For example, if the wall you will be adding a log to is currently 69" tall at one end and is 72" tall at the other (3" of slope), and you would like the wall to be level on top after you finish fitting the next log, then you should select a log to use next that has about 3" of taper.

The map in Figure 17 is an example of this, too. Wall X is currently 69" and 72" tall at its ends—which means it slopes 3" right now. If you want the wall to be level on top after you fit the next log, then the next log should have 3" of taper. The Rules say to use a 12-17 (5" of taper), but if we use a 14-17 instead, then Wall X will be 86" tall at both ends after we are done scribing and fitting the next log. It is level!

Here's a quiz: in Figure 17, to make Wall Y level and the same height as Wall X (86" at both ends), what diameter log would you use next on Wall Y? Instead of the 12-16 suggested for Wall Y by the Rules, use a _____ and it will be 86" tall at both ends, like Wall X *(answer at the bottom of the page)*.

Automatically Corrects Itself

Is there a danger that fiddling with the results of the Selection Rules like this will screw things up? Not too much, because the Rules automatically adjust for shoulders that were left too tall or too low by the last logs, and you don't have to think about it.

Obviously, if you discover that one wall is 6" shorter than the parallel walls, then you should make up the difference in the next two courses, not with just one log, or else you would produce some very tall shoulders with the log you are adding. Better yet, pay closer attention by making the sketch map more often, and then you'll be adjusting diameters by just an inch or two, not half a foot! (Metric readers: "half a foot" is 6", or 150mm.)

If you were a bit low with the last shoulder, my Selection System brings you back up where you belong. Shoulder too high?—The next log will automatically bring you back down just the right amount.

Using the existing shoulder heights to help pick the next log is the key to my method's simplicity and success. The Rules do not require ideal conditions—they work with the actual logs you have in your building yard, and they adapt automatically to how you actually build. The Rules work for all sets of logs—perfect cedar gun-barrels, and gnarly Eastern white pine.

Wall Logs for Interior Walls

For interior log walls, it works best to use logs that have less taper, but have the same average diameter as the exterior log walls.

Here's what this means. Let's say the logs average 12" tips and 18" butts (average diameter 15"). For interior log walls use logs that are close to average size, and avoid using tips smaller than 12" and butts larger than 18".

In this example, good logs for interior walls would be 13-17 or 14-16. These still have an average diameter of 15", but they don't have the extremes of large butts or small tips.

Answer: choose a log that is 16-19 for wall Y.

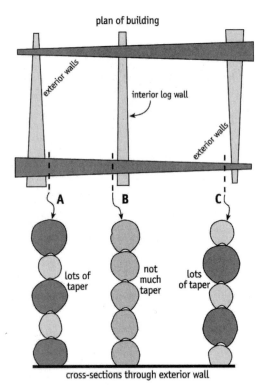

plan of building

exterior walls

interior log wall

exterior walls

A   B   C

lots of taper

not much taper

lots of taper

cross-sections through exterior wall

Figure 19: When selecting logs for the interior wall, use logs that have average diameter, but have less taper than average. For the logs of the interior wall, avoid butts that are much larger than the average butt and avoid tips that are much smaller than the average tip.

Because the logs making up the exterior wall have very little taper near the middle of their length (at B), it is best if the logs of the interior wall have about the same amount of taper as the logs of the exterior wall that they are notched into.

Note that when you are selecting a log for the long exterior walls of this house (the logs that have 3 notches), you measure the shoulder heights only at the ends of the wall.

Figure 20: Looking at "B," in Figure 19. Where an interior wall or a stub wall comes through a long log wall, avoid using small tips or large butts for the interior wall—instead, use logs that are more average in diameter. This is because the long-wall logs are average here, and don't have big butts or small tips.

Definitely avoid larger-than-average butts for the interior wall unless you know that the next log used for the exterior wall will be able to cover it without shoulder-height problems.

To select logs for the long walls in this sort of floorplan *(Figure 19)*, figure the diameter of the next wall log by measuring shoulder heights at the far outside corners. In other words, go to the log deck looking for a butt of a certain diameter and a tip of a certain diameter—don't try to find a log that also has a middle of a certain diameter. But, pay attention to the shoulder left by the interior wall—don't let it get too tall or you'll be cutting a lot of square notches to keep the long logs from breaking in half!

Keep the average diameter of the logs used for the interior wall close to the same as the average of the logs used for exterior walls so that the interior wall and exterior walls gain height at about the same rate *(Figure 19)*. Make wall-height sketch maps often enough that you keep walls on target.

the logs in this wall are all about the same diameter because they are near their mid-span

Log Selection Q&A

Log selection helps get the building started right (the sill logs), and it keeps you out of trouble as you choose which log to use next. Everything about log selection comes from the basics you learned in this chapter, but here are some common questions about log selection.

1— What if my set of logs are a mix of small and big logs?

Logs that are bigger than average usually look best lower on the wall, and logs that are smaller than average look best a bit higher up. If you look at your set of logs and can easily (without a tape measure) say "big" or "small" about each log, then I suggest that you divide your walls logs into two groups.

Find the average butts and tips for the large logs, figure the selection rules, and then use those logs to start the walls. Then, when the large walls logs are gone, measure the tips and butts of all the logs remaining (the smaller logs), figure the new selection rules for these, and continue on building the wall.

2— What about log selection at stub walls?

Where a stub log wall (an interior wall) joins a long log wall near the middle third of the long wall, then I use logs of average diameter for the stub wall. I watch out for the shoulder height that the interior wall log or the stub wall log leaves each time, remembering that it is the middle part of the long logs (and not their tips or butts) that must cross these shoulders.

For example, if a set of logs averages 11-19, then for a stub wall I'd try to use those logs that were more in the range of 13" for tips and 17" for butts. I do not use the smallest tips or the largest butts in a stub wall or interior wall. (Note that 11-19 and 13-17 have the same average mid-span diameter, 15", so the long wall and the short wall will gain height at the same rate, which is important.)

3— Should I use the formulas every time I choose a wall log?

You can, if you want. But it is easier to just write the four numbers for T1, T2, B1, and B2 on a piece of plywood near the building, and use them. Then you just have to add three simple numbers: existing shoulder plus desired shoulder plus 1", and there's no need to refigure the formulas every time.

4— Is there a way to do this without the formulas?

Yes. As always, start with the Rule for T1. Take the average tip diameter and subtract the loss to scribe. Let's say the average tip is 12"—subtract 1" and you get 11", which is the average gain per tip.

One third of 11" is 3⅔"—round this up to 4". Make this the Rule for T1—leave a shoulder on T1 that is about 4" tall.

The Rule for T2 will be equal to 11" (the average wall height gained with a tip) minus the Rule for T1— or 11" minus 4" equals 7". That's the Rule for T2— leave a shoulder that is about 7" tall.

Figure 24: These large logs don't taper much, so log selection is easy. logs with lots of taper are more difficult to use, but my Selection Rules will keep you out of trouble.

Take the average butt diameter and subtract the loss to scribe. Let's say the average butt is 19" in diameter; subtract 1" and you get 18", which is the average gain per butt.

The Rule for B1 will be equal to 18" minus the Rule for T2— or 18" minus 7" equals 11". That's the Rule for B1— leave a shoulder that is about 11" tall.

The Rule for B2 is 18" minus the Rule for B1— or 18" minus 11" equals 7". That's the Rule for B2— leave a shoulder that is about 7" tall.

T1 T2 B1 B2 T1 T2 B1 B2...

Hint   ø is the symbol for *diameter*

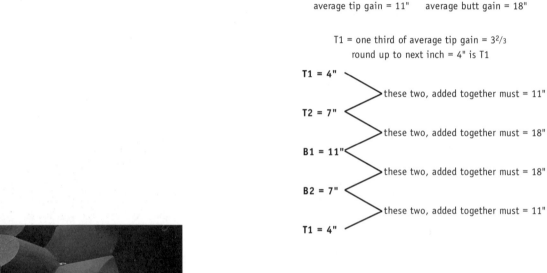

average tip ø = 12"   average butt ø = 19"
loss to scribe is 1"
_____
so
average tip gain = 11"   average butt gain = 18"

T1 = one third of average tip gain = $3\frac{2}{3}$
round up to next inch = 4" is T1

T1 = 4" ⟩ these two, added together must = 11"
T2 = 7" ⟨
⟩ these two, added together must = 18"
B1 = 11" ⟨
⟩ these two, added together must = 18"
B2 = 7" ⟨
⟩ these two, added together must = 11"
T1 = 4"

Figure 25: A round-notch corner in a building from the 1930's. Now that you know what you're looking for: a low shoulder height left by B2, you can study corners that you see in finished homes. Can you see the BBTT, BBTT pattern here? Every scribe-fit log building has it.

5— Should I reshape a log to change its diameter to match the log the Rules say I need?

No. The log selection result is to keep you out of trouble, and you have some flexibility. I never re-shape logs.

Sometimes I sink one end of a log down by rough-notching the end that would leave too high a shoulder. But I do this only very rarely (like one log in 3 buildings). I do this when my log deck is almost empty and I don't have much choice between logs of different diameters and tapers.

6— Are there other sets of Selection Rules?

Yes. The Rules given in this chapter work well—they are based on leaving one-third of a T1 uncut by its notch. You can use them with confidence.

Professional builders sometimes want more flexibility with shoulder heights. I have written about alternative sets of Selection Rules in *Log Building News* #18.

Here are some other sets of Rules that will work for 12-19 logs. Instead of using formulas, I just start with T1, as always, and then work out what the other Rules would be if I started with 4", or with 5", or with 5½" for T1.

Any one of these sets of Rules will work for logs that average 12-19. But you cannot mix-and-match between the sets—you must use all the Rules that are in one box. Changing T1 from 4" to 5" changes all the other Rules below it.

average tip ø = 12"   average butt ø = 19"
loss to scribe is 1"
_____
                    so
_____
average tip gain = 11"   average butt gain = 18"

Set 1	Set 2	Set 3
T1 = 4"	T1 = 5"	T1 = 5½"
T2 = 7"	T2 = 6"	T2 = 5½"
B1 = 11"	B1 = 12"	B1 = 12½"
B2 = 7"	B2 = 6"	B2 = 5½"

Note how B1 plus B2 equals 18" in all 3 sets of Rules. T1 plus T2 always equals 11" in all three sets of Rules. T2 plus B1 always equals 18". Two adjacent Rules, when added together, must equal either the average gain per tip or the average gain per butt.

You can download my professional-builder log selection calculator (Excel spreadsheet) on my www.LogConstructionManual.com website. This will let you plug in your average butt and average tip, and will give you the Rules that I suggest are best for your logs. The Rules may differ slightly from those found in this book because the on-line calculator will use "alternative" sets of Rules.

7— Should I use different loss-to-scribe amounts for different logs?

No. It is true that straight, smooth logs have shallower, and more uniform, long-grooves, and crooked, bumpy logs have curvy, and deeper grooves *(Figure 26)*—and slightly larger loss to scribe.

But of all the buildings I have built, and in all the builder's yards I have taught, the loss-to-scribe has varied only from about ¾" (super smooth and straight logs) to 1¼" (super bumpy and bent logs). So, I always use 1" for the loss-to-scribe—which means I'm never more than ¼" (6mm) off. And because we don't need accuracy any better than ½" (12mm) for shoulder heights, log diameters and log selection, 1" (25mm) works nicely all the time for the loss to scribe.

Figure 26: Straight, smooth logs have long grooves that are quite uniform in width (left). Crooked, bumpy logs can get wildly-shaped long grooves (right). Both can be made to fit perfectly.

Sill Logs

Now that you know the basics of building log walls, and have a handle on selecting the next log, we will tackle the *first* logs—the sill logs.

Sill logs are the most important logs in the walls. If you get started right, then the walls will be easy. If you start the sills wrong, then you'll be scratching your head every day.

How important are sill logs? Log-selection problems that appear at waist height, or higher, are usually caused by wrong shoulder heights in the sill logs. When you are four rounds up, and you have a B2 with a shoulder of 14" (36cm) and there is no tip (T1) in your yard that can possibly cover it (your largest tip being, say, 12½"—32cm), it is natural to blame that particular B2 for being the source of the problem. But it is often the sill logs that are partly, or even completely, to blame—that's how important sill logs are.

Note that several popular log building books have directions for cutting sill logs, also known as the *base round*, that start the building wrong—with bad shoulder heights in the corners.

In this chapter we'll start with sill log basics—like which way to point the butts and tips. We'll study the options for controlling when certain wall logs will be level. Next we'll learn how to cut each sill log—how big to make the sill logs at both ends.

Then, having covered the basics (and probably 95% of all log homes), we'll move on to complicated floorplans that have lots of walls, hexagons, and buildings with odd-numbers of exterior walls, like prows.

Starting a Building Right

Figure 1: This window header is at a poor height—in the long groove. Headers work best in the middle half of a log's diameter because the log is strong there, and the flat produced is wide enough to cover the settling boards. This header will guide water right back into the building.

Sill logs are the key to several important log-wall concerns:
1) Keeping shoulder heights out of notching trouble;
2) Making particular wall logs level on top; and
3) Making door-header logs a good height.

We know about shoulder heights (#1, above) in detail already—that's the Log Selection Rules. We have touched on the fact that it is best to have the top surfaces of certain wall logs level (#2 in list above), like door headers, floor joists, and roof plate logs.

Number three is new to us, but is very important. Doors are available in one standard height: 6'-8". When you add door jambs, framing lumber and settling space, the head of most log wall openings for doors is about 88" or 89" (225cm) off the finished floor. If you cut the sill logs without thinking ahead then you may end up with the situation in Figures 1 and 2—the header log is at a poor height for installing doors.

Windows don't usually have this problem. First, unlike doors, windows can be bought in a wide variety of heights, not just one height. Second, windows can be varied slightly in their vertical position in the wall so that both the window sill log and the window header log have flats that are wide enough, look good, and don't weaken the logs too much *(Figures 1 & 3)* .

How Many Logs Are in a Wall?

First, you must know how many courses of logs your walls will have. In the old days (the 1980's!), we used to just keep adding logs until the walls were tall

Figure 2: Plan ahead so that 88" or 89" (225cm) off the floor comes out in the middle half of the header log. On the left, the door header is in a bad place on its log. On the right, a few inches was left on the sill log and now the door header comes out better.

The logs in both drawings are the exact same logs. The sill log is the *only* difference between these two walls: a few extra inches were left on the bottom of the sill log on the right.

In other cases, depending upon the average wall-gain of your set of logs, you will cut the sill logs normally, or perhaps lower than normal. Plan ahead—sill logs really do matter!

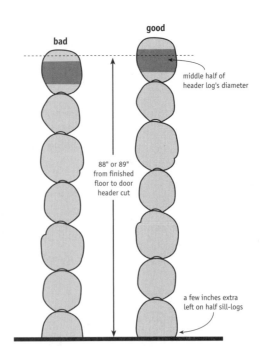

enough. "Too low to put the floor joists in for the loft?—Oh well, maybe the next round will be high enough." This method wastes time, money, and logs—and you also lose control over having logs level when you want them to be.

Let's say you planned for Round #9 to be level and to be the top plate logs, but you get to Round #9 and the walls are only 8'-2" tall (way too low after settling). So you add another course of logs and make Round #10 the plate logs, and now they are 9'-4" off the floor. Problem solved? Well, one problem is solved, but you have caused a new problem! Now the tops of the plate logs are 4" out of level because you planned for Round #9 to be level, not Round #10.

Here's how to avoid this kind of problem. Use your average log diameter to figure how many logs you need to get a log wall that is the minimum height needed *(Figure 5)*.

For example, if your logs average 12" tips and 17" butts, then your average log diameter is 14½" (because 12 + 17 = 29, divided by 2, equals 14½"). Remember that you won't gain 14½" of wall height every time you add a log. You lose about 1" every time you scribe and cut a long groove, so you'll gain about 13½" per round with this set of logs. Now, how many logs do you need to get your walls to 9'-0"? Divide 9'-0" (108") by 13.5" per log and you get 8 logs. With logs this size, you would need 8 rounds to reach a wall height of about 9'-0".

Also figure out which round of logs will have second-floor (loft) log joists. The flooring groove (or "slot") works best if it falls in the upper-middle of the wall log that is the same round as the joists *(Figure 4 and Color Figures C-28 and C-29)*. Also consider any log beams that might be below the floor joists to support them—they must be high enough off the floor, after settling, to avoid headroom problems. Finally, door headers need to be a certain height—and as we have seen, it is best if you are in the middle half of a log when you are at 88" or 89" (225cm) off the finished floor *(Figure 2)*.

So, if your door headers (88") would come out at a long groove, then you could make all the sill logs 3" taller (by using bigger logs to cut the sills from)—this would raise the door headers to be in the middle of a course of logs *(Figure 2)*. Or you could make all the sill logs a bit shorter if needed, depending upon your average gain per wall log.

Figure 3: Photo of header cut in good location. Drawing of a header log that is cut too thin.

It is much better to have 88" to 89" (225cm) fall in the middle half of a header log (photo) instead of the upper one-quarter of the log (drawing). This problem can is easily avoided by planning ahead before cutting the sill logs

Figure 4: Log floor joists and the flooring grooves that are sawn into the wall logs to support flooring, and conceal wane. It is best if the wall logs that get the flooring groove are about level on top—if they are, then there is a constant distance between the top of the floor groove and the top of the wall logs. This requires that the floor joists be in a round that is about level on top.

As you can see in this photo, if the floor groove were any higher it would be too close to the long groove joint in the left wall logs; and if it were any lower it would be too low in the right wall log. If the tops of these particular wall logs are not level, then achieving this balanced location can be very difficult.

See also Color Figures C-28 and C-29 for more floor slot photos.

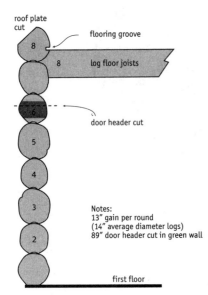

roof plate
cut

flooring groove

8

8   log floor joists

door header cut

6

5

4

Notes:
13" gain per round
(14" average diameter logs)
89" door header cut in green wall

3

2

first floor

Figure 5: In this building, door headers are in Round #6, floor joists are Round #8, and the roof plate log is Round #8. Six and eight are even numbers, so I would cut the sill log to make even-numbered rounds about level on top.

Which Courses Should Be Level?

Plan ahead: know your average log diameter, and then make some simple drawings of the height of door headers, floor joists, support beams, and plate logs to figure out which course, or round, these will be in (*Figure 5*).

It makes sense for the top wall log (the roof *plate* log) to be level, or close to level, on top. After all, it is tough to put a roof on a wall that slopes.

Now, which rounds of logs will have the loft log floor joists, and which round will have door headers cut into them? Are they in Round #7? Round #6? It all depends upon the average diameter of your logs, and how tall you cut your sill logs.

Once you know how many logs will be in the walls of a building, and which courses will have the plates, headers, and joists, then you can plan ahead so that these important rounds of logs will be about level on top. There may be other courses of logs that would be nice to be level, but door headers, floor joists, and plate logs are the three most common rounds to consider.

Even Courses Level

In our example here (*Figure 5*), the three important courses are in Rounds #6 and #8, and since 6 and 8 are both even numbers I would start this building with sill logs cut so that even-numbered courses are about level. This means that Rounds 2, 4, 6, and 8 will be about level on top (*Figure 6a*).

Odd Courses Level

In another building, or with different diameter logs, you might find that headers, joists, and plates are in Rounds #7 & #9 (odd numbers), and then you would probably start the sill logs so that the odd-numbered Rounds 1, 3, 5, 7, and 9 are level on top (*Figure 6b*).

Middles Level

Or you might find that headers, joists, and plates are a mix of even-numbered and odd-numbered rounds (like 7, 8 and 9). I would start a building like that with the horizontal middle of all logs about level (*Figure 6c*). In fact, many professional builders start all their buildings with middles of logs level, just to keep things simple for their crews, and this is often an acceptable compromise.

Figure 6: The three options are: even-numbered rounds about level on top, odd rounds level about on top, or split the taper so the middle of each log is about level. (Dashed lines are level lines.)

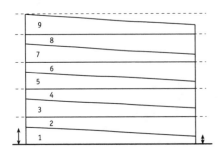

6a: Sill log was cut so that the top of all even courses are about level.

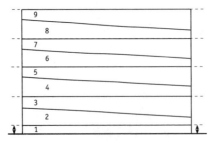

6b: Sill log was cut so that the top of all odd courses are about level.

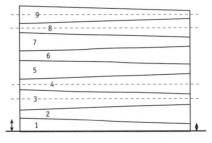

6c: Sill log was cut so that the horizontal centerline of every log is about level.

Summary

Every sill log is in Round #1, and "1" is an odd-number. The sill logs can be level on top, or they can slope with the average taper, or they can "split the taper."

You need to know two things: 1) the height of certain rounds (door header logs, roof plates logs, joists, joist support beams, and so on); and 2) which rounds of logs that door headers, joists, and plates are in (odd, even, or both?); and then you can decide which rounds you want to be about level on top.

Who would have guessed that high wall logs like door headers, floor joists, and roof plate logs would owe so much to the very first round of logs, but they do.

Sill Log Basics

There are many alternative ways to lay out and cut sill logs to accomplish different goals. It can become a complex, but we will keep things simple to begin with. You can use the information in this chapter to lay out and cut the sill logs for virtually all log homes.

Please note that I have been thinking and writing about sill logs and log selection for years. The best, most simple and useful system is the one found in this book.

All log homes have exterior walls that are made of horizontally-stacked and fitted logs. Some homes also have interior log walls. First, we are going to focus on getting the sill logs of exterior walls right *(Figure 7)*, and after this I'll show you what to do about the sill logs of interior walls *(Figure 8)*.

We will begin with floorplans that have 90° corners and are simple combinations of rectangles and squares. Then we'll look at complex combinations of rectangles. Then we will move on to buildings with walls that don't have 90° corners, but still have an even number of exterior walls, like a hexagon, which has 6 walls.

Finally, we'll tackle the toughest floorplans—those that have an odd-number of exterior walls, floorplans with prows, and more. Buildings that have an odd number of exterior walls require special treatment, and a *special* sill log—and they are best left to experts.

Figure 7: All the drawings in this chapter show only the sill logs. So that you can tell the half-log sills from the whole-log sills, the 'wholes' are shown as going over the 'halfs,' just as they do with real sill logs, and the whole-log sills are a darker gray.

Figure 8: Some log walls are exterior and some are interior. First, figure out sill layout for all the exterior walls, and then, when that has been done, decide on sill layout for the interior walls. Do this on paper before you start cutting logs.

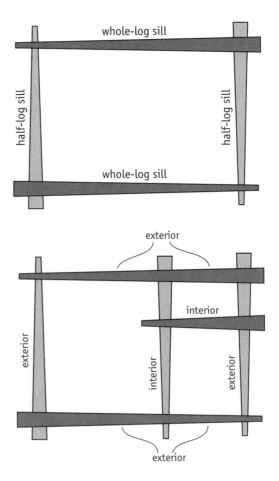

Buildings with an Even Number of Exterior Walls

Floorplans that are made up of combinations of rectangles and squares always have an even number of exterior walls. This is true no matter how many walls there are, and no matter whether the squares and rectangles are large or small, narrow or wide. For such buildings, 50% of the sill logs are 'half-logs' and 50% of the sill logs are 'whole-logs,' or sometimes called *three-quarter* sill logs.

As recently as the 1950's, all sill logs were full logs, and the masonry foundation stepped down on some walls to accommodate them. Now, however, nearly all log buildings are set on a plywood deck (just like conventional 2x4 houses). Since the floor deck is all in one plane, not stepped, we have to cut some sill logs about in half in order to get the corner notches started with the right amount of overlap.

Half & Whole Alternate
Start with any wall, and count around the exterior walls in either direction (clockwise or counterclockwise) and you'll see that the sill logs form a chain that alternates: half-log, whole-log, half-log, whole-log, half-log, and so on *(Figures 7 & 9)*. When there are an even number of exterior walls in a building with 90° corners, then its parallel walls will always be the same: either all half-logs or all whole-logs.

We'll cover much more about these 'half-log' sills soon, all you need to know for now is that the walls that start with the 'half-log' sills are parallel to each other—and you can see this in the drawings that follow.

Start Each Corner with Either Two Butts, or with Two Tips
At every exterior corner, sill logs start with either two tips, or with two butts, and never with one butt and one tip *(Figures 7 & 9)*. While you can start log buildings with tip-butt combinations (and some builders do), there is no question that starting the sill logs in each corner with either two tips or with two butts is superior in important and useful ways to starting corners with a tip-butt combination.

Point the tips "Parallel-Opposite"
When you start the sill logs of a simple rectangular building with either two tips or with two butts in each corner, then the tips on parallel walls will point in opposite directions *(Figure 10, right)*.

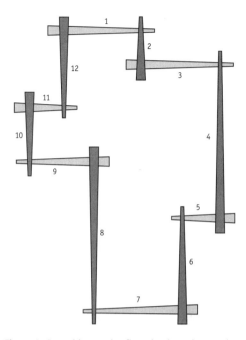

Figure 9: Even this complex floorplan is made up only of rectangles and squares of different sizes. There are always an even number of exterior walls in a floorplan that has all 90° corners. This house has 12 exterior log walls (I am not showing any interior walls in this drawing). The sill logs of exterior walls form a chain that always alternates: half, whole, half, whole, half, and so on, no matter where you start counting, and this means the half-log sills are always on parallel walls. In this building, all east-west walls start with half-log sills (light gray), and all north-south walls start with whole-log sills (dark gray).

The "B2-T1 Problem"

We want each log in the walls to be either a B1-T1 log or a B2-T2 log. If you check the drawings that show sill corners starting with two tips or two butts you'll see this is true. When started like this, every log in the entire building, all the way up all the walls, will be either B1-T1 or B2-T2.

If we start a corner with a butt and a tip, however, then we cannot avoid having other combinations, like B1-T2 and B2-T1 logs. The B2-T1 log is a problem because it does not slope the same as all the other types of logs (B2-T2, B1-T1, and B1-T2).

In fact, I don't use the "parallel-same" sill log layout because it always has a B2-T1 log—and that's why all its walls cannot be level at the same time.

You can start parallel walls pointing the *same* way, and my Selection Rules will still work. But, starting "parallel-same" makes it very difficult to have all the walls in a log building level at the same time. I like to have all the logs about level on top for the door header round; and I also like all the top-plate logs on all the walls to be about level. With "parallel-opposite" this is easy.

Some builders don't like that the top-plate logs of "parallel-opposite" buildings have one butt and one tip pointing out on each gable end. Plates logs are often cantilevered to help support gable-roof overhangs, and the building looks more symmetrical if the plate logs on a gable end are either both butts or both tips. But this is easily fixed by cantilevering out the top two logs on each wall of a "parallel-opposite" building, instead of just the top log on each wall—this fools the eye into thinking both walls are the same.

Parallel-opposite is a pattern that is easy to see in a simple rectangle floorplan like Figure 7. Now, take another look at the complex rectangle floorplan of Figure 9 — the parallel-opposite pattern is not easy to see. But, as long as every corner starts with either two tips or with two butts, then sill layout is correct.

Figure 10: A 4-wall building with the sill logs laid out so that parallel walls have tips pointing the same direction (left), or pointing opposite directions (right). The superior sill layout is "parallel-opposite." Each corner starts with either two tips or with two butts, & not with a tip-butt combination as in parallel-same layout.

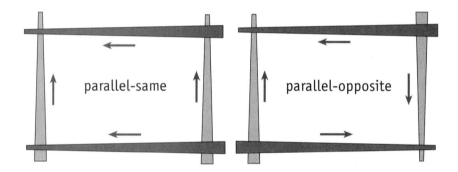

Sill Layout Rules

The basic rules of sill-log layout for buildings with an even number of exterior walls are:

1- *At all corners of exterior log walls, sill logs start with two tips, or with two butts, and never with a tip-butt combination.*
2- *Half-log sills and whole-log sills alternate as you count around the exterior perimeter of the walls.*
3- *Cluster butts together and cluster tips together in the sill layout.*

Rules 1 & 2 we've already been using. As a floorplan gets more complicated or when there are interior log walls, then we'll need Rule #3, too: clustering. In the floorplan in Figure 11, Sill Logs A, B, E, and 2 all have their tips in the same cluster, and Sill Logs C, D, and 2 have their butts clustered together.

Clustering is what protects each log added to Wall #2 from having shoulder-height problems. To visualize this, imagine if the sill log of Wall B were reversed. Then the

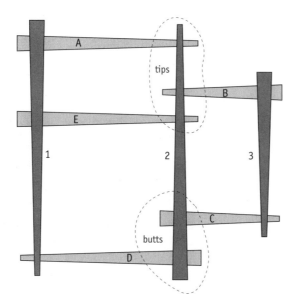

tip end of Log #2 would have to cross a butt (Wall B) and also a tip (Wall A). By laying out the sill logs with the tips clustered as shown, the tip of Log #2 crosses a tip at Walls B & E, and also a tip at Wall A—easy.

The sill logs of every building that has an even number of exterior walls, no matter how complex, and no matter how many walls and corners it has, can be set up by using these three guidelines. But do not pick and choose which guidelines to use, and ignore some of the others—every sill-log layout must use all three guidelines.

Start with the largest rectangle of the structure and work out its sill log directions, and decide which walls start with half-logs and which with whole-logs. Then move on to the smaller rectangles that have exterior log walls. After all the exterior walls are decided, then move on to sill log layout for the interior log walls.

Interior Log Walls

Working on your paper-and-pencil sketch of the sill logs, complete the layout of all the exterior walls before you start the layout of interior walls and stub walls.

If there's an interior log wall in a rectangular floorplan, then orient the butt and tip of the interior log wall so it matches the exterior log wall that it is closest to (*Figure 12*). Clustering butts together and clustering tips together does this for you.

If an interior wall is exactly half-way between the exterior walls, then it doesn't matter which way the tip and butt point for the sill log of the interior wall—either orientation will work.

I use less-tapered logs on interior walls. If your logs average 12-17, then try to use 13-16 logs for interior walls, and avoid using 11-18 logs. All three logs have the same average gain per log, but by using less taper and smaller butt diameters, shoulder heights won't force you to use square-notches that might otherwise be needed to keep the 3-notch logs from being weak.

Stub Walls

Even when an interior wall is not continuous, but is just a stub wall at each end, everything said here about butt-tip orientation and clustering is still true.

Figure 11: In complex floorplans, lay out the sill logs to cluster butts together, and cluster tips together. This is what makes an interior log wall, Wall E, point the same direction as the closest parallel exterior wall (Wall A).

Does this floorplan match the 3 sill-log guidelines? Yes.

1- Every exterior corner starts with either two tips or two butts.

2- Half-log sills and whole-log sills alternate around the outside perimeter.

3- Walls C, D, & 2 have their butts clustered near each other, and Walls A, B, E, & 2 have their tips clustered.

For sill log layout purposes, first work out the biggest rectangle of the floorplan: Walls A, 2, D, & 1. Then work out the smaller rectangle, Walls B, 3, C, & 2.

Wall 2 seems to be both an exterior wall and an interior wall, but starting sill log layout with the largest rectangle before moving to the smaller rectangle solves this.

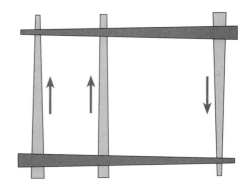

Figure 12: Orient the tip and butt of an interior log wall to be the same as the closest exterior log wall.

If the interior wall is exactly in the center, then you can point the tip and butt of the interior wall either way.

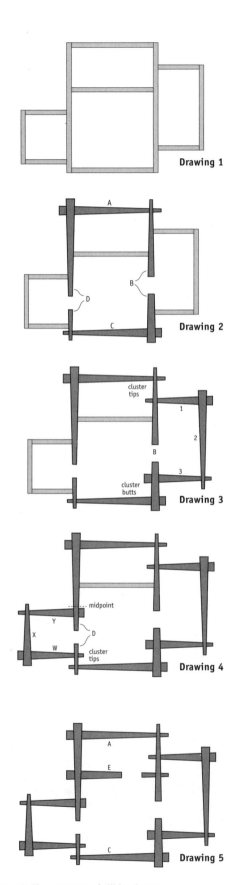

Example Layout

Here is an example of how to use the Sill Guidelines to lay out the sill logs for an actual floorplan, step-by-step *(Figure 13).*

Drawing 1

The concrete foundation of the house we are going to build.

Drawing 2

Start sill-log layout with the largest rectangle of the building (A-B-C-D). Each corner must start with either two tips, or with two butts. I usually start the gable ends (the short walls) of the largest rectangle with the half-log sills. The exterior perimeter walls alternate: whole, half, whole, half. Because this building has an even number of exterior walls, the parallel walls point opposite directions.

It's up to you which corner of the largest rectangle starts with two butts. I have put two butts in the upper-left corner (where Walls A & D meet), but you could have put two tips there—and then you would change all the other corners of the A-B-C-D rectangle to match.

Drawing 3

Lay out one of the smaller rectangles next—we'll do rectangle 1-2-3-B. Wall 1 is a half-log sill, not a whole-log sill, because all the half-logs run east-west in this building. All the north-south walls are whole-log sills in this building.

Wall 1 has its tip end clustered with other tips. Wall 3 has its butt end clustered with the other butt ends. With Walls 1 & 3 set, now we know how to orient the butt and tip of Wall 2: every exterior corner must start with either two tips, or with two butts.

Drawing 4

Lay out the sill logs of the smallest rectangle (W-X-Y-D). Wall W has its tip end clustered with the other tips, and it is a half-log sill because it is east-west, like all the half-logs in this building.

Wall Y is parallel to Wall W, so its tip must point the opposite direction (parallel-opposite). The midpoint of the length of Wall D is shown, and Wall Y is actually a bit closer to the tip of Wall D than it is to the butt of Wall D, so you might think that Log Y should be flipped end-for-end. We can't do that, though, or Walls W and Y would be parallel-same, and we don't want that. Finally, Wall X is oriented so it has two tips in one corner and two butts in the other, and it is a whole-log sill because it is north-south. Now, just one more log to lay out…

Drawing 5

Decide the sill-log layout of any interior log walls last. Wall E is closer to Wall A than it is to Wall C, and so it is oriented to point the same way as Wall A—cluster tips together and cluster butts together. Wall E is east-west, so it is a half-log sill.

Sill layout is complete.

Figure 13: The sequence of sill log layout.

How to Rip the Sill Logs

Next, we need to decide how tall to cut each sill log. In a nutshell, our goal is to have the shoulder heights of the whole-sill logs, after they are notched and in place, match the shoulder heights found in the Selection Rules for our logs.

Each sill log is flattened on its bottom, of course, so that it can sit on the foundation, or more often, on a plywood subfloor. The *Log Building Standards* (see Appendix, Section 2.C.2) recommend that the sill-log flats be sawn so they are 4" wide, or wider, everywhere along their length. This is so the logs are stable (won't rock) and will be easy to weatherproof. Usually the sill logs sit on a foam seal or other gasket that prevents wind and water from getting between the logs and the plywood subfloor.

The next step is actually choosing which logs to use, and ripping them flat at the correct heights. To choose which logs to use, and know how tall to cut them, you need to know the desired shoulders for T1, T2, B1, and B2, and the average tip diameter and average butt for the set of logs. If you haven't done that yet, do it now.

If you use the 3 guidelines for sill-log layout, then no matter how many walls are in your building, there are only two types of sill logs: half-logs (which are B1-T1) and whole-logs (which are B2-T2). You won't need to use formulas for every log, because all the half-logs have the same heights and all the whole-log sills are the same, too *(Figure 14)*. Only the lengths will vary—to match the length of each wall.

As an example, we'll use a set of logs that have 12" average tips and 17" average butts. Also, we'll make the center of each wall-log about level (not even-level or odd-level).

Just plug in your own average tip and average butt diameters into each formula in Figure 14 to come up with a sill-log cutting list.

For Sill Logs A and B (the 'half' logs in Figure 14) the tip end should be cut so it is ½T tall. The average tip (T) is 12", and one-half of 12" is 6", so we would cut the tip ends of Logs A and B to be 6" tall *(Figure 15)*.

The butt ends of the half-log sills should be half an average butt. Our average butt is 17", and half of this is 8½". So the butt-ends of Sill Logs A and B are 8½" tall.

Figure 14: Sill log layout to have the horizontal center of each log about level in every round. There is a lot of information about sill-log layout in this drawing—it's worth studying.

1) Corners start with either two tips or with two butts, and NOT with a tip-butt combination.

2) The top surface of every log slopes from end to end with about half the average taper of your logs. If the average taper is 5" (12" tips and 17" butts), then the top of the average log slopes about 2½".

3) The two half-logs (A & B) come from one straight log, slightly larger than average size, ripped in half.

4) The two whole-log sills (X & Y) are cut the same size. They come from two straight, larger-than-average logs.

5) The shoulders left up at all four corners are the same height (check the Rules and you'll see that T2 and B2 have the same formula, $^2/_3T - 1$).

6) The height formulas for Logs X & Y come from adding the height of the log below ($^1/_2T$ for the tips and $^1/_2B$ for the butts) plus the shoulder height for B2 and T2 from the Rules (which is $^2/_3T - 1$).

Here's what that means: for the tip-end of Logs X & Y, you add $^1/_2T$ plus $^2/_3T - 1$, and that equals $^7/_6T - 1$. For the butt-end of Logs X & Y, you add $^1/_2B$ plus $^2/_3T - 1$, and that equals $^1/_2B + ^2/_3T - 1$.

METRIC USERS: instead of subtracting 1, you should subtract 25mm.

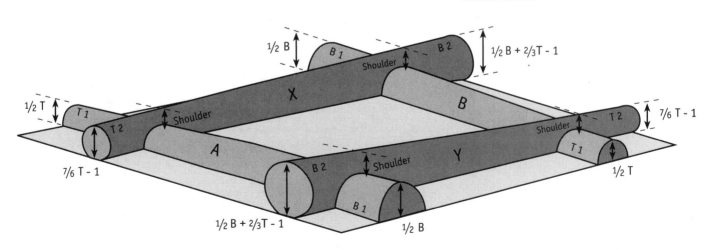

Figure 15: Half-log sills (A & B) for a building with 12" average tips and 17" average butts, and with the middle of every log level. You just plug your average diameters into the formulas shown in Figure 14. For example, for the butt-end of this sill log, the formula is ½B, so, half of 17" equals 8½". You try the tip-end.

Figure 16: Whole-log sills (X & Y) for a building with 12" average tips and 17" average butts, and with the middle of every log level. You just plug your average diameters into the formulas shown in Figure 14. For example, for the butt-end of this sill log, the formula is ½B + ⅔T - 1", or half of 17" plus two-thirds of 12" minus 1", and that equals 15½".

Hint	Here's how to multiply a number by a fraction:

We want to know ⅞ times 12

First multiply 7 times 12, and then divide the answer you get by 6

So, 7 x 12 = 84, then 84 ÷ 6 =14

So, ⅞ times 12 equals 14

If you had a log that was 13" diameter at the tip and 18" at the butt, then you could cut it in half, and end up with two half-log sills each about 6" at the tip and 8½" at the butt. (Allowing 1", 25 mm, that is lost when you rip and flatten them.)

In fact, that's how I pick a log for the half-log sills—I choose a straight log that is about 1" bigger than average at both ends to get two half-logs from it.

Sill Logs X and Y are next. I call them 'whole' log sills, even though they are flattened some, and so are not really whole. (Some builders call these the *three-quarters* sills, though they aren't really ¾ either!)

The tip ends of the whole-log sills are cut so they are ⅞ T - 1" tall. Our average tip is 12", so multiply 12 by ⅞ , and then subtract 1" (the loss to scribe) and you get 13" *(Figure 16)*. The butt ends of the whole-log sills are supposed to be half an average butt plus two-thirds an average tip minus the loss to scribe. In our example this would be 8½" + 8" - 1" = 15½".

So, Sill Logs X and Y would be 13" at their tips and 15½" at their butts measured from round to flat. Choose two logs that are about 1½" or 2" bigger than this (14½" tips and 17" butts) so that when you cut them down to the correct heights, they have a continuous flat that is 4" wide, or wider, where they sit on the subfloor.

Tweaking
If needed to solve wall height problems like door headers, you could add 1½", or some other amount, to all these numbers (half-log sills A and B cut with 7½" tips and 10" butts; and whole-log sills X and Y cut with 14½" tips and 17" butts). Or, you could cut all the sills an inch or more shorter, if that helps solve wall height problems *(Figure 2)*.

But, if you modify the size of one sill log, then you must change ALL the half-log and whole-log sills in the entire building by the same amount. If you don't change them all, then you will affect the shoulder heights the sill logs leave, and that causes log selection problems.

Odd-level & Even-level
We have used an example of middles-of-logs-about-level, but you have two other options. Figure 17 shows the sill-log heights if you want odd-numbered rounds about level on top, and Figure 18 shows sill-log heights to have even-numbered rounds (2, 4, 6, 8, and so on) about level on top.

Which Walls Get the Half logs?
If you are making wall logs odd-level or even-level, then start the gable ends of the building with the half-log sills. The gable ends are usually the short walls of a rectangular floor plan. Use the whole-log sills to start the eave walls.

I do this so that when all the walls have the same number of rounds of logs (say 9 rounds on every wall) then the eave walls will be half-a-log taller than the gable walls, and this makes roof layout easier, as we will see later.

In buildings that have a floorplan that is anything other than a simple rectangle *(like the one in Figure 9)*, then not all gable ends can start with half-logs—some gable ends will start with half-logs and some gable ends start with whole-log sills. In this case, there will be an odd number of rounds of logs on some walls and an

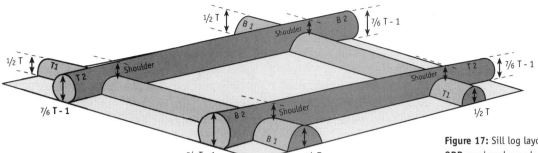

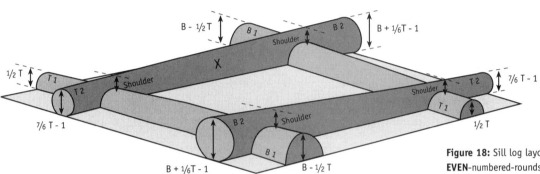

Figure 17: Sill log layout for a building where the top of **ODD**-numbered-rounds are about level. Round #1 is odd (1 is an odd number), and so the sill logs should be about level on top—and they are.

Figure 18: Sill log layout for a building where the top of **EVEN**-numbered-rounds are about level.

Metric users should use the formulas shown in Figures 14, 17, and 18, and simply use their average butt and tip diameters in centimeters, and where needed subtract 2.5cm instead of subtracting 1-inch. The fractions shown work with either metric or imperial units.

even number of rounds on the other walls. If you lay out the sill logs to have middles level then you can have the eave walls end higher than the gable walls (desirable if there is a log truss, as we will see later) without having some walls level and some sloped.

If you are making the center-of-every-log-level *(Figure 14)*, then you can start either the gable walls or the eave walls with the half-logs—it's your choice, though it is usually easier to find a suitable short, straight log to cut in half for two half-logs, than it is to find a long, straight log.

Summary

Draw a sketch of sill log layout showing half-logs and whole-logs, and butt-tip orientations for all the walls. Start with the exterior walls of the largest rectangle, then do the smaller rectangles, and finally lay out the interior log walls and stub walls.

Find the average tip and average butt diameters of your set of wall logs, and figure out which rounds will have roof plates, joists, and door headers. Decide whether you want odd-rounds level, even-rounds level, or the center of every log about level. Plug your average tip and average butt into the right map *(Figure 14, 17, or 18)*, test how tall your log headers for doors will be. Adjust all the sill logs taller or lower if needed for door-header height, and then rip the sill logs.

Sill-Log Layout Rules

FOR BUILDINGS WITH AN EVEN NUMBER OF EXTERIOR WALLS

1- At all corners of exterior walls, sills logs start with two tips, or with two butts, and never with a tip-butt combination.

2- Half-log sills and whole-log sills alternate as you count around the exterior perimeter of the walls

3- Cluster butts together, and cluster tips together

Hexagons

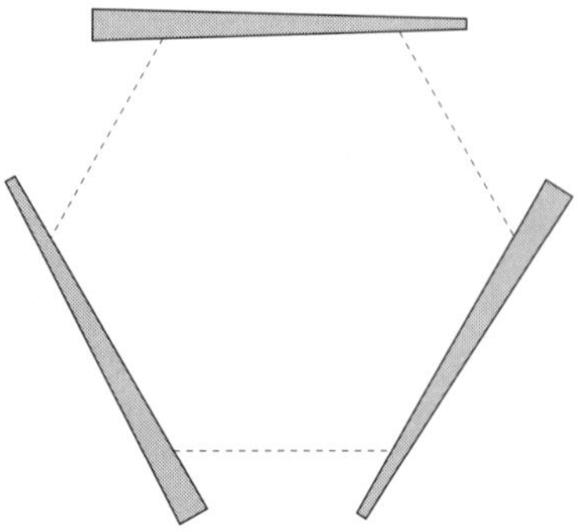

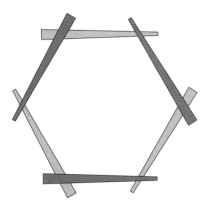

Figure 19: The half-log sills for a hexagon. Note that they all point clockwise and that they skip one wall each time.

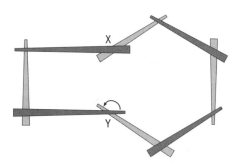

Figure 19a: The whole-log sills cross over the half-logs. They point the opposite way (counterclockwise). Note that every corner starts with either two tips or with two butts.

For professional builders: note that my sill log layout for a hexagon allows *three* active walls at all times, not just two—a benefit that really speeds up their construction.

Figure 20: The sill-log layout for a hexagon with an attached rectangle. Every corner starts with either two tips or with two butts. The perimeter alternates: half, whole, half, whole, half, and so on.

Note that the extreme-angle notches (at Corners X & Y) are more difficult to scribe, cut, & get to fit tightly than 120° hexagon corners or 90° rectangle corners.

Some people are drawn to hexagon floorplans, though most professional log builders are not wild about them. The oblique angle of the notches can make them difficult to fit tightly, especially with green logs that will shrink over time. The sill-log layout is not difficult because hexagons have an even number of exterior walls. Every corner starts with either two tips or with two butts, and half-log sills and whole-log sills alternate as you count around the exterior walls *(Figure 19)*.

Nearly all the floorplans shown here are regular hexagons (that just means that all corners have the same inside angle, 120°, and all walls are the same length), but the same rules work for hexagons that have other angles in the corners, and have different length walls—these are called irregular hexagons.

The tips of the half-log sills all point clockwise, or all point counterclockwise *(Figure 19)*. Hexagon sill layout alternates half, whole, half, whole, as you count around the exterior walls like on the face of a clock. If the half-log sills point clockwise, then the whole-log sills point counterclockwise *(Figure 19a)*. (If the half-logs point counterclockwise, then the whole-logs point clockwise.)

The walls of a hexagon log building can be about level on top of even-numbered rounds, or about level on top of the odd rounds, or have the horizontal center of every log about level—you can choose.

It is interesting that with hexagons, the sills on parallel walls are one half-log and one whole-log sill *(Figure 19a)*, and on parallel walls, the tips and butts point the same way, not the opposite way. But this is correct—the sill log layout for hexagons (and even the complex hexagons to come) follow the three rules of sill log layout.

Complex Hexagons

Now let's try adding some other rooms onto a hexagon. First, we'll add a rectangle—one log of the hexagon is left off *(Figure 20)*, and in that opening we add a simple rectangle. This brings no change to the hexagon sill-log layout except that we left one wall off. All the corners start with two butts or two tips (and not a tip-butt combination). All the sill logs alternate half, whole, half, whole, half, and so on, as you count around the building's perimeter in either direction, starting at any wall.

Now let's try another variation—a hexagon with two walls left off, and a rectangle added *(Figure 21)*. This is easy, too, and follows the rules we have been using. The hexagon (the part of it that is still remaining) has sill logs laid out just like the simple hexagon in Figure 19. The rectangular addition is laid out as we have done many times for rectangles: every corner has either two butts or two tips; and half-logs and whole-logs alternate as you count around the exterior walls of the entire building.

Odd-for-Odd or Even-for-Even

When the home is being designed, if you remove an even number of walls from the hexagon (2 removed in Figure 21), and then add an even number of walls (4 added in Figure 21), then sill layout stays easy. If you add 3 walls on where 1 wall was removed *(Figure 20)*—then we exchanged an odd number of walls for an odd number—still easy.

Hexagon-based floorplans become difficult, and require special sill logs, when they have an odd number of exterior walls. So, to keep complex hexagon floorplans easier, if you remove an odd number of walls, then replace them with an odd number of walls; if you remove an even number of walls, then add an even number of walls in the hole you made.

Any of these buildings can have odd-rounds level, or even-level, or middles-level. The actual sizes from flat to round of all the sill logs follows the rules we have already used in this chapter. Know your average butt diameter and average tip diameter; figure out the selection rules are for T1, T2, B1, and B2; find out which rounds will have plate logs, joists, and door headers, use your four selection numbers in the correct map *(Figures 14, 17, 18)*, test them to see if they should be tweaked higher or lower to make door headers come out at a good height, and then cut the sill logs.

Double Hexagon

Even double hexagons are easy—they follow the same rules we've been using for all buildings that have an even number of exterior walls. As you count around the sill logs (starting anywhere, and going in either direction) the sills of exterior walls alternate: half-log, whole-log, half, whole, half, whole, and so on. Every corner starts with two tips or two butts.

Note that in this floorplan *(Figure 22)* one wall of a hexagon was removed and then five walls of another hexagon were added—we traded an odd (1) for an odd (5), just as we should.

Other Polygons

Some hexagons don't look like hexagons—they are known as *irregular*. Figure 23 is a hexagon with four walls pushed in, but the sill-log layout is the same as the regular hexagon in Figure 19a.

The rules we have been using apply to many other shapes: octagons (8-sides), and more. The corners do not have to be at 90°—and the building can have a mix of different corner angles. As long as the building has an even number of exterior walls, then the sill-layout rules will work.

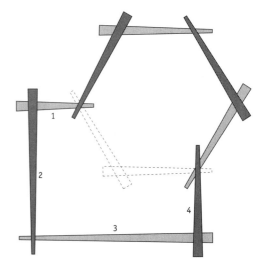

Figure 21: Sill-log layout for a hexagon with a large attached rectangle (a total of 8 exterior walls). Two walls of the hexagon were removed (dashed lines) and replaced with four exterior walls of a rectangle.

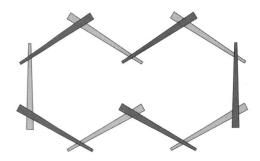

Figure 22: Sill logs for a double-hexagon (10 exterior walls). Half-logs all point clockwise; whole-logs point counterclockwise.

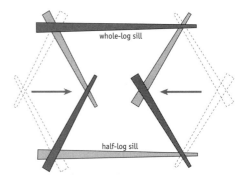

Figure 23: Sill logs of a hexagon with its sides pushed in. Doing this does not change sill-log layout—it is identical to the layout in Figure 19. I have never seen a log building this shape, and I doubt that I ever will. But it shows that the angles of the corners do not influence sill log layout.

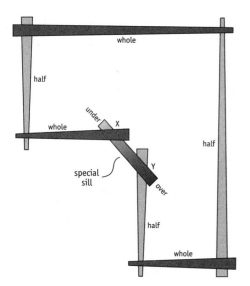

Figure 24: This building has an odd number of exterior log walls— seven —and so it has a difficult sill log layout.

The sill logs in every corner start with either two tips or with two butts. (Even Corners X & Y both start with two butts.) Half-log sills and whole-log sills alternate half-whole-half-whole as you count around the perimeter in either direction.

But we need one special sill log—it is a half-log sill at one end and a whole-log sill at its other end. This special sill has two butt ends. We can do this because we made the shortest wall in the building the special wall, and short logs don't have much taper. At Corner X, the special sill is a B1 (because a butt goes over it), and at Corner Y the special sill log is a B2 (because it goes over a butt).

Figure 25: Adding a clipped corner, Log C, gives this prow building 10 exterior walls, an even number, and it does not need a special sill log.

Buildings with an Odd Number of Exterior Walls

It's time to tackle the tough buildings—those with an odd number of exterior walls. These are for expert log home builders with years of experience, not for beginners. Experts need to know all the tricks of the trade, and that is why I am including this information here.

It is easy to avoid odd-exterior walls when you are designing a home. Don't design a heptagon (7 sides) when a hexagon (6 sides) or octagon (8 sides) works just fine! Or, if you really want a prow, then design the building so it has a prow and still has an even number of exterior walls (*see Figure 25*).

My advice is: avoid log buildings that have an odd number of exterior walls. If you have never built a log home with an odd number of exterior walls, and don't intend to, then I suggest you skip to page 82 of this chapter. Come back to this section in the future, if needed.

'Butts Chasing Tips'

Other methods are sometimes used for odd-wall buildings— like *butts-chasing-tips* sill layout. Sometimes they can be made to work. But the method described here is easier and much more reliable than any other method I have ever heard of, or tried.

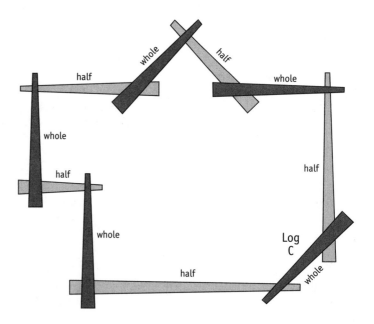

Introduction

Special Sill Needed

Buildings with an odd-number of exterior walls require one *special* sill log. All the other sill logs are laid out the way we have been describing throughout this chapter.

The special sill log is not a half-log and not a whole-log—it is *both*! Think of it this way: the special sill is a half-log at one of its ends, and it is a whole-log at the other end. It goes *under* at one end (like a half-log), and it goes *over* at the other end (like a whole-log) (*Figure 24*). No other log does this.

Special Logs Must Not Taper Much

The special sill log works best if it is in a short exterior wall. This is because each log that you add to the special wall either has two tips, or has two butts. I know this sounds strange—how can one log have two tip ends and no butt end, or two butt ends and no tip end? Well, it can't really, but it can come very close to this if it doesn't taper much. Let me explain.

For example, if the average tip is 12" and the average butt is 18", then a log that is 11" at one end and 12" at the other (an 11-12 log) could be considered a log with two tips, because both of its ends are very close to the average tip diameter. And a log that is 18-19 could be considered a log with two butts.

This is much easier to do if the special wall is short—try to keep the special wall shorter than 14-feet (4.2m) long (or keep taper less than about 2"). The longer the special wall is, the more tapered its logs, and the tougher it will be to have two tips or two butts on each log (*Figure 26*).

Every time you add the next log to a special wall you alternate: two-butt log, two-tip log, two-butt log, two-tip log, and so on—it needs to be a small log with little taper, or a big log with little taper. The wall that starts with the special sill needs these special logs all the way up!

Design To Have a Short Wall

This means that buildings with an odd number of exterior log walls are more difficult to build if no wall is short. It is possible to build odd-wall buildings that have no short wall, but you will have to expect problems as you build. It won't be easy balancing shoulder heights for good notching and keeping walls level and sloped at the proper rounds.

Guidelines

Here are the guidelines for laying out the sill logs of floorplans that have an odd-number of exterior log walls:

1- *At all corners of exterior walls, sill logs start with two tips or with two butts, and not with a tip-butt combination.*

2- *Half-log sills and whole-log sills alternate as you count around the exterior perimeter of the walls.*

3- *Cluster butts together, and cluster tips together, in the sill layout.*

4- *One sill log is special—it is a half-log at one of its ends, and a whole-log at its other end. Make a short exterior log wall the special wall.*

The logs that are placed on the special wall, all the way up to the roof plate log, work best if they have either two tips or two butts—that is, either they are tip-sized without much taper, or are butt-sized without much taper.

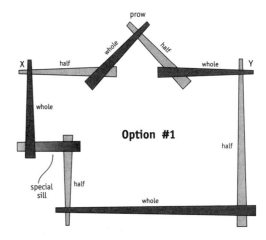

Figure 26: This building has nine exterior log walls—an odd number—and so it needs a special sill log. The pointed corner is called a "prow."

The sill logs in every corner start with either two tips or with two butts. Half-log sills and whole-log sills alternate half-whole-half-whole as you count around in either direction. There must be one special sill log—it is a half-log sill at one end and a whole-log sill at its other end. In the layout above the special sill has two butt ends.

For Option #1 (above), I didn't locate the special sill at the prow, instead I made the shortest exterior wall of the building the special wall. Doing that has implications, however. For example, the walls adjacent to the prows are a half and a whole in the "same" wall. It could be very difficult, or even impossible, to have a header log span from Corner X to Corner Y, if a header is needed.

But note that if you make one of the prow sill-logs the special wall (Option #2, below), then the walls adjacent to the prows can both be half-logs, and a one-log header from "X" to "Y" would be easy. You must consider your choices before you pick the layout that will work best for you.

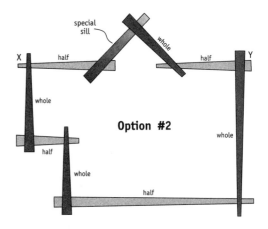

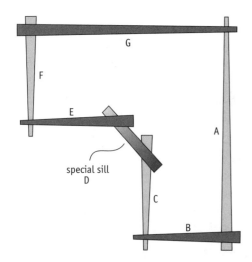

Figure 27: Above is the floorplan and sill layout of a 7-exterior-wall building. Below is the spread-out "chain" of its sill logs, in elevation (the vertical scale is exaggerated). Logs A, C, & F are half-log sills. Logs B, E, & G are whole-log sills. Log D is the Special Sill. Middles-of-all-logs are about level . Note that every sill log below slopes 3", except the Special Sill, which slopes 6" (from 15" to 9").

Further, because we keep adding logs to the special wall that don't have much taper (either a log with two butts or a log with two tips), this means that the special wall *always* slopes.

When Are Walls Level?

The special wall is not easy to get level. Here's why: the special sill-log is either a B1-B2, or it is a T1-T2—and in both cases it is a half-log sill at one end and a whole-log sill at its other end. Think about that and you will realize that the special sill is going to slope a lot *(Figure 28)*. In fact, the special sill slopes an amount that is equal to your shoulder-height Selection Rule for B2 and T2—which is often between 4" and 7".

Furthermore, since you want to keep adding log after log without much taper to the special wall—logs that have either two tips or two butts—the special wall keeps sloping all the way up. This is true whether you start the sill logs for odd-rounds level, even-rounds level, or all middles level.

Adjusting the Special Sill

Still, it is often best to try to reduce the slope of the special sill somewhat. Work out the sill log sizes for all the regular half-logs and whole-logs. Once you have those, then you can figure out the special sill log by applying the Selection Rules you are using for your set of logs. Then, try reducing the slope of the special sill somewhat.

For example, if the special sill is 15" tall at its B2 end and 9" tall at its B1 end *(Figure 28)*, make it 13" and 10" instead—then it slopes 3" overall instead of 6" *(Figure 29)*. If you leave the adjacent sill-log sizes unchanged, then they will have shoulders that are smaller than the Selection Rules recommend.

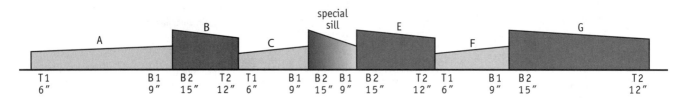

Figure 28: You have never seen a sill log "chain" before, so let me tell you about it. Think of the sill logs in the floorplan in Figure 27 as a chain, or necklace, where each sill log is a "bead." I 'unclasped' the necklace at the corner where Logs A & G meet, and straightened it out. The drawing immediately above is what that straightened chain looks like from the side — an elevation view.

The numbers (6", 9", 12", and 15") come from Figure 14—sill log sizes for middles of each log about level—and using the average tip (12") and butt (18") diameters of this set of logs.

Every shoulder here is either a B2 or a T2, and measures 6" tall—as it should with this set of logs. For example, look where Log G covers Log F—15" going over 9", leaves a 6" shoulder. Log B (a T2 that is 12" tall) covers Log C (a T1 that is 6" tall)—leaving a 6" shoulder height for the tip end of Log B.

Or, if you want, you can change the slope of an adjacent sill log a little in order to keep the shoulder heights closer to the Selection Rules. Or you can adjust both the slope and the shoulder heights a bit *(Figure 29)*.

Cut Sill Logs

To cut the sill logs, you'll need to have the four Selection Rules worked out for your set of average tips and butts. Let's say T1= 5", T2= 6", B1= 11", and B2= 6" (12" average tips and 18" average butts, 6" average taper).

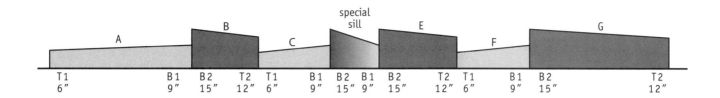

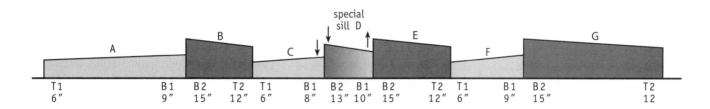

Figure 29: The "chain" of sill logs, in elevation (the vertical scale is exaggerated)—at the top is the unadjusted chain (Figure 28), below is the adjusted sills to reduce the slope of the special sill from 6" to 3".

Only Logs C & D have been adjusted compared with Figure 28. The arrows show that the butt end of Log C has been lowered 1". The butt of Log D has been lowered 2", and the tip end has been raised 1"—it was a 9-15 and now is a 10-13.

For this example we'll make the horizontal centerline of each log about level (that is, split the taper each time we add a log). Starting with the half-log sills (A, C, & F)—choose a log that has average taper and is slightly larger than average (maybe 13-19) and cut it about in half (to make two half-logs— 6-9 & 6-8).

Next, cut the whole-log sills B, E, & G tall enough to produce the shoulder height specified by the Rules. (B2 and T2 both have desired shoulders of about 6" for this set of logs.) We have adjusted the B2 shoulders of Logs D & E down to 5"—which should be fine because having a B2 shoulder slightly smaller than desired is okay—a tip is next.) The Special Sill Log D has been adjusted to 10-13 *(it was a 9-15 in Figure 28)*.

Last, as always, tweak the sill logs all a bit bigger or all a bit smaller, if needed, to get the door-header round at a good height.

Getting Towards Level

As the walls get near the door-header and plate-log rounds, measure walls heights and see how all the walls, and especially the special wall, are doing. You may be able to tweak the walls closer to level at these rounds, if you need to, by selecting logs that have more taper, or less taper, than normal.

You have great flexibility with the roof plate logs because it does not matter if they leave shoulders that are bigger than the Selection Rules recommend. After all, the plate logs are the last wall logs—no log is notched over them.

Interior Walls

The interior log walls of buildings with an odd-number of exterior walls, have their butts clustered and tips clustered just the same way as buildings that have an even number of exterior walls.

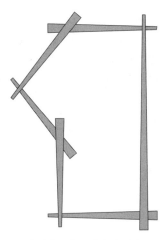

Figure 30: Is this a "prow" design? Does it require a special sill log? Count the number of exterior walls. There are six exterior walls (an even number), and this means that sill layout is easy—no special sill is needed.

Sill-Log Layout Rules

FOR BUILDINGS WITH AN EVEN OR AN ODD
NUMBER OF EXTERIOR WALLS

1- At all corners of exterior walls, sills logs start with two tips, or with two butts, and never with a tip-butt combination.

2- Half-log sills and whole-log sills alternate as you count around the exterior perimeter of the walls

3- Cluster butts together, and cluster tips together

4- If there are an odd-number of exterior log walls, then a special sill is required that is both half and whole. Every log put onto the special wall either has two tips, or has two butts. The special wall should be short.

Learning to See

When you are deciding sill-log layout for a house, find a quiet place to work, and get out some paper and a pencil. Even simple rectangular layouts need to be sketched. Follow the Rules in this book that apply to your floorplan. There are always several different ways to lay out any floorplan—for example, which walls start with half-logs and which with whole-logs. Work out the details of each alternative before you decide which to use.

Some floorplans look like they might be difficult, but are actually easy. This floorplan (*Figure 30*) looks like it might be a difficult prow, and require a special sill log. But, take a look—every corner starts with either two tips or with two butts. Every sill log is either a half-log or a whole-log sill—there is no special sill, because this building has an even number of exterior walls: six.

Learning to see, as I call it, takes practice. Work out the Selection Rules for your logs, and know the average gain you will get with your logs (average gain per log equals average log diameter minus loss to scribe, 1"). Now you can figure out which rounds get door headers, floor joists and plate logs. Knowing this helps you decide whether you want odd-rounds about level on top, even rounds, or middles of all wall logs about level.

Count the number of exterior log walls. If there is an odd number, then try to have at least one short log wall, and make the special wall a short wall.

Start layout with the biggest rectangle in the floorplan. Decide which walls will get the half-log sills. This is more important if you will be having either odd-rounds about level on top, or even-rounds about level on top. It is not as important if you plan on making the horizontal middles of all wall logs about level. Put two butts in one corner and then follow around the rest of this rectangle alternating half-logs and whole-logs, and two tips and two butts.

Move on to the other, smaller rectangles. In buildings with all 90° corners, half-logs all run east-west, or all run north-south, and are parallel to the half-logs in the first rectangle you laid out. Cluster tips together and cluster butts together.

Interior log walls are laid out last, after the exterior walls are known. They match the layout of the closest exterior wall that is parallel.

Using your Log Selection Rules, work out the sizes of all the sill logs at both ends. Check to see if you need to make all the sills a bit higher or a bit lower to get the door headers in a good elevation. Rip the sill logs and you are ready to start building.

Sill and Selection Summary

1) Measure your set of logs and find average tip (T) and butt (B) diameters, and the average log diameter (T + B) ÷ 2. The average height lost to the scribe (L) will be 1" (2.5 cm).

2) Decide which rounds should be level on top. The usual choices are: window and door headers, floor joists, and plate logs. This decision influences whether you cut your sill logs to have the odd-courses about level, the even-courses about level, or the horizontal centerline of every log about level.

3) Adjust the sill logs, if necessary, to be a few inches taller or shorter if you need to get door headers to a better position in their logs. Are the bottoms of the floor joists going to be above head height after settling? Cut and flatten the sill logs so they produce the desired shoulder heights as indicated by the Rules, and to have the walls level where it best suits your building and your logs.

4) Each time you pick the next log, use the Rules to find the desired shoulder height for its place in the BBTT, BBTT sequence. The desired diameter of the next log equals the desired shoulder height you want to leave plus the existing shoulder height you have to cross, plus 1" (the loss to scribe). Do this for each end of every log so that you know the desired butt and desired tip diameters when you looking for logs in the pile.

   Use the Rules more to decide which logs *not* to use rather than to decide which log to use. What I mean by this is if your Selection Result said to use a 12-18, then when you go your log deck you'll know that a 15-22 and a 9-13 should probably not be used this time. Whether you choose a log that is bigger or smaller than the Selection Rules recommend depends upon whether the log you are choosing has a B2 or has a B1.

5) Try to keep walls about the same height as you build by selecting all the logs on parallel walls at the same time, making a map of current wall heights, and then adjusting the selection diameters by using logs that have more, or have less, taper than the log you would have used if you went by Selection Rules alone.

These steps simplify selecting the next log, and eliminate the horrible notching problems with first tips that used to be common. I can usually predict future shoulder heights and wall heights within an inch (25mm).

No more ugly surprises of tips that cannot cross over butts at the corners. No more going to the log deck, round after round, to find "the biggest tip we got!"

Log Selection means avoiding these problems altogether. The Rules adapt automatically to the way you build, the size, taper and shape of your logs, the width of your grooves, and the variation between large and small logs that are actually in your building yard. Thinking ahead makes building with logs more predictable, and it's easy to do. Getting started with the right sill-log layout is the key to successful building.

Figure 31: There are 8 logs in both eave and gable walls, and the builder probably made the center of all logs about level. The door-header cut is in the middle half of its log (a good location), but you can see that the top of the header log is sloping down to the right, and is not level. This works just fine here, but in a longer wall it is definitely best to have the top of the door header log about level. Can you see why? Answer: a door to the far right on this gable wall would have just a thin slab of log left above the door.

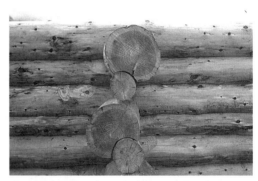

Figure 32: Well, at least the butts and tips alternate! It is difficult to build with crooked logs , but obviously it's not impossible.

Figure 33: A strange sill log. I have never seen another sill log like this one. The builder started this wall with a log that tapers from zero to a half-log. Hmmm.

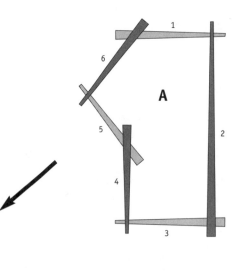

A

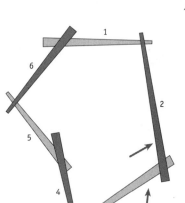

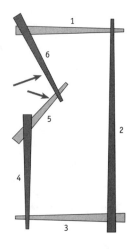

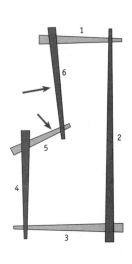

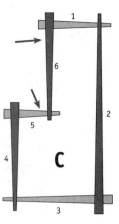

C

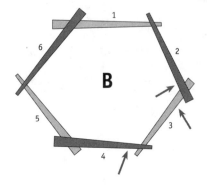

B

Two Final Thoughts

There were two break-through thoughts that helped me understand sill logs. I am including these here in case they might help you, too.

First, when it comes to sill log layout, the shape of the floorplan doesn't matter much. To see this, try to think of the sill logs as a chain or necklace, with flexible joints, instead of as rigid, notched logs, bolted to the foundation. Now we can change the shape—and starting with the floorplan above, we can follow either the track to the left, or the track to the right, and flex the sill-log necklace.

Buildings A, B, and C have very different floorplan shapes—but they have the same sill log layout! The same logs are half-sills and whole-sills, the butts and tips are in the same places and point the same way. All that's changed are the lengths of the walls and the angle of the corners. This helped me see that no matter what the floorplan shape is, the sills logs have just a few simple "rules" that always apply: they alternate half-whole-half-whole-half around the exterior perimeter; and every corner starts with either two tips or two butts.

Second, for buildings with an odd-number of exterior walls, the special sill can go anywhere you want. Yes, try to make the special wall one of the shortest walls, instead of a long wall. But the special sill does not have to be at the prow, or at the clipped corner—it can go somewhere else, if that helps you.

The only way to know where to put the special sill is to try several options on pencil and paper, and compare what happens in each layout. Don't decide sill-log layout in the building yard, half an hour before you start slabbing logs. Do it at home first, on paper—and you have a great chance of making things easy for yourself.

Putting Logs
On The Wall

It's time to put it all together. In this chapter you will learn how to put logs on the wall, step-by-step. Follow the order of the steps carefully: complete each step before moving on to the next. Details for certain steps—sill logs, plates, log selection, cutting notches—are found in other chapters. This chapter is a descriptive summary that puts all the tasks in proper order.

"Scribing" is at the heart of the process of handcrafted, scribe-fit log home construction—that's me scribing on the front cover of this book. The *scriber* is a tool that accurately transfers (scribes) the shape of a log onto the shape of the log above it, and leaves a scribe line on both logs.

When you carefully cut out the wood between the scribe lines: along the length of the log (the long groove or lateral), and at the corners (the notches); then you produce a log that will tightly fit the log below. Scribers feel all the small, unique shapes and bumps of both logs and draw those contours onto both logs.

People have been using scribers for centuries, though the tool has improved from rough-and-ready "scratchers" to accurate and clever tools. An unknown builder added a bubble level to scribers some time before 1945 to help him keep the points plumb; and clever swiveling ends for the pens were added by the Timmerhus staff before 1993.

At first, log building will seem complicated—so many steps! Getting them in the correct order! In fact, log building *is* a complex craft, not a simple one. But each and every wall log goes through the same steps in the same order. With repetition and practice you will begin to feel the rhythm of log building. It is, in the end, just one log after another.

CHAPTER TOPICS
Step by step instructions

1) Choose which log to use next
2) Put the next log on the wall
3) Rough notches
4) Plumb line on one end
5) Lay out the bottom of the saddles
6) Logs that have fewer than 2 notches
7) Mark the center of intersecting walls
8) Bring the new log down
9) Snap the top-center chalkline
10) Lay out & cut the saddles
11) Cut the rough notches
12) Clean up the bottom surface
13) Put the log back on the wall
14) Trim this log's ends
15) Find the widest gap
16) Plumb the scribers
17) Scribe the groove
18) Scribe the notches
19) Scribe the flyways
20) Drill holes
21) Double check that everything is scribed!
22) Bring the log down
23) Score the notches & cut them out
24) Cut the groove, flyways & kerf
25) Test the fit

Step-by-Step Instructions

1) Choose Which Log to Use Next

Size

Use the Log Selection Rules to figure out what size log to use next. Find the desired tip and butt diameters, and the length of the log, and then consult your log list to identify the good candidates.

Be especially careful when you are choosing a log that has a second butt (B2)—remember that it should not leave too tall a shoulder after it has been notched and fitted.

Shape

From the candidates, choose the one that has a shape that best matches the shape of the log below, or at least does not clash with it. If the log below has a dip down, a depression that is a natural part of the shape of that log, then don't choose a log to use next that has a dip up at about the same location, or that will cause a very wide gap (*Figure 1*).

The final scriber-setting you will end up using is determined by the widest-gap-of-all between two logs—not by the average gap, and not by the minimum gap. When choosing a log for its shape, the goal is to create a gap with a uniform shape—a gap that doesn't have both extremely close and extremely distant portions. If you don't or can't reduce the maximum gap, then the long groove will be very wide and deep, and the loss to scribe will be large.

Save your straightest logs for important purposes: door headers, plate logs, and roof systems. Don't waste these precious logs down low on the walls.

Figure 1: When you select the next log consider first whether it is the right size and length, and second if it has a good shape. Avoid logs that have a shape that will create an extreme gap—like the top drawing.

Sometimes it is easy to fix a problem with the shape of the gap. The log, if you slide it to the left (drawing below), becomes a wonderful match for the log below it. The widest gap (X) has moved, has become smaller, and the shape of the gap is much more uniform, which means that less wood will be lost fitting these two logs together.

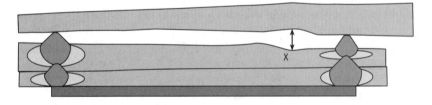

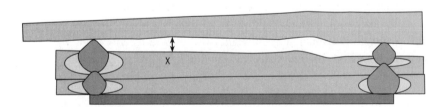

Other Considerations

Remember to also consider that parallel walls should be about the same height; that some courses of logs may need to be level on top (like floor joists, door-header logs, and roof plate logs); that some logs work best if they are at a certain height (door headers); and that some logs must be above a particular height to provide headroom (floor joists and beams that support joists).

It is best to choose all the logs you can at one time—for example, choose all the north-south logs for Round #5. Make a map to help you select the logs you'll use next, and take it with you when you go looking for logs.

If you decide to use pieces instead of a full-wall-length log, then it looks best if all the pieces come from one log that is bucked into lengths and spread apart *(Figure 2)*. This is also a great way to use a log that is too short, or too bowed, to use in one piece.

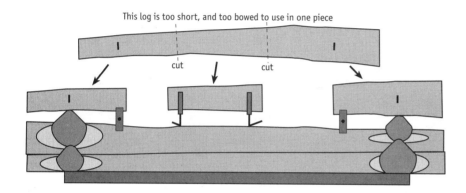

This log is too short, and too bowed to use in one piece

Figure 2: A log can be cut (*bucked*) into pieces and stretched in door or window openings. This is a good way to use a log that has too much bow to use in one piece. The diameters match on both sides of each opening, so the top surface of these pieces will be quite straight, and make it easy to place the next log. Note the marks that indicate an adequate length for the flyways.

2) Put the Next Log on the Wall

While the new log is on the ground, roll it so the bow will be mostly out once it gets on the wall, and the straightest surface is on top. It is easier and safer to roll a log on the deck than when its up on the wall, especially when the walls are tall and you're working from ladders or scaffold.

Make certain that the log will actually bow out when you get it onto the wall, and doesn't bow in. Figure out where the tip and butt will go, and which surface you want to face up on the new log—it is easy to get confused.

Mark one or both ends of the log to indicate where to position it end-for-end. If the finished flyway length is 22" (55cm) from center, then make a clear mark that is, say, 24" (60cm) from the log end *(Figure 2)*. This means that the people who are up on the wall, helping put the log in place, won't have to fumble for a tape measure at the last moment.

My philosophy is that every thing that can be done easily on the ground, should be done on the ground. (I admit that I wasn't so philosophical when I was younger.) It is much safer to make measurements, snap chalklines, and use power tools with both feet on the ground—it's also easier and usually faster, too.

Figure 3: The next log has been selected, rolled so the straightest surface is on top with the bow out, and rigged with lifting straps. A tag line helps a person on the ground keep the log safely under control.

When using straps or tongs to lift logs, here's a good way to get a log to hang level: decide where the log would balance on **one** point like a teeter-totter, then center the straps or tongs on that point.

Figure 4: Place wall logs so the bow is oriented out on the wall, and the straightest surface is on top.

Figure 5: A peavey is used to slide the end of a log into position. The log will roll, and the straightest surface will no longer be on top, unless its other end is held from rolling with a peavey or log frogs.

Lift the log onto the wall *(Figure 3)*. While the crane is still attached to the log, chock it at both ends with log frogs, also called cleats—they help keep the log in place and prevent it from rotating. (Log dogs, which are like big staples, aren't used much anymore.) You don't want the log rolling off the wall: be aware of this danger. Logs that bow are especially likely to roll *(Figure 4)*.

Refine Placement

Stand back and look at the wall and the log you have just set. How well does it shape match the log below? Shift the new log, push it back and forth, and rotate it to find the best match to the shape of the log below, but try to keep the straightest surface on top.

The widest gap of all is usually caused by some small feature on one, or both, of the logs—a knot, flat spot, kink, scar, or cat-face. The widest gap can be caused by full-length sweep, but more often it is the small-scale, unique bits of shape that cause it. Because the little things really do matter, you have an opportunity to tweak the position of the log to overcome some of the extreme uniqueness that is causing the widest gap of all *(Figure 1)*.

Rolling a Log on the Wall

To safely roll a long log on the wall, there should be a person with a peavey at both ends of the log. Professional building yards keep a peavey or cant hook in every corner of the building so that no one has to look for this essential tool.

While one person uses a peavey to roll the log, the other keeps their end in place. Cleats or frogs can also allow the log to roll in place if the top points face away from the log. (When the top points stick into the log they prevent rolling.)

If one end of the log moves while being rolled, and is no longer centered over the wall below, then use a peavey as a lever to slide it back into place *(Figure 5)*. The log will roll instead of sliding unless there is someone at the other end of the log with a peavey to prevent it from rolling. Well-placed frogs, with the tops points jammed into the log, can also keep a log from rolling. Working together, two people (one levering and one hooking to prevent rolling) can make quick work of getting a log in position.

Straight-Face Up

Keep the straightest side of the log on the top, whenever possible, because this makes matching the shape with the next log easier.

Logs sometimes sag and it is difficult to get the straightest surface on top—this happens with small diameter logs that span a long way between notches. But this can actually help. Here's how: if the log is bowed (has a long sweep), then you can put its bow mostly up (instead of mostly out, horizontal), and let it sag down to straighten itself out. I do this all the time. Do not prop up the center of a sagging log unless it bows down no matter which way it's rotated. If a log sags down and in the process becomes straighter, then don't prop it up.

Centered On the Wall

Working just by eye, position the new log so it is centered over the wall below. If the log looks like it is well centered with its mass over the center of the wall below, then that's close enough for now. After we cut the rough notches we'll precisely center the log on the wall, but that accuracy wastes time at this stage. A good deal of a craftsman's efficiency is in knowing when to take the time to be precise, and when close is good enough.

If you just can't get happy with the log no matter how you rotate it, slide it, or push it, then just stop. Perhaps a different log would work here. It's better to take a log down and try another one than it is to force a log to be used where it doesn't belong. Sometimes you just can't tell for certain whether a log will work until it is up on the wall, but your judgement will improve with every mistake you make.

Keep in mind that no log is perfect and you can build great log walls with logs that are not perfect. Your goal is to simply get the best you can from each log. Remember that eventually you'll have to find a place for every log!

3) Rough Notches

The primary purpose of rough notches, also called *pre-notches*, is to improve the shape of the gap between two logs. Rough notches also make a log more stable on the wall, and reduce the final scribe setting—it's easier and more accurate to final-scribe 4" (100mm) than it is to final-scribe with a setting of 10" (250mm).

The log is lowered to its finally-fitted position in two steps. It comes part way down with rough notches; and then later it will come all the way down to touch the log below with final scribing. The purpose of rough notching is to produce a gap with a more uniform shape; and the purpose of final scribing is to get the logs fitted tightly.

Make the Gap More Uniform

Look at the shape of the gap between the log that you just added and the log below it. Learn to see this empty space as an object with a particular shape—a shape that depends upon the unique character of the log above and the log below.

A stone mason can look at a hole in a wall, walk to the stone pile, and choose a good match—he doesn't use arithmetic, or calculus, or make a paper tracing of the hole's shape. You must learn the same visualization process with logs. First, train your eye, then trust your eye.

Instead of seeing two logs, try to see the one space that is between them. This mental trick is very effective at helping you decide how to position and rough-notch the next log (*Figure 6*).

Hint If you are having a hard time picturing the bow or sag, you can tightly stretch a string on the top or bottom surface of the log (stretch the string on the surface that is more concave over its length). Stand back and compare the log to the string. The string is straight, and gives you a good reference to train your eye to see how much bow a log actually has. After you train your eye and your judgement, you won't need the string any more.

Hint A good deal of a craftsman's efficiency is in knowing when to take the time to be precise, and when close is good enough.

See the shape of the gap as an object itself

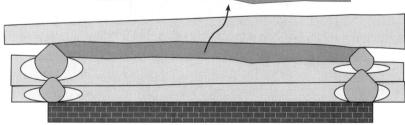

Figure 6: Once the next log is on the wall, and has been rotated and positioned, stand back and look at the space between it and the log below. Is it uniform in shape? Where will the widest gap be?

Try to see the gap as an object itself. Here I have pulled the gap out from between the logs and now it's easy to see its shape, where the widest gap is, and whether the gap is uniformly shaped. This is a useful mental trick to help you understand that the space itself is the object that you need to see.

Figure 7: Identify the 2 or 3 most likely places that could end up being the widest gap of all after rough-notching the log. Do this by eye only, do not measure any of the gaps. For this log, it looks like either A, B, or C could become the widest gap of all. We want to cut rough notches that assure that the widest gap of all will be at B (and not at A or C)—because B is the candidate that is closest to the center of this log's length. Ignore any widest gap candidate that is in a flyway (like C).

If you rough notch so the log becomes 3" (75mm) apart at both ends, then candidate "A" might be the widest gap of all. You can avoid this by bringing the "A" end down a bit closer (say 2¹⁄₂" or 65mm) than the "C" end (3" or 75mm).

The arrows show the end-gaps that we compare and use to help determine the rough-scribe setting. I have shaded the gap a dark gray to help you see it.

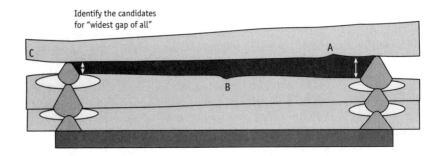

Identify the candidates for "widest gap of all"

Identify the features that you think are the top candidates for widest-gap-of-all after the rough notches are cut *(Figure 7)*. The current widest gap is probably near the tallest shoulder, but I'm not talking about that gap. Imagine the log *after* it's rough-notched: then where will the widest gap be?

Of the candidates, one will be closer to the lengthwise center of the log, and we want that one to "win"—we want it to become the widest-gap-of-all after rough notching. The other candidates gaps will be located closer to the ends of the logs, and you do not want the widest gap to end up there.

So, how do we accomplish this? In practical terms, we'll use rough-notching to bring one end of the log a bit closer to the log below.

Do not measure and compare the candidate gaps. You cannot solve this riddle with arithmetic. I know you'll be tempted to measure the candidate gaps—but don't do it—none of those numbers will help you.

Instead, you must train your eye to: 1) see the shape of the gap; 2) identify likely candidates for widest gap after rough notching; and 3) make a guess at how much closer to rough-notch one of the ends so that the widest gap will end up near the middle of the log's length instead of near its ends. Soon you'll be making fast and accurate judgements by eye.

Caliper the Ends

Use inside calipers to compare the existing gaps very near the notches at the ends *(the arrows in Figure 7)*. We'll try to get the new log down so it is about 3" (75mm) away from the log below in the areas near its end notches, though as we just noted, one end is often brought down a bit closer than the other.

So, for example, if the gap near one notch is 5" (125mm), then set a scriber to 2" (50mm) and scribe a rough notch. If the other notch now has a gap of 8" (200mm), then use a scribe-setting of 5" (125mm) for it. After you cut these rough notches, the gap will be about 3" (75mm) at both ends—you have made the gap more uniform. The gap between these two logs started at 5" and 8" at its ends, and after rough-notching the gaps became 3" and 3" at the ends. You have given the gap a more uniform (less tapered) shape, which is the goal of rough notching.

It is almost always best if the widest-gap-of-all ends up about centered along the length of the new log. Doing this ensures that the long-groove will be as narrow and shallow as possible. Here's another way to see this: the loss to scribe is kept to a minimum when the widest-gap-of-all is near the middle of a log's length.

Some logs have a shape that will tend to make the widest-gap-of-all, after rough notching, close to one end (like Figure 7). You can easily prevent this by rough-notching to bring that end of the log closer than the other end. Perhaps bringing it down to 2½" (65mm) will work, or even closer if necessary, while the target gap at other end-notch is 3" (75mm) *(Figure 8)*.

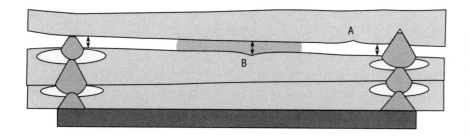

Hint The loss to scribe is kept to a minimum when the widest-gap-of-all is near the middle of a log's length.

Figure 8: After rough notching, the gap has a more uniform shape. Compare with Figure 7, which is the same log, before rough notching. The widest gap of all is now at "B," and to prevent "A" from being the widest gap we had to rough notch the new log so it is closer to the log below at A than it is at the other end of the log. We want the widest gap to end up, if possible, in the middle half of a log's length—which is the shaded area.

If you end up guessing wrong, and if the widest-gap-of-all ends up too close to one end, you can still raise the other end with a log horse to move the widest-gap-of-all closer to the middle half of the log's length. We'll cover this a little later in this chapter *(Step #13, and Figure 28)*.

It is a guessing game at first how much closer to bring one end—but getting good at guessing becomes, with time, having good "log sense."

Just One Rough Notch

Some logs don't get a rough notch at every corner. For example, if you caliper the gaps near the end notches and one gap is 4" (100mm), then don't bother scribing and removing 1" (25mm) to make this gap 3" (75mm) after rough notching. In fact, many scribers won't even close down that small—you may not be able to scribe 1" (25mm).

What can you do? Use a different gap size as your target. For example, make both ends have a 4" (100mm) gap after the rough notches are cut, instead of 3" (75mm). If one end has a 4" (100mm) gap, and the other has a 6" (150mm) gap, then scribe nothing out of the 4" end, and set a scriber to 2" (50mm) and use it to scribe the end that is now 6". After you cut these rough notches, both ends will have about a 4" (100mm) gap—a uniform shape.

As before, one end may need to come closer than the other end in order to make the widest gap of all end up near the middle of the log's length.

More than Two Rough Notches

Floorplans with interior log walls or stub walls will have logs that have more than two notches. And logs that cap over second-floor joists will have a lot of notches *(Figure 9)*.

Figure 9: A log building with a corner notch at each end and 3 floor joists. The plate log comes next, and it will have 5 final notches in it—compare with Figure 10.

For logs that have more than two notches, the rough notch scribe-settings can be determined by making a simple scale drawing. As usual, find the gap at each far end, and decide what scribe-setting to use for each of those two rough notches *(Figure 10)*. You need the actual scriber settings you'll be using—you may have decided to scribe one end with a setting of 2" (50mm) while the other will be scribed down 6" (150mm).

Then make a drawing to help you set the scribers for all the other rough-notches. Draw a line 12" (300mm) long —the baseline. On the left end draw a vertical line that is exactly as tall as the actual left-end rough-notch scriber-setting (6", or 150mm, in our example). On the right end of the baseline, draw a vertical line that is exactly as tall as the right-end rough-notch scriber-setting (2", or 50mm, in our example). Connect the tops of these two end lines with a straight, sloped line that we'll call the scribe-setting line *(Figure 10)*.

Now, along the baseline, map out where the other notches are: floor joists, tie beams, stub walls, interior log walls, and so on. The location of these notches must be to scale along the baseline. At the location of every rough notch, draw a vertical line up from the baseline to meet the sloped scribe-setting line. The scribe setting for each rough notch is equal to the length its vertical line.

Set the scribers for each rough notch by holding the scriber on the vertical line that represents that notch *(Figure 11)*. You don't even have to measure the amount—just set it with the scribers. (I usually add ¼" (6mm) to the rough-notch settings for mid-wall notches, just to be certain that the log won't teeter-totter on them after it's been rough notched.)

Note that sometimes a log you are about to rough-scribe sits on two of its notches and doesn't even get close to other notches—there's an air space between the new log and the notch below (log "A" and two of the joists in Figure 10 are like this). That is okay, and requires no special treatment. This does not in any way change the scribe-settings determined by the drawing.

Leveling a Log

There are times, though not many, when the goal for rough notching is not to give the gap as uniform a shape as possible, and not to put the widest gap in the middle

Figure 10: With the next log on the building, and in position, measure the gap at A and at B. Decide how big a gap you want between these logs after rough-notching, I like a 3" gap after rough-notching. Draw the triangle as shown—the baseline is to scale and shows the location of each corner notch and joist notch. The vertical scale is 1:1, that is, it is actual size. The rough-notch scribe-setting for every notch (corners and joists) can be set from this simple drawing.

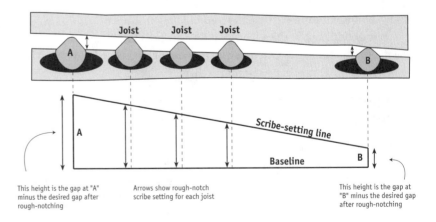

This height is the gap at "A" minus the desired gap after rough-notching

Arrows show rough-notch scribe setting for each joist

This height is the gap at "B" minus the desired gap after rough-notching

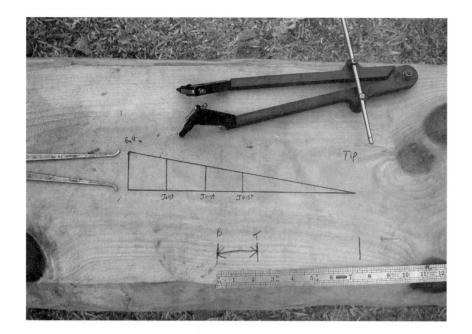

Figure 11: This is the actual drawing used for the building shown in Figure 9. On the bottom is the gap at the butt end and the gap at the tip end of the new log—the arrow shows the difference in gap size (just over 2", 50mm). The triangle-shape drawing has the 2" (50mm) setting at the butt end (left) and the 0" (no rough notch) setting at the tip end (right). The location of each joist is drawn to scale on the horizontal baseline. Every rough-notch scribe-setting for this log was set directly from this drawing.

of a log's length. One example is when you want the new log to be about level on top, even if this means that the gap under it will not be uniform. Floor joists, and top plate logs (the highest logs on the walls) should be about level on top.

Here's how to scribe rough-notches that will make a new log about level on top after the rough notches are cut. First, measure the height of each end of the new log above the temporary stumps or off the floor. Let's say the height of the new log at one end, before rough-notching, is 108" (2740mm) and the other end is 110" (2790mm). Bottom line: the rough-notch scriber-setting for the end that is higher must be 2" (50mm) larger than the rough-notch scriber-setting for the end that is lower.

By removing 2" (50mm) more wood from the rough-notch of the end that is 2" (50mm) taller now, the new log will be level on top after the rough notches are cut.

Continuing this example, if the rough-notch scribe-setting at the end that is 108" tall is 4" (rough notch setting of 100mm at the end that is 2740mm), then the rough-notch scribe-setting for the end that is 110" tall must be 6" (because 108 - 4 = 110 - 6). (In metric, the rough notch setting for the end that is 2790mm must be 150mm, because 2740 - 100 = 2790 - 150.)

Other Reasons

If the new log is going to leave too tall a shoulder, then you could consider using a rough-notch scribe setting that sinks the high end down, even though this means that the gap between the two logs will be tapered and not uniform.

Perhaps you made a mistake with log selection rules, or you're running out of logs to choose from. In either case, if the log you have to use for the next log has a butt diameter that is much larger than desired (and especially if it is going to be a B2), then you can, as a last resort, rough notch that end down close to the log below. As a result, the shoulder it will leave will be lower than it would have been if you had rough-notched it normally to give the gap a uniform shape.

Hint A log that is level on top after being rough notched will be level in the wall. A log that slopes 2" after rough notching will slope 2" in the finished wall.

Final scribing cannot change a log's slope—only rough notching can do that.

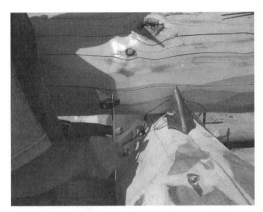

Figure 12: Scribe rough notches quickly, but don't be sloppy. Log frogs are used to prevent the log from rolling or sliding.

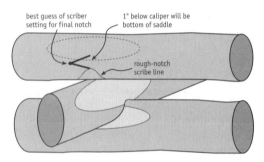

Figure 13: Once you have the rough-notch scribe line, you can lay out the bottom of the saddle. To do this, measure up from the rough-notch scribe a distance that is the final scribe setting you anticipate, and then come down 1" (25mm)—this is the bottom of the saddle.

Figure 14: Two devices that can be used to hold up the free end of one-notch logs. The first, a goal post, fits into a hole drilled into the log below, and its height is adjusted with an arm welded to a nut. The second is called a butterfly or bowtie. The spike on the metal flap can be hammered into the log below. Log height is adjusted with the nuts on the threaded rod that connect the two wooden ends.

In an emergency, this can help you avoid trouble. But, doing this wastes wood and produces are a very wide and deep long groove. *Making a dugout canoe*, log builders call it. It is often better to select another log than to sink a log that is too big.

Start Scribing

Set the scribe distances you need, plumb the scribers *(see Step #16)*, and scribe the rough notches, one after another. Work swiftly and leave just one, smooth scribe line *(Figure 12)*. If you get multiple lines, do not bother to drawknife them off, just indicate the one to cut. I'm not encouraging you to be sloppy, but scribing rough notches is an opportunity for you to practice scribing notches quickly.

4) Plumb Line on One End

Use a 2-foot level to draw a plumb line on the end-grain of one end of the new log. The temporary plumb line does not indicate anything about the center of the log, so keep it well off to one side. And don't mark a plumb line on both of the log's ends—you don't need two lines, and, if they don't agree, then there will be disputes over which line to believe. This line is used only to re-plumb the log when it's back on the ground, before we snap its centerline.

5) Lay Out the Bottom of the Saddles

Make a good guess at the final scribe-setting for this log. This guess is based on two things: how close will the ends be after you cut the rough notches? And second, how smooth and straight are the new log and the log below—that is, how much variation is there in the shape of the gap between the two logs?

For example, if you have you scribed rough notches that will bring the log ends down to about 3" (75mm), and the new log and the log below are fairly straight and not too bumpy, then the final scriber-setting may be about 4" (100mm)—about 1" (25mm) bigger than the gaps near the end-notches.

If you scribed rough notches that will bring the ends down to 4" (100mm) and the logs are a bit lumpy, then the final scribe setting may be 5½" (140mm) which is about 1½" (40mm) bigger than the 4" (100mm) gaps near the end-notches. Unless the log is very short and smooth, then the final scribe setting will be at least ¾" (20mm) larger than the gap near the notches. And even with long, crooked, bumpy logs, the final scribe setting is rarely more than 1½" (40mm) larger than the gap near the notches.

Use your guess for the final scribe-setting of the new log, and measure up that amount from the top of the scribed rough-notches. Make sure that you measure up from the highest point of the rough-notch scribe line, and not from the top of the log below *(Figure 13)*. Come down 1" from there and make a mark—this is the bottom of the saddle. Make this mark on both sides of every notch.

6) Logs that Have Fewer than Two Notches

Some logs don't extend all the way from corner to corner, they may go from one corner to a door or window opening and stop there. In fact, some log pieces have no corner notches at all—like those that are between a window and a door. Back when logs were cheap, and long, straight logs were easy to find, builders often used full-length logs on every wall, and only later cut doors and windows out of nearly solid walls.

Now, however, many of the openings are made as the log walls are being built. This is an opportunity to *straighten* bowed logs by cutting them into shorter pieces, or use a short log to save material. Using log pieces takes a bit more time, however, and requires a way to prop up one-notch logs.

It looks best if the pieces come from one log, and are bucked off that log and put onto the wall in the same order and orientation *(Figure 2)*. You can also use pieces from different logs, but carefully measure the diameters, and try to match the shapes so that the pieces look as if they came from one log instead of looking like scraps from the firewood pile.

One-Notch Logs

Some builders like to cut the rough notch for a one-notch log on the ground, without ever scribing it—they lay it out from measurements made with a ruler and a short level. I have found it to be about as fast, and definitely more accurate, to put a one-notch log onto the wall, scribe its rough notch, and bring it back down to cut the rough notch and saddles.

One end of the log piece will need a temporary support with adjustable height, like a butterfly or bowtie, or a goal post *(Figure 14)*. Set the height of the device before the new log is put on the wall—they're easier to adjust when the weight is off.

Be very careful when setting and positioning a log piece on the wall—pieces are much less stable than two-notch logs. Always have a person at each end, and handle it carefully. For added security, leave the crane attached to the lifting straps.

It is not necessary to adjust the height of the butterfly or goal post to make the gap perfectly uniform. The gap under a one-notch log can be tapered without affecting the rough-notch scribe *(Figure 15)*. Three inches (75mm) of taper, or more, is not a problem.

But, log pieces must be carefully centered over the wall below by eye. If a one-notch log piece is not centered over the wall before you scribe its rough notch, then it will be very difficult to get it centered later *(Figure 16)*.

No-Notch Logs

Sometimes these pieces are called *islands* since they sit between door and window openings, totally separated from corners. No-notch logs do not have to be rough-notched because they have no notch. Wait until you are going to final-scribe, and then put no-notch logs on the wall to scribe their long groove.

These pieces are unstable, and if you build using them you must be very careful. It is corner notches that give a wall stability, and without a corner, a log wall is like a of stack pencils. As an island gets tall you are balancing a tippy log on top of a stack of tippy logs. How many pencils could you stack one on top of another, before they fall over?

Some builders nail 2x6 strongbacks to the island of logs as they go up—with long spikes driven into the logs, and then braced down to the ground. I also use a heavy truck ratchet-strap (sometimes called a *load-binder*) to help keep an island stack in one piece.

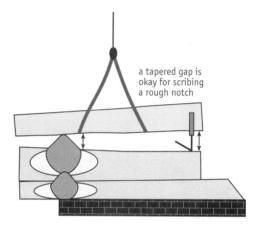

a tapered gap is okay for scribing a rough notch

Figure 15: A one-notch log, with its free end supported by a goal post. Keep the crane hooked up, if you can, because one-notch logs are unstable. The gap can be quite tapered while you scribe the rough notch, so don't waste much time adjusting the height of the goal post or butterfly.

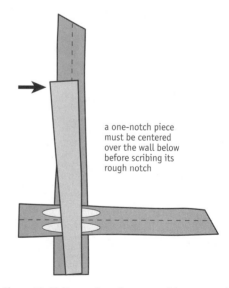

a one-notch piece must be centered over the wall below before scribing its rough notch

Figure 16: Bird's eye view of a corner with a one-notch log piece being positioned. If you scribe it's rough notch without aligning the log on the wall, then it will be very difficult to get the log piece in proper position after its rough notch is cut.

Better yet, as you add more layers, send a log all the way over to one corner, and then in the next layer, send one all the way over to the other corner (*Figure 17*). This avoids no-notch logs—every log has at least one notch.

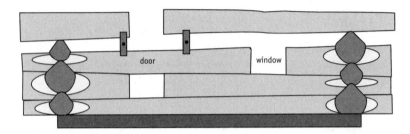

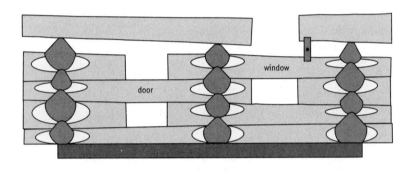

Figure 17: Stagger the log pieces with each new layer—this makes the wall more stable as you build it up. This is a good practice whether there are two corners on a wall (top) or more (bottom drawing).

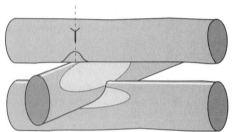

Figure 18: Saddles look best if they are all centered on the corner. Use the chalkline on the log below to transfer a mark to the new log that shows where the center of its saddle will be.

7) Mark the Center of Intersecting Walls

Make marks on the new log that show the location of the chalklined center of the intersecting wall below (*Figure 18*). These marks will be used to lay out the width of the saddles on the new log.

Don't bother laying out the saddles while the new log is still on the wall. It's faster and safer to finish their layout after you bring the log down to the ground, but you'll need these center-marks to do it.

8) Bring the New Log Down

Bring the new log down. Most builders set the log into "V-blocks" which are simply scrap pieces of logs that have a V-shaped piece cut out of either their end grain or the side (depending upon whether you like the V-block to be lying down or standing up, *see Figure 19*). Some building yards build special horses to hold logs. Lining the block with a slippery material (like an empty, plastic bar oil container) makes it easier to rotate the log.

9) Snap the Top-Center Chalkline

Use a peavey to rotate the log in the V-blocks and make the line you drew in Step #4 read plumb. Clean up the top surface of the log with a drawknife, curved planer, or sander to make it smooth, to remove scars, and get it ready for the centerline.

How do you decide what is the center of the log? Think of it this way: if you cut the log in half on its centerline, both halves should weigh about the same. Half the

Figure 19: Logs are held in V-blocks cut from scraps of log (top), or in specially-built log horses (bottom).

log should be to the left of the line, and half should be to the right of the line. This is easy to visualize with straight logs. Bowed and crooked logs require practice.

Do not use a tape measure for this task—you need only a chalkline and your judgement. It can help to have two independent perspectives: get a helper, stretch the line, stand at opposite ends of the log, and adjust the line until you both agree that it divides the log in half. Ignore the ends of the log that are in flyways.

Lately, I've been using a heavy duty mason's chalkline (see the chapter on Tools)—it produces a wide line, but that doesn't matter since ¼" (6mm) accuracy is sufficient for a centerline.

Stretch the string tightly while it's resting on the log. Then, standing near the middle of the log's length, pull the chalkline up at least 2-feet (60cm), pinched between your thumb and index finger, not curled over one finger. Hold the line and wait. When your helper says the chalkline is plumb, then release it with a satisfying ping, and a cloud of chalk dust.

With care, you can get an accurate line on logs up to about 40-feet long. If you're having trouble getting a straight line, here are some things to consider. **1)** Smooth the knots and bumps, and remove all bark—they can deflect a line. **2)** Stretch the line tight. **3)** Be certain to pull back and hold the line plumb—if you're working alone you can use a 2-foot level to help. **4)** Pull the line high so that it is touching the log only at the last inch of the log ends, and nowhere else.

Chalklines are quite sensitive to wind, and especially wind from the side. Try holding the line down at a mid-point (first making certain the line is straight), and then snap it in two halves: on your left side, and then on your right. Always check the line by getting down on one knee and sighting down the line—wobbles show up clearly.

To summarize: centerlines are snapped by eye, not by measuring the log ends and dividing the diameters in half. The goal is to center the mass of the new log on the wall. Ignore the portions of the log that are outside the notches—in the flyways. Straight logs are easy to centerline, but with bowed logs you do your best, and train your eye. The chalkline will not be in the "center" of a bowed log near its end-notches *(see Figures 4 & 20)*.

Logs that bow more than one-half of the small-end-diameter are difficult to use full-length. Stretch a string or a tape measure from the middle of the tip end to the middle of the butt end—if the string comes outside the surface of the log, then it may have too much bow to use in one piece. If a log is so bowed or crooked that you cannot even decide where to hold the chalkline, then it may be best to buck it into shorter pieces *(Figure 2)*.

Do not worry about anything less than ¼" (6mm) when positioning a chalkline. So, if your helper asks you to move your end of the line by ⅛" (3mm), then you know you are ready to snap it.

Once you have snapped the top-center chalkline, do not draw a plumb line on the end-grain of the logs—doing this will only confuse you later. I'll explain why in Step #13.

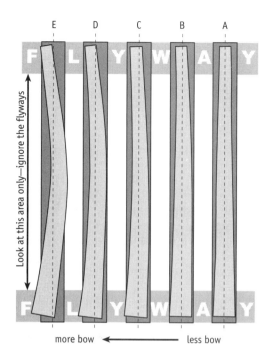

more bow ⟵ ⟶ less bow

Figure 20: Plan view of logs with a variety of bow, and on each log I have shown where I would snap the centerline. Ignore the parts of the log that will be in a flyway (the shaded areas in this drawing). The gray rectangles indicate the mass of the log wall below this log, and illustrate the purpose behind the centerline: to keep the center of each new log plumb over the center of the wall below.

The centerline of a log is not something that can be scientifically determined in a log yard, in fact you shouldn't even use a tape measure. A chalkline and your judgement are all that is needed.

Straight logs ("A" on far right) are easy. Extreme bow is difficult (log "E")—in fact, I would not use log "E" in one piece—it has too much bow. Log "D" has bow that is equal to one-half its small-end diameter—which is the most bow that I will use.

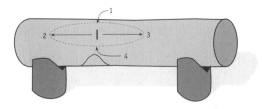

Figure 21: Saddle layout requires 4 points. The top (point 1) is 2" (50mm) down from the high spot of the log. The sides (points 2 and 3) are a set distance out from the center mark you made in Figure 18. The bottom (point 4) is 1" (25mm) below your best guess of the final notch scribe line (Figure 13).

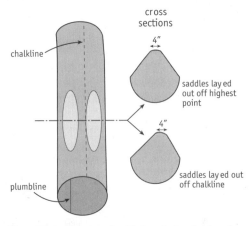

Figure 22: Another view of saddle layout, also showing why you don't use the chalkline to lay out the top point of a saddle. Using the chalkline, on bowed logs, will give the saddles different slopes. Using the highest point gives saddles equal slopes.

Figure 23: Try to cut the rough notch with just two cuts, removing a large wedge of log that looks like a section of an orange.

10) Lay Out & Cut the Saddles

Over the years I have tried several different shapes for saddles: first, basically flat, with ramps leading into the large flat area; then, butterfly, or Beckedorf shrink-to-fit; and finally, concave. About 12 years ago I settled on saddles (also called *scarfs* or *receivers*) that have an even concave shape. This shape has most all the features that I want in a corner notch.

Lay out saddles now, while the log is plumb, and before you roll the log over to cut the rough notches, or else you'll have to plumb the log again.

Some professional log builders like to use a saddle board to help them lay out the saddle size and shape. This is a piece of wood cut to the saddle dimensions, but to me, this is just one more thing to get lost in the sawdust. Concave saddles are easy to lay out using only a tape measure and a felt-tip pen—and these are two things that I always have in my tool pouch. Because my saddles are concave, a board with straight edges wouldn't work anyway.

Each concave saddle requires just 4 layout points *(Figure 21)*—don't try to draw an oval that connects the 4 points. You'll actually get a better saddle if you don't try to draw the entire saddle outline.

Point 1) The top of each saddle is 2" (50mm) down from the highest point of the log. To find the highest point, place a 2-foot level on top of the log, over the center of the notch, and make the bubble read level. The highest point of the log is where the level touches the log—mark this point on the log. In practice, I don't use a 2-foot level to find the highest point, but it is a good aid for learning what to look for.

The highest point is not the same as the chalkline unless the log is very straight. The problem with using the chalkline for saddle layout is that when the chalkline does not follow the highest point, then the saddles will have different slopes, and it is very difficult to get the notch that is scribed over these saddles to fit tightly over time *(Figure 22)*. By laying out the saddles from the highest point instead of the chalkline, the saddles always have the same slope.

The round portion of the log that is between the two saddles can be wider or narrower than 4" (100mm)—and what amount you decide to use depends upon the average width of your long grooves at the notches. If your grooves are often 5" (125mm) wide near the notches, then you should leave 5" of round log between the tops of the saddles. As you scribe a long groove does your scriber fall down onto the saddle? Then you should leave more round log on top for the long groove to rest on.

You can change the distance between the tops of the saddles as you build—your logs will tell you how much room they need for their long grooves. Listen to them.

Points 2) and 3) The corners look best if all the saddles in a building are the same width *(Color C-31)*. I lay out the sides of the saddle by first finding the biggest butt end in the log deck. For example, if the biggest butt is 16" (40cm), then I make all the saddles 26" (66cm) wide—10" (25cm) wider than the biggest butt on any wall log. Measure out 13" (33cm) each way from the center-of-notch mark you made in Figure 18. You could make all the saddles 20" (50cm) wider than the largest butt, or even wider than this, if you want, but you'll probably need extra-long flyways, too.

Point 4) The bottom of each saddle was found in Figure 13 by making an estimate of what the final scribe setting will be, and then measuring up from the rough-notch scribe-line an amount equal to the final scribe minus 1" (25mm).

Cut the saddles, and sand or plane them smooth.

11) Cut the Rough Notches

Rough notches are cut quickly—it is a rough notch, after all. I sometimes cut these on the wall (at least, when the walls are low), but it's much safer and faster to use a chainsaw with your feet solidly on the ground.

The rough notch can be, and usually is, just two cuts to remove a chunk of wood that looks like a section of an orange. Try to make the two cuts meet so the chunk just lifts out, with no carving and no brushing *(Figure 23)*. Do not bother to score the scribe line with a knife or chisel—rough notches don't need this. Cut as closely to the scribe line as you can—sometimes going slightly over, and sometimes leaving little fragments of the line intact *(Cutting chapter, Figure 2)*.

Experts use a few extra tricks when cutting rough notches that make building easier. Overcut the top of the rough notch to create an opening shaped like an upside down "V" *(Figure 24)*. This gives the bottom leg of the scriber room to slip inside when you're scribing the final notch *(Figure 39)*. The point of the V should be right at the bottom of the saddle.

There can be a problem when final-scribing a B2 notch: the bottom leg of the scriber may hit the log, and the upper leg of the scriber can't reach to make a scribe line *(Figure 25)*. One way to fix this is to use a chisel and mallet to chop away the wood that's in the way. Be careful, however, that you don't move the log.

Some professional builders cut the rough notch of B2 logs to remove a little extra wood to help avoid this scribing problem. These little openings are called *notch windows (Figure 26)*. Don't overdo it, though, because if you remove too much wood, then there will be a hole that the final scribe can't close up. You'll have a gap. Notch windows are a trick for experts, not beginners.

12) Clean Up the Bottom Surface

After the rough notches are cut, and while the log is still upside down, look over the bottom surface and clean up the rough spots. It is easier to get accurate long-groove scribe lines if this area is smooth. When you are scribing, small imperfections feel like a crevasse—spending a little time now with a drawknife brings big benefits later.

13) Put the Log Back on the Wall

Lift the log back onto the wall. The log should be quite stable now, and you won't need to use log frogs to keep it in place unless the rough notches are very shallow.

Figure 24: The rough notch has been cut and the log is rolled back right-side up on the wall. Note the rough notch was cut to a point, and this provides a small cave that can be useful: the bottom leg of the scriber can fit into it to scribe the final notch. The point of the V should be right at the bottom of the saddle.

Figure 25: When scribing the final notch, the bottom leg of the scriber sometimes hits the log, preventing any further scribing. This happens most often when scribing the notch of a B2 log.

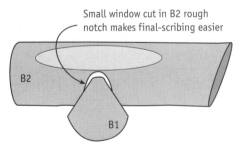

Figure 26: Some professionals cut a little extra wood out of the upper sides of the B2 rough notches. Notch windows provide room for the scriber leg while scribing the final notch.

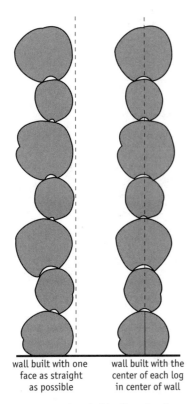

wall built with one
face as straight
as possible

wall built with the
center of each log
in center of wall

Figure 27: I recommend you build walls so that the centerline of each log is plumb above the center of the wall (drawing on right). Trying to make one side of a wall more or less plumb can be difficult, unattractive, and perhaps unstable (drawing on left).

Figure 28: If needed, improve the location of the widest gap after rough notches are cut. If the widest gap ends up near a notch (arrow, top drawing), then you can raise the other end of the log with a butterfly until the widest gap jumps into the target area (shaded area in the bottom drawing). Note that when lifted, the rough notch on the right end of the log doesn't touch the log below.

Approximately Center the Log

Move the new log so that its top centerline is in the center of the log wall—this divides diameter, taper, and bow evenly between the inside wall surface and the outside wall surface. I have seen buildings where the inside surface of the log walls was plumb. This is difficult to do, and it makes the outside surface very irregular *(Figure 27)*. If the logs have a lot of taper or bow it could even make the walls unstable.

Move one end of the log, and then the other end, so that its top-center chalkline is plumb above the center-of-wall marks that are on the sill logs. For now, if the log is centered within about 1" (25mm) that is close enough because you need to find the location of the widest-gap-of-all. If the new log is not approximately centered, however, then this will affect the size and location of the widest gap.

Widest Gap Location

Using inside calipers, find the widest-gap-of-all between the two logs. It should be somewhere in the center half of the wall's length—or at least not near the end notches. If it isn't, then re-cut one rough notch (to lower that end a bit), or even easier, raise the other end with a butterfly *(Figure 28)*.

The widest-gap-of-all will move towards the end you are raising. If both the logs were gun-barrel straight and as smooth as steel, then the widest gap would move slowly, inch by inch, towards the end you are raising. But because the logs have unique, natural shapes, the widest gap will tend to jump from one spot to the next spot. Stop raising the end as soon as the widest gap has jumped to the middle half of the wall's length.

Rarely, you will find that you have exactly the same gap in two places. That is okay, and you do not need to adjust the gap if those two wide spots are far apart.

To level the top of a log, ignore the location of the widest gap for now, and instead raise the low end of the log with a butterfly as you measure the height of the top of the log off the stumps or the floor.

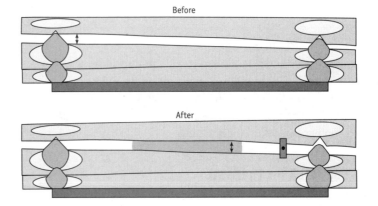

Before

After

Positioning Log Pieces

You want to position log pieces on a wall as if they were one log that is as straight as possible. Even some professional builders do not do this correctly, and so I want to take some time explaining how to set up and position log pieces for final scribing.

As you know, a full-length log should have its widest gap somewhere near the middle half of the wall's length. This is true for pieces, too: it is usually best if the widest gaps are close to the middle half of the wall (*Figure 29*). Adjust each goal post or butterfly so that the widest gaps are close to the middle half of the wall. Often this requires that the widest gap for a piece will be very close to, and sometimes even right at, its free end.

Don't put the widest gap near the middle of a piece, unless that piece is very long. If you put the widest gap of each piece near the middle of its span, then it will look like it's pointing down at the opening (*top drawing in Figure 29*). Repeat this mistake for a couple layers of logs and the effect accumulates and becomes unattractive. It is not unusual to see handcrafted log houses that appear to droop down at door openings. But they haven't sagged over time—they were actually built that way when the log pieces were set up incorrectly.

This is avoidable. Sight down the top of the pieces from one end—they should look like a one-piece log (*Figure 30*), that is, they should be straight on top. If they aren't, then adjust each butterfly up or down as needed, and then check to find the widest gaps.

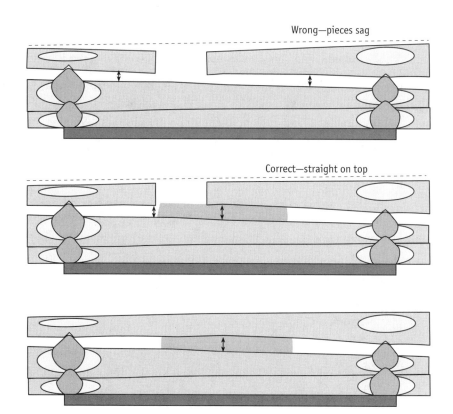

Wrong—pieces sag

Correct—straight on top

Figure 29: It is important to set up log pieces correctly, or the wall will appear to sag at openings. The key is to adjust log pieces so that they look as if they are one straight log.

The top drawing shows a common mistake: if you put the widest gap near the center of each piece, then the wall soon starts to look like it's sagging at the opening.

Instead, put the widest gaps near the center of the **entire wall** (middle drawing), which is how we treat the widest gap in a full-length log (bottom drawing). That's why I say you should treat log pieces as if they were a one-piece log.

The dashed lines are not level (though they could be), they are straight. The tops of logs may or may not be level, but it is best when they are straight. Using pieces lets us straighten out bent logs, so we gain nothing if we put pieces on the wall and make them crooked again (top drawing).

Note that goal posts or butterflies are not shown in these drawings to make it easier to see the shape of the gap and the location of the widest gaps (indicated by the arrows).

Center the Log Accurately

We started Step #13 by centering the log just enough to find the location of the widest gap. After you have confirmed that the widest gap is in a good place, or have raised one end of the log to move the widest gap into the target area, then you need to center the log accurately (*Figure 30*).

Figure 30: Three pieces of log were used to make up this next layer. Each piece was given its own chalkline, and then each piece was centered over the chalklines below. Also, sight down the pieces from the end, like this, to make sure they line up. Here the center of the wall is clearly marked on the sill log, and a chalkbox is used as a plumb bob.

Do not draw plumb lines on the log ends—I have often seen them used, but I have rarely seen them work well. There are two reasons for this. First, a level is not accurate for log buildings—when you hold a level on a sheetrocked wall you can quickly draw an accurate plumb line, but hold the same level on the end of a log, and it is unreliable. Second, the plumb line will become unplumb when the log is pushed one way or the other on its saddle. The details follow right now.

As strange as it sounds, a level must be held on a flat and plumb surface in order to read plumb. A level can have its bubbles precisely centered, and yet the level can be way out of plumb if it leans and is twisted—try it. When a level leans towards you or away from you, then its bubbles are thrown off by even a small amount of twisting—so little, in fact, that you might not notice that it's happening. This means that a high quality, and accurate level is not reliable when held on a log wall.

The second reason I don't draw plumb lines on the log ends (except for the sill logs) is that it is inevitable that when you set some rough-notched logs back on the wall, their plumb lines will no longer be plumb. Then you'll waste a lot of time trying to shim one side of the rough notches to bring the plumb line back to plumb when it doesn't matter. With concave saddles (the kind described in this book), when a log is pushed to the left or right to get it centered on the wall, it rotates slightly on the curves (or ramps) of the saddles. This means that log-end plumb lines waste time. (As a business owner I know first-hand that this is true.)

Keep it simple—first, we need a top-center chalkline that divides each new log in half: and second, we need to move a new log so its chalkline is plumb above the center of the sill log. You don't need end-grain plumb-lines to do this.

Using a Plumb Bob
I like a plumb bob for centering logs, and I make them myself from a 2- or 3-pound (1kg) fishing downrigger (a ball of lead), tied to a length of brightly-colored mason's line. They are cheap (less than $4 each), so I keep a plumb bob in every corner of the building, and I never have to go looking for one.

Figure 31: It's also easy to use a plumb bob off the side of the wall, as shown. Screw a piece of plywood to the sill log so it has a straight edge that is an exact distance from the sill log centerline. In this drawing, the new log needs to move to the left about one inch (25mm) because its chalkline is 11" from center, and it's supposed to be 10".

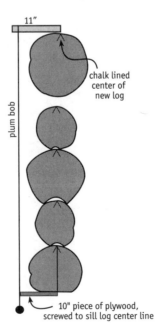

11"

plumb bob

chalk lined center of new log

10" piece of plywood, screwed to sill log center line

Sometimes you can drop the plumb bob off the end of the new log, holding it on the top chalkline *(Figure 30)*. Or, you can hold the plumb bob off the side of the new log *(Figure 31)*. Just don't let the string ever touch any of the wall logs or it won't hang plumb.

How Close?

It is not really necessary to center a chalkline over the sill log any closer than about ¼" (6mm). After all, you might have easily decided to hold the chalkline one way or the other by ¼" before it was snapped, so the line itself is only accurate within ¼". As long as you center every new log by comparing it to the sill log, then ¼" is close enough.

If you transfer plumb lines up the log ends, one layer after another, however, and never refer back down to the sill log, then even ⅛" (3mm) accuracy may not be enough—the top log could be 1½" (40mm) off center because error will accumulate from log to log to log. But, if you always compare each new log all the way down to the sill log, then no log will ever be more than ¼" (6mm) off center.

Using a Level

Instead of a plumb bob, each new log can be centered with a long level, but only if use the following procedure. First, screw a piece of plywood to the bottom, sawn surface of the sill log so that its outer edge is exactly 12" (30cm) from, and parallel to, the centerline on the bottom of the sill log. Hold a level flat on the plywood edge *(Figure 33)*. With the level held plumb, measure 12" back towards the top of the new log—you can easily tell whether the new log needs to move in or out. When the wall gets taller than your longest level, screw small plates into convenient locations to act as level extenders.

Square the Building

Every several courses of logs, measure from one chalkline to the chalkline on the parallel wall to make sure the walls are the right distance apart. You should also check diagonal lengths from corner to corner to make sure the building is staying square. If the diagonal measurements are equal, then the corners are 90°.

14) Trim this Log's Ends

Cut the log's end clean and straight, and about even with the flyways below. This makes it easier to scribe the end grain of the flyway's long groove.

The log ends should still be several inches (50mm+) longer than their final lengths so you can trim them all later, after the walls are completed. It is difficult to trim ½" (12mm) from a log end because chainsaws don't cut straight when there is wood on only one side of the bar.

Some flyway styles must be trimmed and sanded to their final lengths as you add each log and before you final-scribe it—they cannot be trimmed later. The classic chiseled log ends and the flyway style in which butts and tips have staggered lengths are like this.

15) Find the Widest Gap

Find the one spot that is the widest gap anywhere between the new log and the log below. Starting near one notch, and holding a pair of inside calipers so the

Figure 33: A level can be used to center each log on the wall, if you set the building up to do this right. A piece of plywood is screwed to the flattened sill log so that its edge is, say, exactly 12" from the centerline of the sill log. The level is held against the plywood and an extender, and is plumb. (The level is accurate because it is resting on flat and plumb surfaces—notice that it's not touching a log.) Measure back towards the chalkline of the log you are trying to center on the wall. As walls get taller, use simple extenders (top photo)—a flat piece of steel brazed to an Olylog screw. I learned this trick from Jeff Westhoff.

Figure 34: Once the log is in position and ready to be final scribed, find the widest-gap-of-all. Always ignore the gaps in flyways. An inside caliper is a fast and easy way to find the widest gap.

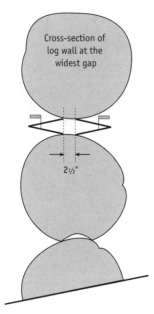

Cross-section of log wall at the widest gap

2½"

Figure 35: Working at the widest gap area, scribe a foot of the inside groove edge and the outside groove edge and measure the width of the groove. You want the groove to be about 2½" (65mm) wide at this location. Open or close the scriber setting, if necessary, to make the groove close to this width. Note that this has nothing to do with any chalkline.

points are about plumb, drag them along between the two logs, adjusting them wider as you go so that they always barely fit in-between *(Figure 34)*. The last place the calipers fall through is the widest-gap-of-all. Don't check the flyways for the widest gap because we don't care if we have a continuous long groove in the flyways.

With practice, you'll be able to stand back and look at the wall and pick out the two or three possible widest gaps, and then check them with calipers. And, as you know by now, the widest gap should be located somewhere in the middle half of the log's length.

If you are drilling electrical or through-bolt holes on the way up as you build the wall, then check every hole to see if it is the widest gap—it often is. If the caliper falls into a hole, even though it does pass through to the other side, then you have to open the caliper up far enough to stay out of the hole.

Set the Scribers
Working at the widest gap area, separate the tips of a pair of scribers so that the groove will be about 2½" (65mm) wide here. This does not mean the tips of the scribers are 2½" apart. This means that measuring from the inside edge of the long groove to the outside edge of the long groove is about 2½" *(Figure 35)*.

Ignore the chalkline—it has nothing to do with scribing—it's only purpose was to position the log on the wall. The long-groove scribe-lines will not be centered off the chalkline, and you do not make any measurements from the chalkline to the scribe lines.

Photos of a variety of scribers can be found on page 227, Figure 6. The four photos on pages 106 and 107 show scribers being held and used to scribe grooves and notches.

With the scribe distance locked, plumb the scribers on a plumb board *(see Step #16)*, then go back to the widest gap area and scribe a foot of the inside edge of the long groove and a foot of the outside groove. Make sure that the groove will be about 2½" wide, or so, here. If it is narrower than 2" (50mm) then it is difficult to keep the groove weathertight, so open the scriber tips apart slightly, re-plumb the scriber, and try scribing this area again.

If the groove is wider than 3" (75mm), then you'll have to remove a lot more wood from the long groove—remember that it will be wider and deeper *everywhere*, not just at the widest gap. To reduce your cutting time, and also reduce the loss to scribe, close the scribers points slightly closer together, re-plumb the scriber bubbles, and test again at the widest gap area. There are other ways to set the final scribe distance, but the method described here is the only way to guarantee that the groove will never be narrower or wider than necessary, and that every electrical hole will be fully covered by a long groove.

16) Plumb the Scribers
Most wall logs gets scribed with three scriber settings. First, the long-groove setting—all scribing starts with the groove; second, the notch setting (if you are underscribing the notches); and third, the flyway (overhang) setting.

There are two types of scribers (also known as *scribes*), those that are always plumb, and those that must be re-set to plumb when you change the scribe distance, or sharpen a pencil, or replace or adjust a pen. The self-plumbing scribers that I like are Timmerhus *Ultrascribes* and Grieb *Gearheads*. These tools save time, and are easy to use, but they cost more than basic scribers.

Other scribers must be plumbed each time before they are used. First, you need a plumb board—a very flat, stiff, and smooth piece of clean wood about 3" wide by 14" long (75mm by 350mm). The exact measurements do not matter *(Figure 36)*. Use screws to attach the plumb board to something that won't move or be bumped: a telephone pole, the wall of a building, or a stiff post.

Tapered wood shims are placed behind the plumb board so that the face of the board is plumb (hold a level on its edge against the board). Tighten the screws slowly until the board is tight and is plumb. Make certain you have not twisted or bent the board with the screws, and double-check the plumb board every morning.

The final step is to lay the level flat (not on edge) on the plumb board, adjust the bubble until it is plumb, and carefully draw a vertical line. This line is plumb both left-to-right and also back-and-forth. The plumb board is ready to use.

Before you plumb scribers, the scribers must be sharp (if you use pencils), the pencil or pen holders must be tightened firmly, and the scribe distance must have been decided, set, and firmly locked. Gently hold the tips of the scriber on the plumb line drawn on the plumb board—that makes the tips plumb (since they are resting on a plumb surface and a plumb line).

Now, adjust the bubble (if your scriber has a single bull's-eye) or levels (if your scriber has two vials) to read level. When the tips are on the plumb line and the bubbles read exactly level, lock the bubbles so they won't move. The scribers are ready to use.

If a pencil wears down or a point breaks off or you sharpen it, or if a pen dries out and is replaced, or if the scriber is bumped, then you have to re-set the scribe distance and also re-plumb the scriber. Repeat this process each time you change the scribe distance setting. Accuracy is very important, and scribers are precision instruments.

A scriber that is always plumb costs about $150 more than an a scriber that requires re-plumbing. But in just one house you will save that cost ten times over. Each wall log needs three scribe settings, an average house has maybe 120 log pieces to be scribed. In one house, you can eliminate 360 trips to the plumb board by using always-plumb scribers.

I am often asked which is better: scribers with a single bull's-eye bubble, or with two bubbles. I own both kinds, but I learned with two bubbles, so I prefer two bubbles. If you learn with the bull's-eye bubble, then that's what you'll prefer. They are equally accurate. Some fans of bull's-eyes say they have only one bubble to watch, so it's easier for them to scribe, but I haven't found that to be a problem. Use whichever scriber you feel comfortable with—the difference between them is not very important to how well your logs will fit together.

Figure 36: The plumb board is attached to a stiff pole or wall and plumbed with shims. Then a plumb line is drawn on the face of the board. Scribers are set by holding the tips on the line, and then adjusting the scriber's bubbles until they read plumb, and locking them there.

Figure 37: Drag the scriber towards you and let your hands and fingers feel the contours of the two logs. The scribers are leading, and you are following their lead. Leave a clear and continuous line, and do not push the scriber into the log too hard. It's okay if the scribers are not 90° to the log (above).

Below, focus on the bubbles, and know where the points of the scriber are by *feeling* them through your hands, not by watching them.

17) Scribe the Groove

Make a record of your scribe setting—you can refer to it if your scribe setting ever changes (when a pencil wears down or breaks off). Do not bump the log or the building until all the scribing is finished. If the log moves before scribing is complete, the lines won't be accurate and the log won't fit.

Hold the scribers like you're holding a live bird—just tight enough that it won't get away, but not so tight that you squish it. The scribers should touch the logs gently—just enough to leave clearly visible lines *(Figure 37)*. If you press too hard the scribers won't slide smoothly, they'll grab and release, and the line will be jagged. You also risk breaking the pencil lead or flexing the pens. Always drag the scribers away from their points (so the points are trailing). Note that the scribers do not have to be 90° to the wall—they can be at any angle that is comfortable.

There are times when you'll need to let the scribers go in towards the log (when the groove becomes narrow) and also times when you'll have to let the scribes come out towards you (when the groove gets wide). A common mistake is to push the scribers too tightly against the log, and not let them float in and out, as they must. The unique shapes of the two logs determine where the scribes go. You cannot control this—in this dance, the scribers are leading, and you are following.

To get the tightest fits possible, the scribers must be plumb whenever they are moving. Watch the bubbles. When they go out of plumb, stop—pull the scriber points ½" (10mm) away from the logs and re-plumb them, and then move in. Don't try to plumb the points when the scriber tips are touching the logs!

Take time to get it right because the quality of the fit depends entirely on accurate scribe lines. You must have smooth, continuous scribe lines. If you get double lines remove both of them and scribe that area again. It doesn't matter how accurately you can cut, after all, if you're cutting to a line that isn't right!

Scribe the inside and the outside edges of the long groove, that is, scribe from both the inside and the outside surfaces of the log wall. Double check to make certain that every inch of the long groove is scribed—on both sides of the wall. If you get a log on the ground and discover that some of the line is missing, then you'll have to put the log back on the building and try to get it back in the same precise location. It's sometimes easier to simply drawknife off *all* the scribe lines and start scribing the log again from scratch.

If the long groove scribe falls onto a saddle you can re-shape the bottom of the upper log somewhat to move the line upwards. If this happens frequently, then you'll want to increase the distance between the saddles at the top of the log to give you some more round log for the long groove to fit *(see Step #10)*.

18) Scribe the Notches

All the notches on a log get the same scribe setting. Starting with the setting for the long-groove, close the scribers down the correct amount *(see the chapter on Underscribing)*, and re-plumb the scribers. Start at the bottom of a notch and gently drag up about halfway, and then start at the top of the notch and gently drag down to meet that line *(Figure 38)*.

The scriber will often be swiveled at angle to the log, not at 90°—use an angle that keeps the tips of the pencils or pens against both the saddle and against the log above. Make certain you are always leaving a line on both logs, and keep the bubbles plumb at *all* times. Repeat for the other three faces, or quarters, of this notch, and then move on to the other notches.

You will not be able to slide the scribers *under* the bottom of a notch—the scribers stop at, or very close to, the long-groove scribe lines. Also, the notch scribe line will not exactly meet the groove scribe line because of the underscribe difference (and you can see this in the top photo of Figure 38).

Most people find that scribing grooves is easier to learn than scribing notches. Scribing grooves is mostly a side-to-side, horizontal motion, while scribing notches requires that you keep the scriber points plumb while they move in three dimensions (left-right, up-down, and in-out).

Here are some more tips. Watch only the bubbles while you are scribing—if your eyes wander to the lines you're leaving, then the bubbles will go out of plumb and you'll have lines to remove. To check out your lines, stop scribing and *then* look. Use two hands on the scribers, if you like, it might make the lines smoother. With your fingers, feel the pen or pencil points through the scriber arms—it will take some time to develop this feel, but it is essential. You need to feel both the points writing, and how softly they are touching, almost floating, and without ever looking at the points.

You need one clear and accurate scribe line for the notch. If you end up with several lines, or a tangled web of lines, then remove them all, and start again—a sharp chisel with the bevel against the log, or a 4" disc sander with fine grit, works well.

You will get better with practice, though some people improve faster than others. And, in my experience, a few never get completely comfortable scribing notches—though they may be great at grooves.

Can't Reach

There are several common problems you'll probably encounter when scribing notches. First, sometimes the scribers won't reach the saddle at the top of the notch, unless you tip the scribe out of plumb (and don't ever do that). This problem is often caused by saddles that were cut below their bottom point—the cure is simple: lay out each saddle's bottom point with care and then do not cut, plane, or sand past it.

This situation can also be caused by not cutting extra wood out of the top of the rough notch *(Figure 39)*. We want to leave a small open cave at the top of the rough notch *(see Figures 24 & 25)* so that the bottom leg of the scriber can reach inside the rough notch and still stay plumb while you scribe the top of the final notch on its saddle (this can be seen in the bottom photo of Figure 38). If your scriber leg won't fit into the cave, then use a chisel and mallet to enlarge it slightly. But before you do that, try swiveling the scriber and entering the cave at an angle (not 90° to the log), though still plumb, of course. Some scriber models, the Grieb Gearheads for example, have slim pens and can reach in more easily than other scribers.

Figure 38: Start at the bottom of the notch and gently lift the scribers, all the while making sure that both scribers legs are constantly touching the logs, and the bubbles are always plumb. (The scribe lines above have been digitally enhanced to make them easier to see in the photo.)

The scriber can be swiveled at any angle to the notch—as long as both points are touching and the bubble reads plumb—it does not have to be 90° to the log, and often isn't (photo below).

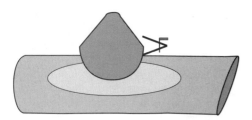

Figure 39: A common problem comes when the bottom leg of the scriber hits the log, and the top leg can't reach without tipping the scriber out of plumb (and don't do that!)—drawing above.

The little cave that is cut out of the top of the rough notch allows the bottom leg to reach inside, and then the top leg of the scriber can touch the saddle of the log above—drawing below.

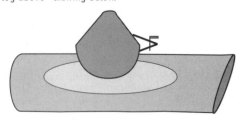

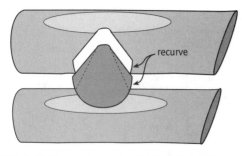

Figure 40: Recurve is a serious problem that must be cured *before* you scribe and cut the notch. This shows a severely recurved notch that is too late to fix.

The bottom edges of the saddles of the log below were not cut low enough. The next log was final scribed and no one noticed that the bottom leg of the scriber had fallen off the bottom of the saddle below. The notch was cut, but it's too narrow at its bottom to even fit over the log below.

Avoid recurve by re-shaping the saddles on the log below before the final notch is scribed. The dashed lines show where the saddles should have been cut to avoid recurve.

Recurve

A second problem is when the bottom leg of the scriber falls off the bottom of the saddle below—this is called *recurve (Figure 40)*. If the bottom of the saddle was either not laid out right or not cut low enough, then you'll have recurve. The second cause is when the final scribe setting for the log you're working on turned out to be much wider than you originally expected. In both cases, the cure is to re-shape the saddles of the log below, remove the scribe line from the upper log, and then re-scribe this portion of the notch. The best tool for reshaping a saddle in place is probably a *slick*—a large and wide chisel used by pushing, not striking.

You might be able to cure small amounts of recurve by removing some wood from the bottom of the upper log—use a slick or small broadaxe to slice off the wood. But if you take off more than ½" (12mm) here it will look like a beaver has been chewing at the notch.

If you do not remove the recurve, then you could have a notch that is too narrow at the bottom to fit over the log below. This cannot be fixed—you'll probably have to throw away the top log and start with a new one.

Underscribe Tabs

There are two ways to hold up the free end of a one-notch log piece that has had its notch underscribed: wedges, or scribed tabs. If you are going to use tabs, then use the scriber (with the notch setting), and scribe 2" (50mm) of long groove at the free end of every log piece. Scribe from both the inside and the outside of the wall. These very short, new scribe lines will be slightly inside the scribes lines that run the full length of the log. When you cut the groove, follow the scribe lines up until you get to the end of the piece, and then switch over and use the underscribed tabs for the last 2" of the log.

19) Scribe the Flyways

Open the scribers ½" (12mm) wider than the long-groove setting for the log, and scribe the flyways, including the end-grain. We don't want the flyways to bear on each other (see *Log Building Standards* Section 2.E.3), and opening the scribe setting ensures that they have a cove that conforms to the right shape, and looks nice, but does not quite touch the log below *(Figure 41)*.

Some builders save a little time by using the same scribe setting for the flyways and for the groove (so they don't have to reset or re-plumb their scribers). Then, when they cut the flyway cove, they simply cut to the outside of the scribe line instead of the inside, as usual. This provides the desired ½" gap.

When flyways are supporting weight—an example is a cantilevered plate log supporting deep roof overhangs—then they can be scribed with the same setting as the long grooves. This makes the flyway cove touch the flyway below, and help distribute the load down the corner.

20) Drill Holes

Log walls need vertical holes for electric wiring and through-bolts. If you are building on the foundation, and the home will not be disassembled and moved, then you must drill holes at this stage. If you are building off-site, then you can drill holes now, or wait until you take a shell apart to load it onto trucks to move it—something that many professional yards do.

I try to use as few different auger bits as possible (for one thing, good bits are expensive). I drill electric-service holes 1¼" (30mm) in diameter. I could go a little smaller for electric holes, but using this size let's me use the same bit for through-bolt holes *(Figure 42)*.

I drill vertical through-bolt holes from the sill log to the plate log—the entire height of the wall. When the log walls are assembled on the foundation I drop a long, ⅝" (16mm) diameter all-thread rod into each hole, and use washers and nuts at the top and bottom. This holds the walls together, and can be tightened in the basement as the logs settle. Dowels and lag screws are used by some builders instead of through-bolts, but they can't be tightened in the future because there is no access to them.

Make sure the holes don't come outside the long-groove scribe-lines—if one does, then it will be visible. The only way to fix this is to scribe the entire log again with a wider scriber setting (or use caulk to fill the gap). To avoid this, use a smaller diameter auger bit, or move the hole left or right to a place where the groove is wider (if you can), or drill the holes before you scribe the log. Back in Step #15 I reminded you to check all holes while searching for the widest gap.

If your auger bit is shorter than the depth of the hole, then pull the auger out (with the drill running forward, not reverse, while you do it) to clear the chips.

Figure 41: Flyways look good when they are scribed to match the shape of the log below, but provide about ½" (12mm) space.

See also color photos C-17 and C-30.

Figure 42: If you are building on your foundation, then you should drill holes while the log is still rough-notched, and just before you bring it down to cut it out. At this stage you can still see where the hole is in the log below, and this helps you line up every hole. A bubble glued to the drill's housing helps you keep it plumb (see Tools chapter).

If you are building off-site and are going to move the shell to its foundation, then you can either drill holes now, or when you take the shell apart.

Figure 43: Cutting out the notches. In this photo, the notch has been scored, cut, and carved, and is being brushed to finish it off. Note. the saw is held against the hip to steady it.

Figure 44: Cutting the long groove. Get the log at a comfortable height for this work. See also Color Figures C-23 and C-24.

21) Double Check that Everything is Scribed!

Check, then check again before you lift the log to ground to cut the notches and groove. If you take a log off the building, turn it over and find that a little bit of the groove didn't get scribed, or one-quarter of a notch was missed, then you are in BIG trouble—you'll probably have to drawknife all the scribe lines off, and start this log again. Of course, if this happens to you once, you won't forget to double check the scribe lines in the future.

As you gain experience, you'll be able to add a missing inch (25mm) by hand without having to scribe it, but anything more than that really should be scribed.

22) Bring the Log Down

Put the straps back on the log, use a tag line to control long logs, bring the log down and set it in V-blocks or horses.

23) Score the Notches & Cut them Out

I like to use a 1" chisel to score the notch scribe line. Scoring prevents splintering, and also make the notch quicker to cut with the chainsaw. Cut large V's out first, then carve the center of the notch (but not the perimeter) to create a slight concavity, and finally, brush to the line *(Figure 43)*. You'll find plenty more about cutting techniques in the next chapter.

24) Cut the Groove, Flyways & Kerf

No need to score the grooves, because the chainsaw is cutting *with* the grain, not across it, so just cut them out with the chainsaw *(Figure 44)*. Trim close to the scribe lines with a chisel or sander *(Color Figures C-23, C-24, and C-25)*.

Cut the kerf on the top of the log below, staying inside the scribe lines. If you are using the double-cut groove, then instead of a kerf you will remove a "V" of wood.

If a log ends at an arched opening, then stop the kerf before the opening, and carefully cove the long groove so it closely matches the shape of the log below. This makes the exposed ends look neat and tidy *(Color Figure C-27)*.

25) Test the Fit

Lift the log back onto the building and line up the notches on both ends so they are exactly on the scribe lines of the saddles below. You cannot tell if the log fits until both of its end notches are positioned exactly right *(Figure 45)*. The chalklines and plumblines are not accurate enough—that's why I double-scribe everything (the scriber has pens or pencils in *both* legs). Put the notches of the new log right on the scribe lines on the saddle below *(Figure 46)*. A crowbar is great for gently moving a log—reach into the flyway and pry *down* to make it slide easily. (If you pry *up* you'll make a dent in the edge of the groove.) The best crowbars I've found are made by Gränsfors Bruks *(see Tools chapter, Figure 14, page 229)*.

The fit is good if all the notches are completely tight and without gaps, and the groove has a long, small, and even gap (the underscribe amount). Your first few logs will probably have hang-ups—areas where not enough wood was removed from inside the groove, notch, or flyway. A piece of wood as small as a couple matchsticks can prevent a log from fitting.

Just because you find one hang-up, don't stop there—find *all* the problems before you take the log off to fix them. This saves time. A 12" General® flexible ruler is a great tool for feeling inside the notches and grooves—probing for hang-ups that you can't see. It's the tool I always have in my back pocket.

Here Are Some Common Hang-ups:

1) The long groove edges came down onto the saddles of the log below. Fix this by sanding off the sharp edge where the top of the saddles meet the round on the top of the log. Never sand below or past the scribe lines or you'll make a gap!

2) Knots inside the groove—bumpy and curly-que grooves can hang-up—especially if there are bumps on the top surface of the log below. I usually smooth out these portions before I scribe the groove, because I know that they can cause hang-ups.

3) Flyway hang-ups—look inside the flyway from its end. The groove can get wide in the flyways, and so it needs to be deep, too. If it's not deep enough, then you have a hang-up. Flyways tend to have wide and deep grooves because the butt ends of logs sometimes flare out and get lumpy.

4) A knot inside the corner notch on the top of the log below, between its saddles. Just sand the knot off the log below—it's easier than turning the log over to make the notch deeper. Next time, when you are cutting the saddles, smooth the area on the top of the log between the saddles and eliminate this hang-up before it ever happens.

Figure 45: Lower the log into position, and at both ends match it up exactly on the notch scribe lines on the saddles below. Then check the whole log for the quality of the fit.

This is a square notch, also called a lock notch, and is sometimes used for plate logs and for support beams where you want to maintain strength in the log, and can afford to spare some wood out of the log below.

Once it is fitted it is impossible to tell it from the standard saddle notch.

Figure 46: A tightly-fitting round notch. Note that there are no saddles, so it is not a saddle notch. Also see Color Figure C-35.

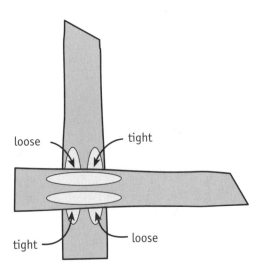

Figure 47: Gaps in the notch? If two diagonal quarters of a notch are tight and the other two-quarters are loose, then the bubble was set wrong when you scribed the notch—it was tipped either to the left or to the right.

Figure 48: The rhythm of log building: it's just one log after another.

5) Areas where you did not remove all the wood right down to the scribe line. The last wood that is left on the log should be colored by the pen or pencil—*no uncolored wood should remain near the scribe line.* But, do not remove a scribe-line from the log or you'll have a gap.

6) Still can't find the hang-up? Gently wobble the log with a peavey to see where it's pivoting—that's where you should look for the hang-up. Probe there with a thin, flexible ruler. As a last resort, put the log in position (as best you can) and give the top a couple good whacks with a sledgehammer. Lift the log off and look carefully for shiny spots—places where the wood has been compressed and rubbed. Sometimes the hang-ups are very small—the size of a single kernel of corn.

7) If a notch fits on two sides, but is loose on the other two other, then check to see if the tight fits are diagonally opposite, and the loose fits are diagonally opposite *(Figure 47)*. If they are, then there is no easy way to improve this fit: your bubble was set wrong when you plumbed the scribers. Unfortunately, you have to decide whether you'll keep the log, or replace it with a new one. There are almost no good ways to improve the fit of a log that was mis-scribed. Accurate scribing is essential for good fits.

8) If you cut or sand over the scribe line, then you'll have a gap. You cannot close up gaps that were caused by the scribe line being accidentally removed. The only way to close such a gap is to remove wood from all the places that now are fitting great. It's a losing proposition to fiddle with all the spots that you like the looks of, in hopes that the spots with gaps might get better. (That's a little like trying to fix a leak in a dam by removing all the parts of the dam that are strong and holding water!) Therefore, it is futile to try to close gaps that were caused by cutting over the scribe line. The lesson? Accurate cutting is also essential.

Once the fit is good, congratulations.

And on to the next log!

Cutting Notches, Saddles, & Grooves

It's time to cut rough notches, saddles, final corner notches, long grooves, and flyways. This chapter will take you, step-by-step, through the process of using a chisel and a chainsaw (and more) to make these cuts safely, accurately, and efficiently.

Guys love to whittle. Given the chance, a man can spend hours with a sharp knife and a piece of wood not much bigger than his thumb. Now imagine how long it could take to whittle a 40-foot log! Cutting notches is not whittling. There is no sense in learning to cut notches like an amateur. You need to cut notches like a professional, and the way to do this is to learn the steps and techniques, and practice them again and again. Demand efficiency from your work—you'll need it because there are hundreds of notches in even a small log cabin. Keep reminding yourself to *not whittle*.

It is also important to work comfortably in order to work safely. If something feels awkward it is probably also dangerous. Evaluate every chainsaw technique and position—it must be both comfortable and safe.

Rough Notches

Figure 1: Rough notch. Even with large logs, the rough notch is just two cuts to remove a block of wood that is shaped like a section of an orange.

As we have learned, the rough notch has many useful purposes. But, a tight fit is *not* a goal for rough notches. The rough notch should be completely cut out with just two cuts, one on each side of the rough-scribe line *(Figure 1)*. While the final notch is scored, cut, carved, and then brushed closely to scribe lines, the rough notch is just cut—no scoring, no carving, no brushing.

The chainsaw should follow the scribe line closely, and not leave too much wood. It is okay if you cut slightly over the scribe line here and there *(Figure 2)*.

To cut rough notches, hold the bar of the saw at 90° to the length of the log and then rotate the bar and saw, and sight down the side of the bar until you can see that a straight cut will follow the scribed line. When you get the bar rotated to the proper angle you'll see that this is actually a straight cut, not a curved cut! The bar should extend right through the log and out the other side. Make the second cut. Remove the V-shaped block of wood and you're done.

It is important to over-cut the top of the rough notch—to cut past the scribe line that represents the top of the notch (it looks like the bottom of the notch, when the log is rolled upside down while it is being cut) *(Figure 2)*. We'll need this open "V" at the top of the rough notch when we scribe the final notch.

Do not, under any circumstances, cut the rough notch into the saddle. If the two rough-notch cuts do not meet at a point, and sometimes they don't (rough notches in B2 logs are often like this), then you'll have to make another cut or two to remove the wedge of wood.

Figure 2: The rough notch was scribed with the log on the wall. Two quick chainsaw cuts complete the rough notch. The saw should closely follow the lines on the side of the notch. But at the top of the notch, the two cuts meet well beyond the scribe line. This creates a gap that is useful when you are scribing the final notch. The gap is removed by the final notch, unless the rough notch cut exceeds the final scribe setting.

Saddles

Over the years, I have used several different shapes, or styles, of saddles, and have come to think that a simple concave saddle is ideal for most circumstances. Log builders use many different names for this cut—*scarf*, *blaze*, and *receiver*. Timber framers have been using *scarf* for centuries to mean something entirely different. It does look like a saddle after both cuts are finished, and when you straddle the log it's like a horse saddle, so *saddle* it is.

Cutting the Saddle

Mark the four layout points with a black felt marker that you can easily see. I do not draw the oval that goes through these points before I make the ripping cut. I strongly recommend that you make this cut without trying to draw the saddle.

The saddle is cut in three steps. First you rip a concave, curved slab off the side of the log. Then you use the chainsaw to quickly brush it smooth and fine-tune its shape. Finally, you remove the chainsaw marks with a sander-grinder, slick, or curved-bottom planer.

Place the log in V-blocks and roll it slightly so that you can see all four points. Stand on the side of the log opposite the saddle you are about to cut. Hold the saw left-handed—the trigger is in your left-hand, no matter whether you are left-handed or right-handed. Your right hand is holding the top handle right over the top of the saw's motor *(Figure 3)*.

Hint Don't try to draw the oval shape of the saddle – you don't really know where it will be, especially on lumpy logs. If you use the four points, then you won't be wrong.

Figure 3: Ripping the saddle. The ripping cut forms the basic shape of the saddle as a long, gentle, concave surface. Note that the trigger is in my left hand and the right hand is on the top handle and in line with the pulling force. This makes it easy to rotate the saw as it enters and leaves the saddle smoothly. The saddle is not drawn out—I use only the four points to guide my cut. I recommend that saddles be cut while the log is on the ground, not on the building like this.

Figure 4: Brushing the saddle. The ripping cut formed the basic shape of the saddle, and brushing fine-tunes the shape. Note that for brushing, the bar is about 90° to the surface of the saddle and that the bar is parallel to the slope of the saddle—the entire bar is cutting. To keep the saddle flat from top to bottom, brush with the straight part of the chainsaw bar, not the curved tip of the bar. The slab of wood that was ripped off of the saddle can be seen balancing on the log below.

Rest the bar flat on the log over the center of the saddle, and hold it at an angle that matches the angle that the saddle will have once it has been cut. Move the bar to the left side of the saddle (as you are looking at it), rotate the bar into the log slightly, and start cutting. You target is a gentle concave cut that will pass just slightly above the four points that mark the saddle.

Flatten the concave as you approach the center of the saddle, and then after you pass through the center, start guiding the bar in a gentle concave cut that will let you exit the log at the last point of the saddle.

Here are some of the keys to making the saddle rip-cut easy and accurate:

1— Hold the chainsaw 'wrong-handed'; this lets you pull the bar in a smooth curve throughout the cut. The gas cap is up, not down—this puts the handle in the right place to pull the saw instead of push it.

2— Keep the bar at the same angle to the ground throughout the cut. If you try to lift the motor up or push it down (to get closer to a mark) then the chain will bind in the cut—you are bending the bar. If the bar or chain get hot, or you see smoke, or the chain isn't moving fast, even at full throttle, then you are probably bending the bar. Do not try to make the top edge of the saddle cut less deep and the bottom edge more deep at the same time—that bends the bar.

3—The way to make the rip cut go near your marks is to find the correct bar angle first and then maintain that same angle throughout the cut. You can make small corrections as you cut by slightly rotating the saw around its bar. If you rotate upwards, then the cut will get less deep at both the top and the bottom of the saddle. If you rotate downward gently, the cut gets deeper at the top and bottom.

Figure 5: Sanding the saddle. Use a powerful sander-grinder or curved-bottom planer to remove chainsaw marks and make the saddle smooth and attractive. This work can be done on the wall, or on the ground. Working on the ground is safer.

Figure 6: A completed saddle—smooth concave left to right, and flat from top to bottom. Note how the bottom of this saddle is just about 1" below the final notch scribe line, as desired. The right side of this saddle was finished with a sander, the left side (the part you see inside the building) was smoothed with an axe. Also note the 4" of round log left between the two saddles of the bottom log.

Brushing the Saddle

Once the slab has been ripped off, brush the saddle with the chainsaw to smooth it off and correct any areas that still have too much wood *(Figure 4)*. To brush, the bar is about 90° to the surface of the saddle. Because the saddle changes shape throughout its concave, this means that the saw must rotate a bit as you brush.

Also, you must brush the entire saddle from top edge to bottom edge (that is, maintain the angle). If you change the angle, then the saddle becomes convex (high in the center), and you want it to be straight from top to bottom. Use the straight portion of the bar and the saddle will stay flat from top to bottom—do not brush the saddle with the nose of the bar.

Finally, remove the chainsaw marks with a 7" sander-grinder or a curved-bottom planer *(Figure 5)*. These tools are only to make the saddle smooth—do not use them to reshape the saddle, it's too slow, and the saddle will end up wavy, not flat. The chainsaw gets the shape of the saddle right, and the sander makes it smooth.

Corner Notches

There are many styles of corner notches: round notch, saddle notch, dovetail, and others. And there are many styles of long grooves (also called *lateral grooves*): deep V, shallow cove, double-cut.

During 20 years of building log homes, and 15 years of teaching log home construction, I have visited the construction yards of many professional log builders. Whenever I saw someone cutting faster, safer, or more accurately, I watched carefully to see what techniques I could use.

The styles and techniques that I will describe in this chapter are not the only ways to cut logs—but they are the best methods that I have found. I am confident that if you follow my directions, study the photos, and practice, practice, practice, you will get the hang of it.

The goal for cutting notches and grooves is to produce a shape that fits tightly to the log below without gaps or hang-ups. We also want notches with sharp, strong edges that bite into the log below without crushing. And we want cutting to be efficient, comfortable, and safe.

Four Steps: Score, Cut, Carve, & Brush

Notches are made in four steps: score, cut, carve, and brush. Scoring keeps the wood fibers from tearing out and helps make a strong, sharp edge. Cutting quickly removes the biggest blocks of wood. Carving produces a slight concave shape on the inside of the notch so that the notch rests on the log below only on its scribed edges. Brushing slowly finishes the edges of the notch to the scored line (*Figure 7*).

I have seen plenty of other notching styles, tools, and techniques. I have tried pneumatic (air) and electric chisels. I have used electric chainsaws to do the final brushing. I have sliced the notch with a chainsaw like I was slicing bread. I have used chainsaw-tooth disk grinders. None of these is as fast, safe, or accurate, in my opinion, as the method that I will teach you here.

For corner notches I recommend a chainsaw about 44 to 60 cc with a 16" bar. Even a giant 28" (71cm) diameter log can be notched with a 16" bar—I've done it. This is because you can notch halfway through from each side of the log. The chain that works best has a curved cutter tooth sometimes called a *chipper* style (see the chapter on Tools for brand-names). For a notching saw you want finesse, speed, and easy handling.

Brushing a notch with a 72 cc saw and 24" bar with full-chisel chain is a miserable job. It's like trying to parallel park an 18-wheeler semi on Main Street. It can be done, but why would you bother doing it? It's the wrong equipment for the job.

Some professional builders use full-chisel chain and get good results—only after years of practice. Chisel chain cuts quickly, but when brushing it is much more difficult to control than chipper chain.

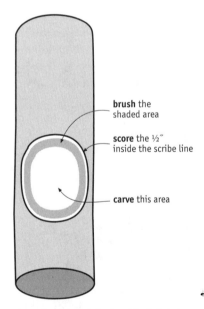

Hint   **Notching Steps**
1) Score
2) Cut
3) Carve
4) Brush

brush the shaded area

score the ½" inside the scribe line

carve this area

Figure 7: Top view of notch as it looks while it is being cut. The gray area is where you should use the brushing technique to remove extra wood. Carve the center of the notch concave—brushing is too slow for that big area. The narrow white edge just inside the scribe line should not be carved or brushed because it was completely finished back when you scored it with a chisel.

1 - Score

The first step is to score the edges of the notch. Scoring does three things: it prevents the grain from tearing out (look at the rough-notch tear-outs in Figure 2); it keeps the edges of the notch sharp and strong; and it makes it easier and faster to cut the notch.

I use a razor-sharp 1" chisel to score the notch-scribe lines. A chisel is better than a razor knife for this job: it is safer and easier to control, and a chisel scores deeply with ease *(Figure 8)*. (Some professional builders learned with a knife, and continue to use one, but that doesn't make a knife better.)

How can a chisel follow the edges of the notch? Isn't the notch scribe-line curved and the chisel straight? No, the sides of a saddle notch are actually straight, not curved—as we will see, and they can be easily worked with tools that work in straight lines. (There is a video on my website you can download.)

If you decide to try a razor knife, then keep both hands on the tool at all times. Don't rest one hand on the log, or eventually you'll cut your fingers.

A chisel scores deeply very easily—I usually drive it in nearly ½" (12mm) in green wood. You just cannot get that deep with a razor knife. By scoring in ½", that part of the notch is finished, and you can stay ½" away from the scribed edge when you are brushing with the chainsaw. If you score only ⅛" (3mm) deep, then you have to come to within ⅛" of the notch edges with a chainsaw, and there is much greater chance of cutting over the scribe line and making a gap.

The flat side of the chisel goes next to the scribe line, the bevel side of the chisel faces the center of the notch. Get close enough to the line so that after scoring there is *no* extra wood left between the scored edge and the scribe line. That is, the very first wood outside the chisel-score should be colored wood—it should have scriber ink on it. If you have any uncolored wood remaining you're not scoring close enough.

I use a 1½ to 2½ pound (1 kg) rawhide-faced cast-iron, or lead-shot dead-blow, mallet with the chisel. I strike the chisel once to set it in place, and then drive it in ½" with one more blow, and then move along. Tap, strike, move. Tap, strike, move. Tap, strike, move. It goes quickly.

Angle the chisel so that the scored part of the notch has a sharp edge that is also very strong. The weight of all the logs will rest on this notch, and if the edge is too concave it will crush. We want the edge of the notch to bite into the saddles below and not crush, even under thousands of pounds of weight *(Figure 14)*.

2 - Cut

The second step to making a saddle notch is to cut out the bulk of the wood. (Some notches will already have a rough-notch that removed some wood.) The secret to making this cut is getting your chainsaw lined up right before you start cutting. You won't be able to turn or twist the bar to change directions while you are in this cut. So, make certain you're pointed in the right direction before you even start cutting. Sight along the side of the chainsaw bar to see where you are headed with each cut.

Figure 8: Score the notch scribe line using a chisel and heavy mallet.The flat side of the chisel goes against the scribe line and inside the notch (the bevel of the chisel faces into the notch). The scribe line will be left on the log and not removed.

Scoring with a razor-knife is just not as good as scoring with a chisel because a knife rarely goes in deeper than ⅛"—and that means you have to use the chainsaw to brush very close to the scribe line. ⅛" (3mm) accuracy with a chainsaw is more difficult than ½" (12mm) accuracy.

The chisel probably takes an extra minute when scoring, but it saves more than that when you are cutting and brushing the notch, so using a chisel saves time and increases accuracy.

Hint   Score as close to the scribe line as you can, because then that part of the notch is finished and needs no extra work to clean it up.

Hint   There is a short video on my website (www.LogConstructionManual.com) that shows cutting a notch with chisel and chainsaw.

Figure 9: Cutting the notch. The saw bar is 90° to the length of the log and tipped at an angle so that the edge of the notch being cut is a straight line. Note that in both photos I am looking down the side of the bar closest to the notch edge so I can get the rotation right, and not cut over the line. Note the angle of feet, hips and shoulders to the log. And the saw is being held against my hip or thigh, not held in mid air. I am cutting only halfway through the notch.

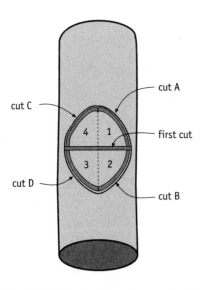

Figure 10: Think of each notch as being made of four quarters (top view). There are four cuts, each halfway through along the scribe lines, and two cuts right down the center.

Wear chainsaw chaps and steel-toe boots. A helmet with face screen is a great way to improve your notching—you'll be surprised at how easy it is to concentrate if you are not getting sprayed in the face with sharp chips of wood.

Roll the log so that it is upside down, in V-blocks on the ground.
- *Your toes should be pointing at an angle towards the log, not straight at the log.*
- *Your hips should be pointing at an angle to the log, not parallel to the log.*
- *Your shoulders should be in line with your hips and toes* (Figure 9).

Do not squat down, or hold the saw between your thighs or knees. The saw should be resting comfortably on your hip or thigh, and held close to your body. Don't hold the saw with just your hands: it is difficult to control, and your lower back will suffer.

The first cuts are right down the center of the notch *(Figure 10)*. The next four cuts are actually *straight* cuts—not curved. To see that they are straight, you need to position the chainsaw and get it lined up before you start cutting—sight down the bar as you rotate it from horizontal until it points down the line, then cut.

Hold the chainsaw on the log on the left side of the notch. The bar should be 90° to the length of the log, and only barely past halfway through the notch. Then tip the bar (rotate the saw around its long axis) until the flat side of the bar lines up with the cut on the side of the notch. Staying about ¼" away from the scribe line, cut down. Repeat on the right side of the notch. Step around to the other side of the log and repeat these two cuts. Break out the large pieces of wood and you'll find that probably 95% of the notch has been removed.

As your skill improves, try cutting all the way through the notch, instead of halfway through. Don't cut over the scribe line or you'll make a gap that can be closed only with caulk.

3 - Carve

After the notch has been cut, and most of the wood removed, the next step is to carve the inside of the notch to produce a shallow concave bowl so that when completed and in place, the notch touches the log below only on its edges.

The goal of carving is to remove any extra wood from the notch that might cause an internal hang-up.

I usually take two or three carving strokes on each quarter of each notch—a total of less than 12 carving strokes per notch.

The carving stroke is like no other in log building. The trick is to have one side of the bar free, and not caught in a kerf. As you can see in the photos *(Figure 11)*, one side of the bar is free and open to the inside of the notch. If the bar has wood on both sides of it, then it is trapped inside a kerf and it cannot rotate.

The second trick to carving is that the saw must rotate around its long axis throughout the cut. You have to hold the saw against the side of the notch as you push down and roll the saw.

4 - Brush

Using the chainsaw to brush has many applications. You brush a saddle to smooth it and make it flat, and you brush flats on sill logs and floor joists. For notches, brushing is done with the bottom of the nose of the bar, and is quite accurate—with just a few hours practice you should be able to brush with ¼" accuracy.

Cutting and carving has removed 99% of the wood from the notch. Brushing is used to make the notch slightly concave in the last inch or two closest to the scored scribe-line. That is, you brush just the perimeter of the notch *(see Figure 12)*.

Hint   Chainsaw chains that have safety links will brush quite slowly.

Figure 11: Carving the notch. The bulk of the wood has already been removed by cutting. Carving produces a notch that is slightly concave on the inside so that it will rest only on its edges. Note the angle of my feet, hips and shoulders to the log. The saw is held against my hip or thigh, not in mid air. Note the position of my hands when carving on the left and on the right. For carving, I roll the saw as I carve down the side of the notch; and there is wood on only one side of the bar. In these photos, I have already carved the near side of the notch and am carving the far side. You can see the white edge next to the scribe line—that's where the chisel scored the notch.

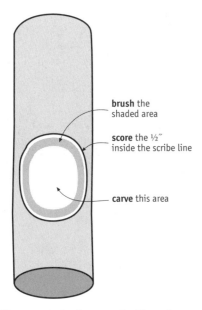

brush the shaded area

score the ½" inside the scribe line

carve this area

Figure 12: The gray area is where you should use the brushing technique to remove extra wood. Do not brush the narrow white edge just inside the scribe line it was finished back when you scored it with a chisel.

Three warnings before you start: first, brush as little as needed to complete the notch; second, don't run the saw for long without much load; and finally, make sure the oiler of your saw is turned down as far as it will go (see the owner's manual). If your oiler isn't turned down, brushing will leave an ugly spray of bar oil on the finished notch.

When you are finished brushing, the chiseled score around the entire notch should still be completely intact and uncut by the chainsaw. That part of the notch was done as soon as it was scored—no more wood needed to be removed *(Figure 16)*. If you have chainsaw brushing marks that extend right out to the scribe line, then you've brushed too much and too far.

Be kind to your saw. Just as you wouldn't put a brick on your car's gas petal and park it at the curb, don't run a chainsaw too long at high *rpms* with no or little load.

Brushing a notch is done with the nose of the bar *(Figure 13)*. When you are brushing areas that you want to be flat you use the straight part of the bar. You want a slight concave surface inside the notch, so use the part of the bar that is shaped to do this—the curve on the bottom of the tip of the bar.

There is little kickback danger in using the bottom of the tip of the bar. It is the top of the bar that can have dangerous kickback, so be careful that you don't let the top of the bar touch wood. At first you will hold back, for fear of brushing over the scribe line. As you get better at brushing you can push the nose of the bar more firmly into the wood.

Figure 13: Brushing the notch is accurate, but slow, so I only brush the last 2" or so that is closest to the scored edge of the notch. Note the angle of feet, hips and shoulders to the log. The saw is being supported on my hip or thigh, not held in mid air. Keep the bar at about 90° to the surface of the wood inside the notch, which means you roll, or rotate, the saw as you brush different parts of the notch, using different parts of the handle to hold.

Concentrate on brushing small portions of the notch—about the size of your palm. Your attention will be caught by little bits of wood here and there, and you'll be tempted to wander. But stay in a small area (just a few square inches) and force yourself to completely finish it before moving on. This is the best way to keep yourself from "whittling."

I have seen students re-fill their gas tank and go back to brushing—that's way too much time spent brushing. I try to keep the total brushing time per notch down to as little as 3 or 4 minutes. Time yourself to see how long it takes you—and know when to walk away.

The Top of the Notch

The top of the notch is handled specially. I make it extra concave from edge to edge, and cut to the scribe lines, but not over them. My goal is for this part of the notch to have a thin, sharp, crushable edge. The rest of the notch edges (where it rests on the saddles below) should be strong and sharp. But at the top of the notch I want the edges of the notch to be sharp and weak. This is part of having a compression-fit saddle notch. You must either remove this wood completely (*Figure 15*), or make it crushable (*Figure 14 and also Color-Figure C-22*).

Figure 14: Saddle notch. This notch is finished. Note that the scribe line is still visible. The sides of the notch are slightly concave, but the edges of the notch are sharp and very strong. The top of the notch, however, is very concave and the edges are sharp and thin so they can crush over time. This log is upside down, of course, while it is being cut, so the top of the notch is at the bottom of the photo.

Figure 15: The top of the notch must allow the notch to compress onto the saddle below. Here that wood is removed, leaving a gap. I prefer to cut to the scribe line, but make the edges thin.

this area is enlarged to the right

Figure 16: Close-up of the scored edge of notch. The notch has been scored with a chisel, cut, and carved with a chainsaw, but not yet brushed. You can clearly see that the 1/2" (12mm) perimeter closest to the scribe line was chiseled and finished—even before you started the chainsaw. All that remains is to brush the ridge that starts 1/2" in from the edge. Scoring with a chisel produces a strong, sharp edge that needs no more work. You cannot score this deeply with a razor knife. See Figure 12.

Test the Notch

Finally, test to see how the notch looks. You are looking for any possible internal hang-ups that would prevent the log from fitting when it is put back on the logs below *(Figure 17 and Color Figure C-22)*.

The notch should touch the saddles of the log below along its edges, and nowhere else. There should be a shallow, smooth concave all the way across. You should just be able to fit your finger between the straight edge and the inside of the notch—about ½", or slightly more, is all you need.

Slide the straightedge slowly throughout all portions of the notch, looking for problems. Find and mark *all* the problems before you pick up your chainsaw to brush any high spots. Over the noise of the saw, and the power of genetic pre-disposition, I hope you can you hear me telling you: "Don't whittle! Fix the high spots and get on to that next log."

You could probably cut the notch straight across from edge to edge, and it would still only touch on its edges when it is placed on the saddles below—because the saddles are concave. So, you should not hollow a saddle notch out—keep those notch edges strong and sturdy.

Figure 17: Testing the notch is the last step. Place a straightedge across the notch. The edge should touch only the scribe line on each side, and no where else. The center of the notch should be concave only by the thickness of one finger. "Do not turn notches into salad bowls," as log builders tell beginners.

Note how gently the concave starts near the scored lines at the edges of the notch—that keeps the edges of the notch strong. Slide the straightedge throughout all portions of the notch to find hang-ups, if any. See also Color-Figure C-22.

As you can see in this photo, the ¹/₂" chisel-score around the notch's perimeter is slightly darker than the middle of the notch that was brushed with a chainsaw. The chiseled surface is smooth and appears dark, while the brushed area is rough and looks lighter Compare this with Figure 12 on page 122.

Long Grooves

The long groove, also called the *lateral groove* by some log builders, is the scribed connection along the length of logs. It is at the heart of handcrafted, chinkless construction. Chinked buildings have scribed corner notches but no long grooves or only partial grooves. Full-scribe buildings have scribed notches and full-length scribed grooves, and this is what really sets them apart.

Several styles of long groove are used. I prefer the Beckedorf *double-cut* groove for wall logs because it is fast to cut. Where the groove becomes visible, as in flyways and arched openings, all builders use a shallow-cove groove because it looks better.

Whatever shape groove you use, its shape, especially near the scribe line, should be adapted to match the gasket you are using. The shape and size of the gasket you use influences the shape and angle of the groove near its edges.

Long groove edges, unlike notch edges, do not need to be scored before they are cut. This is because the chainsaw is cutting with the grain, instead of at 90° to the grain fiber, and so tear-outs are not common.

With dead-standing logs, however, sometimes splinters get raised when ripping long grooves. One way to avoid these tear-outs is to walk forward with the chainsaw, instead of backwards. This means the teeth are cutting down into the wood, instead of pulling out, and though it takes a while to learn the feel (the bar is always trying to climb out of its cut, or kerf), it is effective.

Cove

The *coved* groove, also known as *coped* or *shallow-cove*, is quite common, though it is a bit slow to cut and finish *(Figure 19 & Color Figure C-27)*. I have tried many tools and techniques to cut and finish the shallow-cove groove including a drawgouge *(scorp)* and an axe (without a chainsaw). The easiest method I have found, so far, though no method is truly easy, is to make four cuts with the chainsaw *(Figure 18)* that removes the bulk of the wood between the scribe lines.

Roll the log so the long groove is facing up (so the log is upside down). Follow the scribe line with a chainsaw, cutting only ¾" deep, or so. If there are very tight curves in the scribed line (maybe there was a knot dimple in the log below)—skip those with the chainsaw, and clean them out later with a sander or chisel.

Keep the chainsaw motor held fairly high (though not uncomfortably high) so there is very little of the nose of the bar in the wood—this allows you to turn on a dime to follow the scribed edge. Walk backwards while you watch the scribe line carefully *(similar to Figure 24)*. Repeat this shallow cut on the scribe line on the other side of the log. Also see Color Figures C-23 and C-24 for similar cuts.

Now change your grip on the saw, holding it almost flat, with the gas tank facing up. It will probably be more comfortable to run the trigger with your thumb instead of your index finger *(similar to Figure 25)*. Walking backwards, cut at a low angle and deep enough to come to the center of the width of the groove. You won't be able to see the scribe line because it is below the bar—that's why you

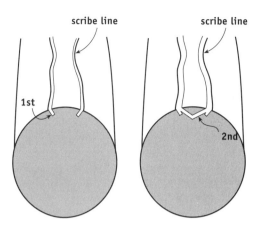

Figure 18: Cutting sequence for a shallow cove long groove. First, make a very shallow cut near each scribe line. Second, cut to the center of the groove, but at a lower angle to remove the piece of wood.

Figure 19: The shallow cove long groove. On the left the groove as seen in a door opening. On the right, a log that has been cut and sanded.

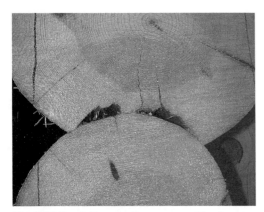

Figure 20 The shallow cove long groove, with a hang-up. Not enough wood was removed in the center of the groove—it is hitting the log below. If more wood than this had been left, the edges of the grooves would not be touching the log below. Use a log "cookie" to test the depth of a shallow cove groove. This groove was cut entirely with an axe.

Figure 21: Cut most of the wood out of the flyway with two low-angle cuts that meet in the middle of the groove's width. Then, standing at the end of the log, brush with the whole bar (not just the nose) to make the shallow cove the shape it needs to be.

Clean the cooling fins of your saw to remove the long strips, and keep it from overheating and seizing.

made the shallow cuts first because they let you stay far enough away from the scribe line that you don't cut over it. The groove, of course, changes width throughout the log, so you'll be pushing the saw nose in deeply in some places, and pulling out in other places. And the wider the cove-groove, the deeper it must be, too—that means angling the saw bar to point down a bit deeper where the groove is wide. Repeat on the other side of the log to remove a long piece of wood *(Figure 18)*.

Use a chainsaw nose to brush the ridges and lumps remaining in the groove, and bring the groove into a coved shape *(Figure 19)*. If you prefer hard work and handtools, this is when you would pick up a drawgouge, adze, or scorp.

To see if a groove is deep enough you can cut cookies from the end of a scrap log and use them to test the depth by holding it in the groove *(Figure 20)*. To be accurate, the cookies should be about the same diameter as the log below.

Finish to the scribe lines with a 5" sander and 24-grit disks *(Figure 26 and C-25)*. I have tried an axe, a sharp chisel, and a scorp to trim to the lines—but I think that nothing beats a small sander for accuracy, speed, safety, and ease.

Flyways

The long groove in a *flyway* (the log extension outside of a corner notch) is almost always cut in a shallow-cove shape *(Figure 27)*. Make two cuts with a chainsaw that meet in the middle of the width of the groove, making sure that both cuts are well above the scribed cove shape on the end grain of the flyway. With the majority of the wood removed in a shallow "V" shape, you then can brush with the grain *(Figure 21)* to get the smooth, coved shape that you want. See also Color-Figures C-30 and C-31 for photos of finished flyways.

This lengthwise brushing produces long strands of wood that get caught in the cooling fins of the saw, and this can cause serious overheating. After brushing, take the cover off your saw and make sure the fins are clear.

Arched Openings

Arched openings are popular in interior log walls—they are a doorway without a door, or a window opening without a window. As a result, the end grain of all the logs is visible, and so are the ends of the long grooves. It looks best if the cove of every groove exactly matches the shape of the log below, and this means they must be scribed *(Color Figure C-27)*.

To scribe the end grain of the logs in an arched opening, you may need to cut small holes in the log wall before you scribe *(Figure 22)*, so that you can reach in with the scribers to mark the end grain carefully.

When cutting the groove in these logs, remember to switch from the four-cut shallow cove to the flyway, two-cut style, so that the shapes will match precisely.

Double-cut Groove

Lloyd Beckedorf, a builder in Alberta, and the man who taught me to build, invented the "double-cut" long groove, for which he holds patents. This is the groove style that I prefer because it is fast and easy to cut. It also securely holds the rubber and foam gaskets that I like to use in the grooves *(Color Figure C-26)*. The double-cut groove requires that you use a scriber that has a pen or pencil in both legs of the scriber, not just the top leg.

It is a groove style that I do not recommend for beginners or do-it-yourselfers—it is for professional builders. Here's why: the deep "V" on the top of each log must be cut by walking backwards, on top of the wall, on each log with a chainsaw. This can be dangerous at low heights, and the problems only increase the higher the wall gets, unless the entire building is safely scaffolded.

If you use my patented Accelerated Log Building Method, then all cutting can be done with your feet on the ground. You'll need a license to use the Accelerated Method, and I recommend Accelerated Method only for professional log builders.

The double-cut groove is cut as follows. First, roll the log upside down so that one of the two long-groove scribe lines is, as close as possible, on the top center of the log. Second, cut out one of the two small "V's" by making a shallow (1½", 40mm, deep) cut near the scribe line *(Figure 23)*. Do not roll the log yet. Remove a long strip of wood by making a second cut to meet the first cut *(Figure 24)*. Note the position of the log, and the two different stances for these two cuts. The strip of wood that is released for each small "V" is about 2" (50mm) wide for it's entire length *(Figure 23, photo on right, & Color Figure C-23)*.

You can feel when the bar is deep enough on the second cut—the chain falls into the first cut you made, and cutting becomes easier and faster. Ride that kerf, and always be aware of not cutting too deeply with the second cut. Also watch out for the narrowest part of the groove so you don't cut over the scribe line that is below your saw bar, out of sight.

Now, roll the log so that the other scribe line is on the top center of the log, and repeat the two cuts to make the second small "V" *(Figure 25)*. This is why the double-cut groove is fast—with 4 shallow cuts the groove is done, except for sanding to the lines.

Figure 22: To get a tight fit in the grooves of an arched opening, first use a chainsaw to cut a small hole in the log above that is very near to, but will be inside, the final width of the opening (you can see the arrow that marks the width of the final opening just below my left hand—the arrowhead points into the arched opening). The scriber reaches in to mark the end grain so that it can be cut with a profile that exactly matches the log below. This makes a nice, tight fit in the groove.

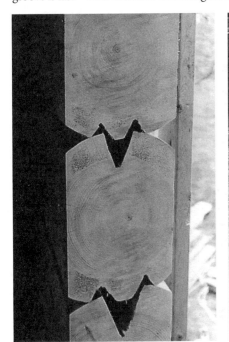

Figure 23: The double-cut long groove. On the left, a finished wall showing the two small "V's" in the bottom of each log, and the deep "V" on the top of each log.

On the right, the small "V's" of a long groove being finished. Note the 2 pieces of wood removed for the small "V's" are lying on the ground—they have a constant width for their entire length.

See also Color Figures C-13, C-23, C-24, C-25, and C-26.

Figure 24: The double-cut long groove. Roll the log so a scribe line is on top. Make a shallow cut close to the line, with the stance and bar angle shown above. Note the scribe line marking the edge of the long groove.

Figure 25: The double-cut long groove. Make a shallow cut to just barely meet the first cut. This strip of wood is a constant width, and the small V is a constant depth. Notice the stance and low bar angle. Use your thumb on the trigger to help you keep the saw at a low angle. Also see Color-Figure C-23.

The purpose behind this specific setup of rolling the log, and then cutting with particular stances is to help you keep the groove the right size and shape. This will help you avoid internal hang-ups caused by not removing enough wood, or by cutting at incorrect angles (*see Color Figure C-13*).

The deep "V" is cut out of the top of the log below. To know where to cut, you need a scribe line on the log below, too. The deep "V" is removed with two cuts, each is 1" inside the scribe-line (*see Figure 23, photo on left*). This means that the deep "V" varies both in width and in depth—and when the groove is wide, the deep "V" can get quite wide and deep.

Double-cut long grooves are not used for flyways because the deep "V" is ugly and would catch rainwater, promoting decay.

To the Edge

To get tight fits, the edges of long grooves must be finished close to the scribe lines. The last wood remaining on the log must be the wood that has scriber ink on it—anything more will cause a poor fit—a *hang-up*.

For years I used a sharp 1" chisel to pare off the last ⅛" (3mm) of wood that remained after the chainsaw cuts. It made a clean, sharp edge, and was an opportunity to put down the noisy tools and work by hand. Some pro builders still use an adze, scorp, or other edge tool for this. But these tools are slow, clumsy, and can lift splinters that cross over the scribe line, causing gaps. Some builders use an axe, which just seems like really the wrong tool to me: good for chopping down big trees and for…trimming ⅛" off to the scribe line?

Since 1992, I have been using a 4½" or 5" sander-grinder with 24-grit disks on a flexible backing wheel (*Figure 26 and Color Figure C-25*). It is fast and accurate. I learned this trick from a past student of mine, Graeme Mould, now a professional log home builder in New Zealand. Try it, you'll like it.

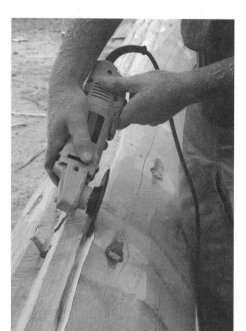

Figure 26: A 5" sander trims to the scribe line easily and quickly. Find a way to work comfortably—get the log at the right height, and rolled to a good angle so that you can see the scribe line without bending over and twisting too much.

Use the flat part of the disk as shown here, not its edge—that is, you are sanding quite a wide area, not a small spot.

A big benefit of a sander over any edge tool (axe, adze, chisel, gouge, scorp, etc) is that the sander never grabs the grain or pulls out a splinter past the scribe line, which is a concern especially in drier or dead-standing logs. Sanding is also very fast.

See also Color Figure C-25.

Kerfing

As logs dry, they shrink, and as they shrink they crack, or *check*. Almost always, a log develops one major check, and several smaller ones.

We want to encourage wall logs to check on their tops, or at the very least, we don't want them to have their major check only on their bottoms, in the grooves. A major check in the groove causes poor fits and slumping.

It's easy to encourage a green log to check—they usually check the closest distance to their heart, the *pith (Figure 27)*. A kerf is simply a way to get closer to the pith, and is usually just a single chainsaw ripping cut *(Figure 19, left)*.

For a kerf to be effective, it needs to be about the same depth (about as close to the pith) as the long groove in the bottom of that log. The log does not need to be kerfed all the way to the pith, and in fact, I recommend against it. The kerf just needs to be about the same, or slightly closer to the pith than any other cut in the log.

In the past, builders would pound oak wedges into the kerf, but this is not needed. Take a look at the checks in the logs of Figure 27—they never had wedges, and yet every time they checked in the place you would expect: the cut that is closest to the pith.

Flyways are not kerfed because the kerf could catch rainwater and promote decay—and also, kerfs are not attractive. For the same reason, stop any kerf before it gets into an arched opening: you don't want to look at a kerf.

Figure 27: Logs check the closest distance to their centers, in this case, starting in the flyway groove.

Note that flyways, as in this picture, are never kerfed—in fact, we *want* them to check in their grooves. If they were to check from the top down, the kerf would catch rainwater, and could accelerate decay.

But this photo clearly illustrates that if logs are not kerfed, then they will check in their grooves nearly every time. That's why we kerf the top of logs—to move the location of the check from the log's bottom to its top.

Underscribing

It is pretty easy to get notches and grooves to fit tightly the day they are scribed—all it takes is a steady hand and careful cutting. The real challenge is to find ways of keeping fits tight over time—over the life of the building, which can be hundreds of years.

In a nutshell, the problem is that as the logs dry out and shrink in diameter, the long grooves stay tight, while the corner notches get a bit loose. This is caused by the fact that trees shrink in diameter, but not much in length.

Then, when the notches get loose, the fits can become even worse because the logs can twist—without tight corner notches to restrain them and keep them in place. As a result, gaps get wider. Logs dry, shrink, check, and twist, and all of these tend to make gaps bigger.

There have been two approaches tried by handcrafted log home builders to keep notches tight over time:

1) Some have tried altering the shape of the notch itself, so that it tends to tighten up as the logs shrink—the *shrink-fit notch*.

2) The other approach has been to make the notches extra-tight at first, and this is called *underscribing*.

Underscribing Keeps Fits Tight

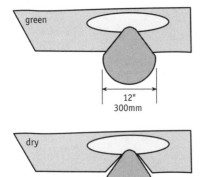

Figure 1: When the logs are green the notch fits perfectly (top); but then the log below shrinks in diameter, while the hole in the log above (the notch) stays the same width. ***The log below is no longer big enough to fill the hole!*** This happens because trees shrink a lot in diameter, but not much in length.

Figure 2: As logs dry, grooves stay tight, while notches get gaps (as can be seen in this photo). Underscribing helps solve that problem—underscribing helps keep both notches and grooves tight over time.

Starting in 1982, Del Radomske, a log builder and teacher in British Columbia, began experimenting with ways to keep notches tight over time. He noticed that logs fit perfectly when they were green, but did not fit as well when the logs eventually dried. In every building, he noticed, the grooves stayed tight and most of the corner notches got loose *(Figures 1 & 2)*.

Radomske imagined that if he could take a dry building apart, remove a fraction of an inch from all the grooves, and put the walls back together, then both the notches and the grooves would be tight.

Obviously, taking homes apart in 3 or 4 years is not practical, and Del's insight was that if he took a little extra wood off the grooves at the start, then when the logs eventually dried this would accomplish the same goal. As he saw it, the grooves were too tight in log homes. In fact, they were so tight that they would not let the notches slide down the saddles and make a tight compression-fit.

Underscribing is a technique that helps solve this problem: the notches and grooves are scribed using slightly different scribe settings. The result is significantly tighter notch fits over time.

In the log building yard, underscribing and overscribing are the same thing. But because of the way Del first visualized the problem— "the long grooves are too tight"— he calls his technique *overscribing* (referring to the grooves, he "opens up" his scribers). Because I optimize my long-groove scribe-setting, I scribe the grooves first and then I close down my scribes so the notches are a bit tight, and so I call the technique *underscribing (Figure 4)*. They are one and the same.

Scribe Settings for Notch & Groove are Different

Radomske started out small at first—using one scribe setting for the grooves and a slightly smaller setting for the notches—this produced a ⅛" (3mm) gap in the grooves. By the time he had two rounds on the building, the ⅛" gap in the grooves had already closed—the logs were visually tight because the notches had compressed into the saddles enough to close the small gap in the grooves. He found that ⅛" of overscribing was not enough to keep the notches tight.

He built another house with ¼" (6mm) gap, and watched it for several years. Del finally tried ⅜" (10mm) and was getting better results.

All the Weight Starts on the Notches

At first, in an underscribed building, all the weight of the building is on the notches, and not much is on the grooves. As the logs dry and shrink, weight is slowly and gradually transferred from the notches to the grooves.

As weight is transferred off the corners, the wood of the saddle rebounds, keeping the notches tight. In order for underscribing to work, the corner notches must withstand great weight without crushing. Notch edges must be so strong that the wood of the saddle is compressed instead of having the edges of the notch crushed (*Figure 5*). Compressed wood will rebound, while crushed wood is forever crushed.

Underscribing also requires that the notch slide down the saddle to a new position that is as wide as its original position, before it shrank. In order to slide, there must not be a hang-up at the top of the notch. This means that either we leave a gap at the top of the notches, or we leave a wood seal there that is so thin that it can easily crush and never cause the notch to hang up (*compare these two methods—they are shown in Color Figure C-22 and in Figure 15 on page 123*).

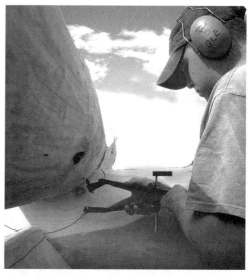

Figure 3: The groove is scribed, then the scribers are closed down slightly (see the Chart on next page), and the notch is scribed. Perhaps you can see that the groove scribe line is a little below the notch scribe line—the difference in height is the underscribe.

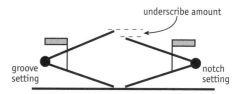

Figure 4: Find the widest gap between two logs and set your scribers at that point—this is the groove scribe-setting. Scribe the groove. Then close the scribers a bit and scribe the notches. The chart on the next page helps you decide how much to close down the scribers.

Figure 5: In the building yard, about 100% of the weight of the building is on the notches, and little weight is on the grooves. In this photo you can see that the notch has compressed the saddle, and this is the result that we want.

Notch edges must be so strong that the wood of the saddle is compressed instead of having the edges of the notch crushed. If the edges of the notch crush, then we have failed.

How Much to Underscribe

Hint Notches that have a lot of weight are underscribed more than notches with very little weight.

Figure 6: This log caps over 8 floor joists and 2 walls. This means that each of the 10 notches gets very little weight and so must be underscribed less than if the log had only two notches.

The amount of underscribe varies with several factors: the distance between notches, and the height of the log on the wall (that is, the round number of the log). In addition, you should consider the climate, elevation, and the initial moisture content of the logs. The chart below will help you get started.

Notches that have a lot of weight are underscribed more than notches with very little weight. The trick is to estimate how much weight is on a notch. The amount of weight on a log is determined partly by the location of the log in the building: logs that are low in the wall have more weight than those that are high on the wall. Logs that support floor joists and roof systems have extra weight on them.

Considering the distance between notches, for example, a 36' (11m) log with just 2 notches will compress its notches more than a 36' log with 4 notches. So, the log with two notches will be underscribed more than the log with four notches. The more weight per notch, the more the underscribe.

An extreme example of this is the log that is scribed over floor joists—it may have as many as 8 or 10 notches *(Figure 6)*. If the "basic" underscribe amount for this log is ⅜" (10mm) (at the lowest rounds), then the scribe setting for the floor joist notches would be about $1/16$" (1.5mm)— half the usual underscribe for its span and its round number (the height on the wall).

Distance between notches (in feet)	5 to 10	10 to 15	15 to 20	20 to 25	25 to 30	more than 30
Underscribe for bottom 3 courses	3/16″	1/4″	5/16″	3/8″	7/16″	1/2″

Walls with 'Un-equal' Notch Spans

If a 35' wall has notches that are 9' and 26' apart, use just one underscribe amount—and split the difference (from the Chart: ³⁄₁₆" for the 9' distance, and ⁷⁄₁₆" for the 26' distance), so use ⁵⁄₁₆" for the notches on this wall. Each individual log uses only one underscribe amount.

Sill Logs

Even the sill logs must be underscribed. Although they have no grooves, you still need to underscribe the notches of the whole-log sills where they cross over the half-log sills. I underscribe the whole-log sills only half of the amount shown in the Chart.

Until compression and shrinkage have occurred, the whole-log sills will not be bearing fully on the foundation. If you are building *off-site* (not on the final foundation), then prop up the center of the whole-log sills with a foam wedge that will compress as the walls gets taller. This will keep the sill logs from sagging, or bowing down, in the meantime.

Reduce the Amount of Underscribe as the Walls Get Taller

Each wall starts with the *basic* underscribe found in the Chart on the previous page, and this amount is used for the first three rounds of logwork. Then, as you get higher in the wall, and there is less weight bearing on the corner notches, reduce the *basic* underscribe amount in the next three rounds by ⅛" (3mm). The underscribes of higher rounds are reduced by another ⅛", and so on.

You use less underscribe as you get higher because there is less weight on the upper logs than there is on the lower logs in a wall. Remember: the more weight on a notch the larger the underscribe; and the less weight on a notch, the smaller the underscribe.

Special Logs Need Special Treatment

Some logs have less than two notches—for example, logs that extend from a corner to a door opening; and some logs have no notches at all, like those between a door and a window. You must use special techniques to underscribe logs like these.

One effective technique is to scribe small *tabs* at the end, or ends, that have no notch. These tabs are portions of the long groove that are scribed with the same scribe setting as the notch *(Figures 7A & 7B)*. The tabs prop-up these special logs, and keep an even gap in the groove, and the tabs are sometimes cut off later when the door or window opening is cut to its final width.

The tabs can usually be about 1" to 2" (25-50mm) long, though you could make them shorter for short logs that don't have much weight on them.

Instead of tabs, you can use wooden shims at the log ends that have no notches *(Figure 7C)*. You can pull the shims out over time to keep the gap in the long groove even, but they usually just conveniently crush on their own as needed.

Climate & Moisture Content Affect Underscribe Amounts

There are other conditions that affect the amount you should underscribe, but it is difficult to quantify exactly how much effect they have.

Reduce the figures in the Chart if you are using dry logs—but test them with a moisture meter to determine how much drying they still have left. Never use the weight of wood to guess at moisture content (MC). The amount of underscribe is proportional to moisture content as long as the wood has a moisture content that is less than fiber saturation (about 28% MC). The drier the log, the smaller the underscribe amount that is needed.

Climate and elevation also affect drying and the underscribe amounts. A log home in a damp, coastal climate requires less underscribing than a log home in a desert. Finally, log homes at high elevations may need slightly less underscribing than the amounts in the Chart.

But keep in mind that underscribing is not an exact science, at least not yet. The basic underscribes shown in the Chart come from trial and error, from experience and observation. Underscribing definitely improves the fits over time, but it is difficult to predict how much a log should be underscribed when you consider all the variables: climate, elevation, moisture content of the tree, distance

Hint You use less underscribe as you get higher on the wall because there is less weight on the higher logs.

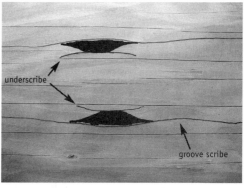

Figure 7A: Tabs can be scribed to hold up underscribed logs at the right gap. The short scribe lines inside the groove were scribed with the notch scribe setting (underscribe). I marked the wood that is to be left as tabs with a black felt pen to make them easy to see.

Figure 7B: Tabs in a long groove that has been cut and sanded. (This is not the same log as Figure 7A.) These tabs are about 1 inch (25mm) long.

Figure 7C: Wooden shims or wedges can hold up the end of the log that has no notch—this keeps an even gap in the groove from end to end. This is a substitute for tabs.

between notches, species, and height on the wall. All we can do is make an educated guess.

Make Sure There Are No Future Hang-ups

You must be precise when scribing and cutting underscribed logs because there is no good way to test for the eventual fit, or find possible future hang-ups.

When you scribe logs with equal scribe settings (that is, no underscribing) you can see exactly how it fits when you put the log back on the wall. If there is a problem, you know it immediately. But, when you put an underscribed log on the wall, the grooves are loose, and there is no easy way of knowing whether there is an extra bit of wood somewhere that will be in the way at a later date.

You should use a scriber that marks both the top and the bottom scribe lines, not a scriber that holds only one pen or one pencil *(Figure 8)*. You need both lines in order to get the log back in exactly the right place. With both sets of lines, just match up the notch with the scribe lines on the saddles at both ends.

Figure 8: Top row: "Mackie-style" scribers; ITT scribers; Sellman scribers.

Bottom row: Lee Valley Veritas; Starrett scribers, Timmerhus Ultrascribers.

Some hold pencils only, some hold pens, and some use computer-plotter pens. All but one scriber (the Starrett) holds two pens or two pencils.

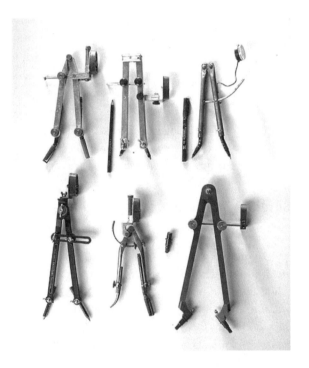

How Long Will There Be Gaps?

The most common question about underscribing is: what do you do with the gaps!? The answer is simple—there are no visible gaps by the time the roof is on the building.

That's the way it's been working for me. There are some builders who use more radical underscribe amounts than I do. But for me, the gap in the grooves are closing by the time there are 3 rounds of logs above them. The grooves are visually tight when there are about 4 or 5 rounds of logs in the walls. The top several rounds in a building are underscribed less, and these smaller gaps appear tight as soon as the roof is on.

Underscribed log walls do not need caulk, chinking, or extra-thick gaskets. The logs all appear tight before the owners ever move in. The reason is that the gaps get closed by compression, not by shrinkage. While shrinkage takes years, compression takes just days or weeks.

Hint For more information on underscribing, see Del Radomske's article in *Log Building News* #11 available from the ILBA for $6, call 1-250-547-8776

Settling

The shrinkage and compression in an underscribed building is identical to shrinkage and compression in a building that is not underscribed.

This means that settling will also be the same—allow about ¾" per vertical foot of logwork (6%). There is no extra weight to cause more compression, and there is no extra shrinkage, and so the settling allowance is the same.

Shrink-fit Notches

The other method that has been tried to keep notches tight over time has been to change the shape of the saddle—creating a high spot near its center.

Invented by Lloyd Beckedorf, the *butterfly* notch, has not been adopted by professional builders as widely as underscribing has been. My own experiences so far with it have shown me that with my logs, in my climate, and using my building techniques, underscribing is more effective for me than shrink-to-fit notches. The 4% or 5% shrinkage in log diameter has not seemed to bring the log back to a fat-enough part of the saddle to keep the notch tight over time.

Figure 9: The shrink-to-fit, or Beckedorf "butterfly" notch. The saddle, instead of being flat or slightly concave, is convex—that is, the saddle is higher in the center.

Settling

All homes that are made of wood settle. Even 2x4 homes made of kiln-dried lumber settle: just look at the flashing between the chimney and the roofing in a 5-year-old stick-frame house and you will see the flashing has bent: the building has shrunk in height. Settling is normal for *all* wooden homes.

Wood shrinks and wood compresses—these two components make up what logbuilders call settling. Wood shrinks in diameter as it dries; and it compresses (gets a bit squished) where it supports weight.

But wood does not shrink much in length, only in diameter. Since most of the walls of a frame home are composed of vertical wooden studs, frame walls don't settle much. Log homes have more settling than frame homes because log walls are entirely horizontal wood.

In this chapter you will learn how wood dries, how much settling to expect in a log home, and what precautions you must take to avoid settling problems.

Settling will happen, whether you build a home with kiln-dried lumber, dead-standing logs, or green logs—you simply must know how to deal with it. It's not difficult or scary, but it is important.

Settling

Figure 1: The classic settling problem: putting a window in a log wall. The window is nailed to these 2x4's, which are called *bucks*, but the 2x4 bucks cannot be nailed to the logs! The space above the framing and below the log header must not be filled with any material that would prevent settling. See also Color-Figures C-12 and C-14.

Note that the spline used here is a piece of angle iron, so the keyway cut is just a single chainsaw kerf. The 2″ x 2″ angle iron is attached to the back of the buck. You can see that the keyway kerf extends above the spline, to allow for settling.

To prevent air infiltration between the buck and the log ends, rubber gaskets are installed on both sides of the angle iron, and they can be seen barely poking out from behind the top of the buck.

Hint A log can lose half its weight before it even *starts* to shrink!

Drying

Logs are made of billions of wood fibers that are like microscopic straws: they are long, thin, and hollow. In a living tree, these cells, and the spaces between them, are full of water and sugars that are needed for growth. When a tree dies, or is cut down, the wood begins a long process of losing this water—the straws empty out over time.

The wood loses *free* water quite quickly—this is the water between the cells (in-between the straws). After the free water is gone, then the wood starts to slowly lose the *bound* water that is trapped inside the cells. Losing bound water causes the cell walls to dry out, and as they lose their plump, round shapes they collapse. The collapse of cell walls during drying is what causes wood to shrink. Logs shrink quite a bit in diameter, but lose almost nothing in length, as they dry.

It is important to note that losing the free water does not cause shrinkage—a log has to lose its bound water before it shrinks. This means that a log may weigh half of what it used to, but hasn't even begun to shrink yet. The moisture content (MC) when free water is gone and wood starts to lose its bound water is about 28%—this is called the *fiber saturation point*.

The moisture content of wood is measured as a percent of oven-dry weight. It is common for living trees to have 100% MC, or more, but this does not mean the tree is 100% water and 0% wood, it means that, by weight, the tree is half water and half wood. In some species, green moisture content can be as high as 200%, which means the tree has 2 pounds of water for every 1 pound of wood fiber.

Eventually, the moisture content of wood falls to the point where it will not lose any more water—it is in equilibrium with local conditions, which is called the *equilibrium moisture content* (EMC). EMC varies from about 4% to as much as 14% depending upon humidity, elevation, and temperature *(Figure 2)*. To get wood below EMC you have to put it into a kiln to artificially dry it out, and once it is back in the natural environment it will rise to the local EMC again.

Shrinkage

As the bound water is lost, wood cells get smaller in diameter, but not much shorter in length. This means that logs stay the same length, but they get smaller in diameter as they dry. This is why we are concerned only with horizontal wood when we allow for settling. Posts, columns, and studs (which are wood placed vertically), don't get much shorter over time, but wall logs, purlins, and joists, (which are wood placed horizontally) do get smaller *(Figure 1)*. Wood shrinks only about 0.1% to 0.2% in length—this is about ¼" to ½" in 20'-0" (10mm or 20mm in 10meters).

The amount of shrinkage varies with tree species and EMC. In general, dense, heavy species (like oak, hickory, and other hardwoods) shrink more than lighter woods like pine, spruce, and cedar. It is the lighter woods that are commonly used for log homes, and these species shrink about 4% in diameter as they dry from living trees (green) to equilibrium moisture content (dry).

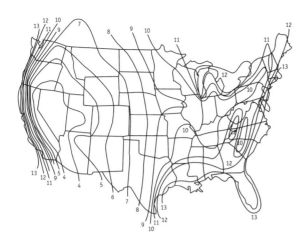

Figure 2: The dark lines show equilibrium moisture content in July. In the US, EMC varies from about 4% to about 14%.

In North America, log builders use 6%, or the equivalent ¾" per foot, as the settling allowance. This means that every vertical foot of log wall should be expected to settle about ¾". Why do we allow 6% instead of 4%? Simple: settling is made up of shrinkage and compression, and logs compress about 1% under the weight of the building. Plus, we must allow a bit extra as a safety factor, and this is built in to the 6% allowance. Shrinkage (4%) + compression (1%) + safety (1%).

Hint Log builders use 6% as the settling allowance. This is the same as 3/4" per foot.

Dry or Green?

Green logs are well suited to building log homes, and in some ways are even preferable to "dry" logs. And as a practical matter, it is difficult and expensive to kiln-dry or air-dry whole logs and large timbers.

A number of log builders have told me that they buy green logs, peel them, store them for a year, and that they build with "dry wood." This is a misunderstanding of how logs dry. Their logs may weigh only half as much as they did when they were green, but the logs are not dry, and they may not even have started to shrink. The free water is quickly gone, but shrinkage doesn't begin until the bound water is lost. Shrinkage starts when a log finally gets below 28% MC, and this can take years.

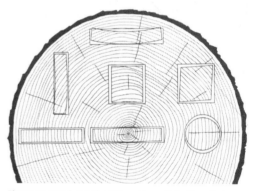

Figure 3: Wood shrinks differently radially than it does tangentially, and this is why lumber changes shape as it dries.

Log Building Standards allow builders to use less than ¾" per foot if the logs can be shown to be drier than 20%MC. Here's how to figure out how much shrinkage to allow for dead logs. First, find the radial shrinkage of the species of wood you will be using (there is a list in the *Wood Handbook*). Next, find the current moisture content of your logs with an electronic meter, and find the equilibrium moisture content for your area on the map *(Figure 2)*. Finally, apply this formula for determining how much your logs will shrink by the time they reach EMC:

$$\Delta D = D \times S \ (\Delta MC \div fsp)$$

Translation: the amount of shrinkage (ΔD) equals the starting dimension (D) times the total shrinkage (S, from the *Wood Handbook*) times the change in moisture content (ΔMC) divided by the fiber saturation point (usually about 28%). A good example of using this formula can be found in *Understanding Wood*, by Bruce Hoadley (pp. 76-79).

Hint It's a lot safer to just allow for 3/4" per foot !

It's a lot easier and safer to just allow ¾" per foot, no matter how "dry" you think your logs are—but don't make a guess because guessing wrong can be a disaster.

Compression

Just as logs shrink more in diameter than in length, wood also compresses (squishes) more easily in diameter than it does in length. This means that logs are stronger as posts (parallel to the grain) than as walls logs (perpendicular, or 90°, to the grain).

Hint Logs do not shrink much in length, so door and window opening widths do not change as the logs dry.

When you take a building apart to truck it to its foundation you can see that the notches have dented into the saddles—not much, perhaps 1/8" per log, but in a wall made of 10 logs, this adds up to a total of 1¼" (30mm) of compression. The settling allowance of ¾" per foot accounts for both compression and shrinkage so nothing more needs to be added when you figure how much settling to allow for.

Underscribing (described in detail in a previous chapter) does not change the settling allowance. Underscribing does not change shrinkage of logs—the same height of wood is involved as with a wall that is not underscribed. And it does not change compression because the weight of the log walls is the same—it's just that some weight has been shifted from long grooves to notches. As long as the edges of the notches do not crush or collapse, underscribing does not change settling.

Slumping

In the past, before kerfing, logs checked in their long grooves and there was a third component of settling called *slumping*. When logs check in their grooves and the check widens, then the edges of the groove separate from each other, come off the location they were scribed to fit on the log below, and slump to a new, lower level. Slumping increased settling substantially, but now that kerfing is widely used, slumping has been nearly eliminated.

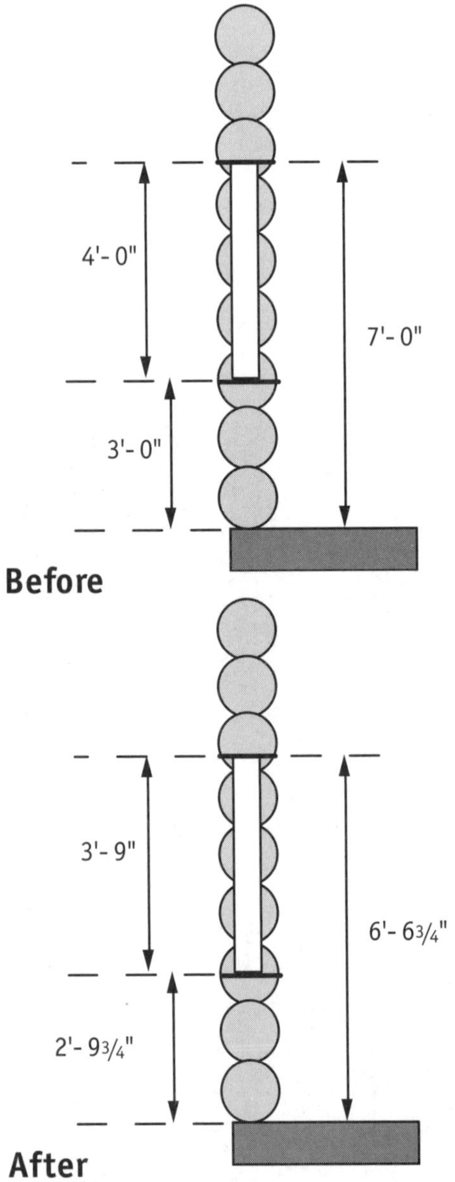
4'- 0"
7'- 0"
3'- 0"
Before
3'- 9"
6'- 6¾"
2'- 9¾"
After

Figure 4: Involved height helps you determine how much settling to expect—the window opening shrinks in height by 3" *and* the window sill also gets closer to the floor.

Involved Height

Before we can know how much settling to expect, we need to understand the concept of *involved height*—which is the log wall height that is involved in settling. Let's start with an example of a window opening in a green log wall in which the log sill is 3-feet off the floor, and the opening cut in the low wall is 4-feet tall *(Figure 4)*. There are three different involved heights in this example (3', 4', and 7')—and which height we use depends upon what part of the opening we want to calculate settling for.

If we want to know how high off the floor the window sill will be after settling, then the involved height is 3-feet. To calculate settling allowance, multiply 3 times ¾" per foot and you get 2¼"—it'll be 2¼" lower than where it started. Then, 3'-0" minus 2¼" equals 2'-9¾". So, a window sill that starts at 3'-0" will end up at about 2'-9¾" off the floor.

If we want to know how high off the floor the log header will be after settling, then the involved height is 7-feet. To calculate settling allowance, multiply 7 x ¾" = (21/4)" = 5¼" of settling. The log header will be 5¼" lower when settling has ended, or it will be at 7'-0" - 5¼" = 6'-6¾" off the floor.

Now, what about the window opening itself—how tall will the opening be after settling? The involved height is 4', because the horizontal logs that are involved in making the window opening smaller are those between the window sill and the log header. Multiply 4 x ¾" = 3". The opening will be 3" smaller than it started, and 4'-0" - 3" = 3'-9" opening height after settling. Yes, it is also closer to the floor, but that fact does not make the opening smaller—only the involved log height of 4' makes the opening itself smaller.

Consider this: if this 4' tall window opening were 1' off the floor it would still be 3'-9" tall after settling. A window opening 4' tall in the second story, 12' off the floor, would also be 3'-9" after settling. The amount that the opening will shrink is the same whether it is high or low on the wall—the involved height here is the height of the opening, not the height off the floor. Before you figure setting allowance you must first determine what part of the height of the wall is involved.

To compute window and door openings it is more useful to work in the other direction—we typically want to know how tall to cut the opening in a green wall, not how tall an opening we will end up with in a dry log wall. For example, if the catalogue says the carpentry rough opening height of a window is 4'-9" and inside the log opening you need 3½" of framing lumber, then you will be putting something that is 5'-0½" tall into this opening. How tall should the log opening be so that when settling is done the opening is not shorter than 5'-0½"?

Here is the easy formula:

Height before settling in inches = height needed after settling in inches x 1.06

For example, if the height after setting must be 5'-0½", first convert this to inches (60.5"); and then multiply this times 1.06 to get 64.13". To convert .13" to sixteenths of an inch, multiply it by 16 = 2.08 , and round this to ²⁄₁₆ths, which is ⅛". The height of the log opening when the wall logs are green is 5'-4⅛". Rounding off, I would make the green log opening height 5'-4".

Non-settling Parts of Log Homes

Settling affects the log portions of homes like door and window openings and plate heights, but many non-settling structures in buildings are also involved in settling.

A classic example is an interior 2x frame wall built so that it sits on the first floor and tall enough to be nailed to the bottom of the second floor *(Figure 5)*. Although this would be typical in a frame house, in a log house it is a problem. In a log house, the ceiling is getting closer to the first floor, and so frame partition walls must be built shorter than the floor-to-ceiling height in the green log building.

Many common parts of a building must be examined for potential settling problems: the chimney, plumbing, staircases, posts, heating ducts, downspouts, flashings, and kitchen countertops (will a log, or a window sill, come down and hit the backsplash?). Posts, frame walls, and staircases are three of the most common settling problems, and as you build you will need to find creative ways of accommodating settling.

Hint   Before you figure setting allowance you must first determine what part of the height of the wall is "involved."

Metric users will be shaking their heads in disbelief at "imperial users." With metric, you simply multiply the involved height by 6% (0.06) to find the settling allowance.

Figure 5: An interior frame wall must be built shorter than the floor-to-ceiling height of the green building. This wall is okay only because it was built after all settling was over.

Figure 6A: This screwjack is bolted to the concrete pillar under it, and the screw rod extends up into the bottom of the log post in a hole that was pre-drilled for that purpose. Turning the nut lowers the log post.

Figure 6B: On the right, the screwjack assembled, on the left it is taken apart. The L-shaped flange welded to the top plate prevents a high wind from lifting the post off the screwjack by trapping the nut between two plates (the nut is never welded to either plate). The top plate is bolted to the bottom of the log post, and the bottom plate is bolted to the concrete base (note bolt holes pre-drilled in both plates).

Hint Have a place in each building where you can measure unrestricted settling.

Use this to then calculate how much to let down each different screwjack in your house.

Screwjacks

Posts are usually outfitted with large diameter metal screws that can be let down gradually over time. These devices, called *screwjacks*, can be fabricated, or purchased.

Sometimes the screwjack is at the top of a post and sometimes at the bottom. In general, if the post is freestanding then it is more convenient to have the screwjack where it is easy to reach—at the bottom *(Figures 6A & 6B)*. However, if things are attached to the post (windows or walls, for example), then it can be better to have the screwjack at the top of the post. In this way the windows probably won't need slip joints where they are attached to the sides of the post.

A screwjack must be strong enough to hold its calculated load forever, so consult your structural engineer. Screwjacks are not removed after settling is done—they are permanently in the building. And, screwjacks must be relatively easy to adjust, and should not require two people, or a power tool, to turn them safely.

Keep a Record of Settling

Keep track of the amount of settling that has happened. Here's how to do this. In a part of the building that has no screwjacks (staircase openings can be good because you can measure from the finished second floor to the finished first floor), measure and write down the height before settling, and then every 3 months measure again to determine how much settling has occurred. Keep a written record of the changes to this height, and the dates.

If your test site started 10' feet tall and has come down ⅜", then all the screwjacks in the building should be let down *in proportion* to this amount. For example, a post that has 20' of involved height would need to be let down ¾" because (20/10) x ⅜" = ¾". A post with 12' of involved height would be let down ⁷⁄₁₆" because (12/10) x ⅜" = ⁷⁄₁₆".

It is easy to wrongly let a screwjack down too little or too much, and you must know how much to lower it each time. For the first 4 years of the building check the settling every 3 months, and make the small adjustments that are needed each time. This is easier and safer than waiting too long and having to let screwjacks down 1" (25mm) or more at one time.

Keep notes on heights and dates so you can clock the progress of settling. So, for example, if the original measurement at the test site was 10'-0" (120") and you measure it now at 9'-7½" (115.5"), then the shrinkage has been 115.5 ÷ 120 = .96, which is 4% shrinkage so far (100% - 96% = 4%). There might be more to come: we allowed for 6% settling. But there is a safety factor built in to the 6% allowance—we don't expect to get all of the 6% of settling we allowed for because then an additional ⅛" of settling would definitely cause problems for our building!

Frame Walls

Frame partition walls that are not bearing any loads are often used in log homes to reduce the need for expensive interior log walls. They are also used for wet walls (plumbing) or walls to hide ductwork. They are usually 2x4, unless they are constructed with plumbing or ventilating in mind, and then can be 2x6, or even larger *(Also see Color Figures C-14 and C-15.)*

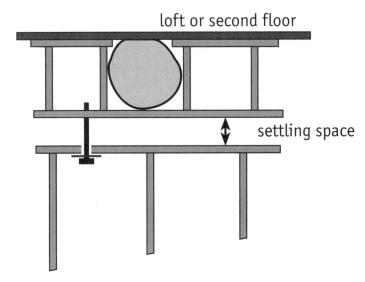

loft or second floor

settling space

Figure 7: A short frame wall is built around log floor joists and is attached only to the ceiling. A taller frame wall is built off the floor, but is short enough to provide open settling space. Lag screws stabilize the lower wall, but allow for settling. Compare with Figure 12.

Most of these walls can be constructed in the same way—a short frame wall is built to the ceiling that comes low enough to get below all floor joists. Then a frame wall is built off the floor that is shorter than the ceiling height by the settling allowance that is expected. The lower frame wall is stood up and nailed to the floor *(Figure 7)*. It is prevented from swaying at the top with lag screws drilled through slightly oversize holes in the frame-wall top-plate and into the ceiling wall—this stiffens the wall and keeps it in place. The walls are finished and sheathed separately—no board or sheathing bridges the settling space. The gap between the two walls is covered with settling boards that allow slippage by being fastened to either the ceiling wall or to the floor wall, but never to *both* walls.

Staircases

Staircases are perhaps the greatest settling challenge for log builders. There have been a number of solutions attempted, and some perform better than others, but none of these is really elegant. When it comes to stairs in log homes, there is still room for improvement, I think.

The most popular stairs for log homes start out with tilted treads that don't get close to level until settling is over. If you guess right, the treads end up almost level. If not, the homeowners permanently get a low landing, or unequal risers, or some other kluge.

Here's the problem in a nutshell. When settling is completed, in four or five years, the second floor will be the height it will always be above the first floor. But in the first five years in the life of a log home, the second floor will be higher than this. The stair stringers must be laid out and cut to the total rise you expect to have after settling is done, which means that the stairs do not fit well in the time before settling is completed—they're too short. The stairs will fit some day, but in the meantime they are not quite right.

To make up the settling difference, it is possible to hang the stairs at a steeper angle than they were intended, letting the stringers hinge from the top, hung from a floor joist or beam.

Hint The stairs will fit some day, but in the meantime they are not quite right.

Figure 8: Half-log treads spiral around a center log post by Caribou Creek Log Homes.

Photo courtesy: Beaver Creek Log Homes

A temporary platform can be built at the bottom of the stairs that is the same height as the total amount of expected settling. This means that the stairs are at their proper angle (treads are level and risers plumb), but it means that there is a landing at the bottom with a riser that may not be equal to the rise of the stair treads, and this can be dangerous. Building codes have good reasons to specify that all risers must be the same height—it's safer.

Stair risers are usually about 7", while the settling allowance between the 1st and 2nd story is usually about 7". As times goes on, in order to keep the stair treads level, the landing would have to be reduced to 6", then 5", then 4", and finally 1", before, you hope, the bottom of the stairs can rest on the finished first floor.

Another problem with this is that while ¾" per foot, or 6%, is a good guess at the settling, the actual settling you get will be somewhat less than this (you certainly hope it will not be *more* than this). This means that the stair treads will not come out exactly level without a shim of some size at the bottom.

There are other solutions to the problem that are quite mechanical, expensive, and complex. These solutions include circular staircases with risers that can be let down over time so that the risers stay all equal height. Or circular wooden stairs cases that have a center threaded rod and solid log treads that have the same amount of green, horizontal wood as the walls, and so should settle the same amount *(Figure 8)*. But circular stairs made entirely of green material are less than satisfactory. And both of these circular stairs have difficult problems with the handrails and balusters.

Perhaps the best solution, in some ways, would be to have a temporary stairs for four or five years and then measure the final floor heights to get the total rise, and make the final, permanent stairs only after all settling is done. Unfortunately, very few homeowners or builders are this patient—of course, that would be asking a lot!

Self-adjusting Staircase

Here's another idea that's still in the works: a self-adjusting staircase. The top of every tread is always level, and the rises between all the treads are always equal—before, during, and after settling, no matter how much settling happens.

The keys features of these stairs are split-stringers, pivoting treads, and open risers. Each side of the staircase needs two stringers, not one. This means four stringers for a simple, straight run of stairs. The stringers can be lumber or log slabs.

The treads *(in Figure 9 they're half-logs)* are attached to the stringers by pins that allow the treads to pivot. The nose of every tread is pinned to the top stringer, and the toe of every tread is pinned to the bottom stringer.

Like most stairs in log homes, as the first and second floors get closer together the angle of the stringers changes—the stairs get less steep. Also, the bottom of the stairs slowly slide out across the floor. This is not new, either.

But in these stairs, the vertical distance between all the treads gets smaller as the building settles. That is, the unit rise changes from, say, 8" before settling to 7⅜" after. And all the rises change at the same rate, and stay equal, all the time.

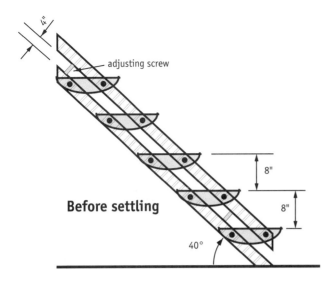

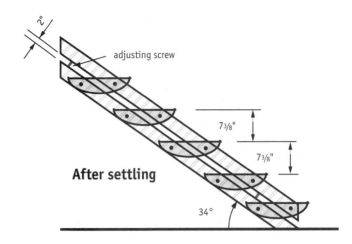

Figure 9: A staircase that self-adjusts for settling over time.

To do this, the top and bottom stringers on each side must get closer together—here they start 4" apart, and end up 2" apart.

I have shown adjustment screws between the stringers, but they might not be necessary. I won't know for certain until a set is built, but perhaps by hanging the top of the staircase only from the top stringers, and laying the bottom of the stairs only on the bottom stringers, no screws will be needed to keep the stringers separated.

I had two inspirations for this stair design. First, the stairs used by cruise ships: the wharf stays the same elevation as ships rise and fall with tides. Thank you to a student for telling me about these gangplanks—I live a long way from salt water.

The second inspiration has been on my desk for years. Parallel rules (a drafting tool I use in roof design) have two parallel "stringers" (the rulers) joined by pivoting "treads."

Chimneys

Chimneys are columns that do not change height. But the roof that surrounds a chimney is slowly getting lower until settling is complete. This has several important consequences *(Figure 10)*.

First, though you will certainly be tempted, do not use chimneys to support roof beams. Normally, a chimney comes out somewhere near the ridge of a house, and so it seems logical that here is a way to reduce the span of a long ridge log—just support it in pockets the masons could build in the stonework of the chimney. Unless the roof is not settling (and this happens only in some piece-en-piece homes), this is a bad way to build.

In general, it is best to have chimneys and masonry columns set slightly to one side or the other of the ridgeline so that the log ridge can continue uninterrupted by the chimney.

Figure 10: A big mistake: the mason cut the stones of the chimney to fit the contours of the logs. There will be gaps between the logs where the stones prevent natural settling. The logs may even break some of the mortar joints as they settle, and stones could fall out.

Figure 11: The trim was cut to match the shape of the log wall—but settling exposes the problem with this method. Compare this with the slots cut in the log wall shown in Color Figures C-14 and C-15.

Figure 12: Frame walls are built to allow space for settling. Slots are cut into the log walls so that trim can appear neat, and allow for the wall to settle *(see Figure 7, and also compare with Color Figures C-14 & C-15).*

Second, the roof flashings that keep water out of chimney openings must have two parts: flashings and counterflashings that allow slipping and still keep the weather out. The flashing and counterflashing are built so that when the house is green they are overlapped to their maximum extent, and they have enough overlap (measured vertically) so that when settling is completed there is still at least 1" (25mm), and preferably much more, overlap remaining. *(See Figure 10 of Log Building Standards.)* The flashing and counterflashing must not be attached one to the other—they must be free to slide past each other.

All framing in the roof—the 2x's and sheathing, must also be kept away from the masonry column. If you are using rough stones of variable size, be certain that there are no lumps that are in the settling path that will hang-up in the future.

Cutting Slots in Log Walls

There are several places where straight, flat surfaces meet log walls. The difficulty is how to make these meet so that the gaps are filled and settling is allowed.

This happens where an interior 2x frame wall meets a log wall. It is possible to cut trim boards that are scribed to the profile of the logs, but this profile changes continuously for the first four or five years. Any scribed trim-board you cut will, in just a short time, no longer fit *(Figure 11)*, and will have to be re-done.

The other way of sealing this connection is to cut vertical slots, or dadoes, in the log wall and slip trim boards into the slots *(Figures 7 & 12, also Color Figures C-14 & C-15)*. These are permanently installed and do not need to be changed, removed, or altered. But they do require that the log walls be cut, and this means that remodeling—moving interior partition walls—will leave unsightly slots.

If cut too deeply, these cuts can also weaken the log wall, so *Log Building Standards* specifies their depth and proximity. The back of the dado (slot) should be plumb or contoured, because if they lean out at the top, then as the logs come down they will push the trim board over, into the room.

Frame walls start with a stud attached to the log wall. This first stud should either sit on a flattened portion of the log wall, or it should be shimmed out so that it is plumb. The bottom of this stud can be permanently nailed or lagged to the bottom log of the wall. But everywhere else higher on the wall, the stud must be attached so that the log wall can settle. This usually means cutting vertical slots in the stud at the midpoint and near the top, and then attaching the stud to the logs with lag screws and washers. The lag screws are put through the slots near the top of the slots, and they are tightened so the stud is secure but so that the washer will slide down the stud.

To have enough width to attach the wood trim board and the drywall or other wall covering, you should add one or even two more studs to this to create a double or triple stud at the start of the wall *(Figure 12)*.

Some builders slide drywall into the slots (also known as *dadoes*), but this is not a method I usually like: it usually leaves unpainted areas of drywall exposed as the logs settle, and sometimes tears the drywall paper. There is also the problem of the drywall edge. It is often preferable to use a piece of wood in the wall slot.

How Long Does Settling Take?

The time it takes for settling to be completed varies with climate, elevation, how the building is used, and the original moisture content of the logs. You should count on four to five years from the date the owners move in.

If the logs start drier than 60% MC, then it may reduce the time by perhaps a year. Higher elevation tends to speed drying, while lower elevation can slow it down.

Once settling is over, the logs have dried to equilibrium with the climate, and have compressed as much as they will, then settling is complete—you do not need to think about screwjacks and settling boards again.

I advise log home owners that after settling is done they should caulk door and window trim just as they would be caulked in a frame house. If you caulk before settling is done the caulk will stretch and rip, so wait to do this. The record that you have been keeping of settling will indicate when it is complete.

Checking

Logs check, or crack, as they shrink—this is an unavoidable, natural consequence of the drying of wood. Trees tend to develop just one major check, and several smaller ones. We can't stop checking, but we can have a lot of control over where the check happens in green logs *(Figure 13)*.

Some long groove styles *(Figure 14)* cause checks in the groove. In fact, checks normally occur from the log's center out to the closest exterior surface. In this case the cove made for the long groove makes an exterior surface that is closer to the log center than anywhere else, and so the log checks there, 100% of the time. Checks virtually never split a log in half unless something else is wrong.

Kerfing is very old technique in which a log or timber has a saw kerf cut lengthwise to encourage checking to happen at that location *(Figure 14, left)*. Kerfs do not need wedges in order to be effective. The double cut groove, shown below on the right, does not need to be kerfed—it has a "V" shape cut that encourages checking on the top of each log.

Hint Settling takes 4 to 5 years, and sometimes longer. It's over only when it's over.

Figure 13: Logs usually develop just one major check. The most likely place for the check is where the log has been cut. Here it is the flyway long grooves.

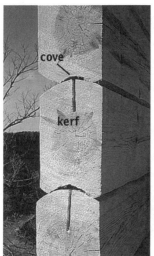

Figure 14:

Far left: Shallow-cove long groove and kerf. The check in green logs will likely form in the chainsaw kerf—a good place for it.

Near left: The Beckedorf double-cut long groove. The check in green logs will likely form in the deep-V on the top of each log—a good place for it.

In log buildings, half-log sills do not contain the *pith* (the center of the growth rings of a log) and will not check, and so do not need to be kerfed. (In fact, that's why timber framers like to order timbers that are *free of heart center* (fohc)—because they don't crack.) Lots more on kerfing can be found in the chapter on Cutting Notches.

Settling & Roofs

The log gable end roof *(Figure 15)* has the potential for severe problems caused by differential settling. I have built only one log gable roof in my career because they have serious structural concerns, most of them caused by settling.

One problem is that the pitch of a roof supported by log gables changes over time. A roof that starts at, say, 10:12 pitch can become 9½:12 after settling. To allow for this change of pitch, log builders must find a way for the 2x common rafters to slide downhill on the plate logs and yet still be attached to them so they don't blow off in a wind. Special care must be taken with chimneys because the roof framing is moving not just straight down but also *downhill* over time and could, if not allowed for, come in contact with the chimney and cause problems.

Figure 15: A simple log gable end on a rectangular floor plan can be okay, but watch out for special settling problems!

If the floorplan is anything other than a simple rectangular box, that is, if the roof has hips or valleys, then settling problems with log gable roofs become severe. And I advise against mixing log gable ends with a log truss—the roof at the gable ends will change pitch, but there is no way to adjust the roof pitch of a truss over time to match.

Post-and-purlin log roofs are generally free of settling concerns if all the posts bear on log walls that are all the same height. If a post must be supported somewhere else, then a screwjack on the odd-length post can usually solve the problem.

In Figure 16, the roof is supported by timber-frame trusses on top of log walls. The plate log extends (cantilevers) several feet to support a wide gable end overhang, but it was not strong enough without being braced. A diagonal brace down to the log corner would have had settling problems because of the involved height between the plate log and the bottom of the brace.

We solved this by using a *suspended brace*. The short vertical post is attached only to the plate log, and is free to slide down the log ends. The diagonal brace is attached to the short vertical post instead of to the log walls, so there is no settling problem.

Summary

You need to keep your wits about you when it comes to planning for settling. When necessary, seek expert assistance. Screwjacks should not automatically be your first choice. Having a house full of screwjacks can be a headache for several years—if you forget to adjust them, or adjust them incorrectly, you'll cause more problems than you solve. In Figure 17, you see the result of not adjusting screwjacks (in this case the builder provided wooden wedges at the tops of the posts, but they were installed improperly and could not be adjusted). When it comes to designing for settling, it is best to keep things simple and effective, and avoid complex or expensive solutions.

Figure 16: A suspended brace supports the plate log of a heavy roof overhang. As the log wall settles, the brace slides down the log ends—an elegant and simple settling solution that needs no screwjacks or adjustment.

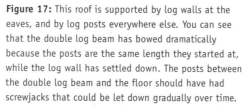

Figure 17: This roof is supported by log walls at the eaves, and by log posts everywhere else. You can see that the double log beam has bowed dramatically because the posts are the same length they started at, while the log wall has settled down. The posts between the double log beam and the floor should have had screwjacks that could be let down gradually over time.

Settling Quiz

Figure out both:

1) the involved height, and
2) the settling allowance for each of these log home situations.

All logs shown here are green; and all framing lumber (including walls and joists) is dry—it is at EMC.

The answers are on the next page.

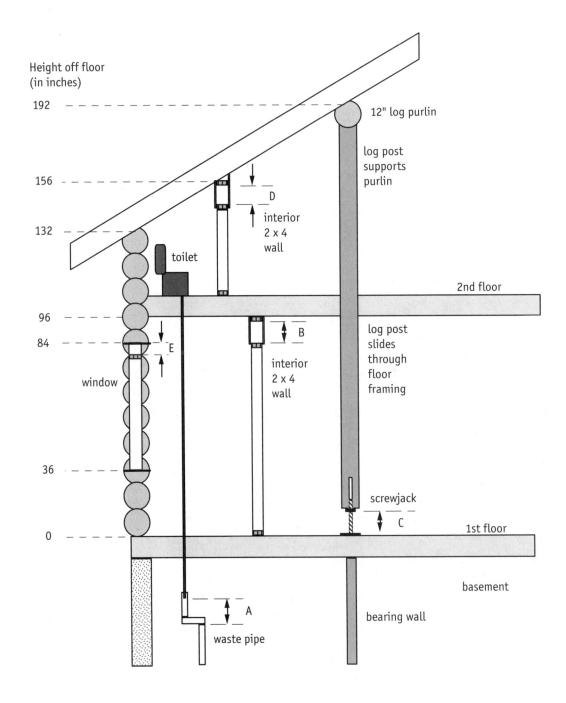

Height off floor (in inches)

192 — 12" log purlin

log post supports purlin

156 — D interior 2 x 4 wall

132 —

toilet

2nd floor

96 — B

84 — E interior 2 x 4 wall

window

log post slides through floor framing

36 —

screwjack C

0 — 1st floor

basement

bearing wall

waste pipe

Answers to the "Settling Quiz"

"A" The second-floor toilet has a waste pipe that extends into the basement. The waste pipe must be allowed to settle into another, larger-diameter pipe with a flexible neoprene collar to keep it safely sealed. Involved wall height is 96" (8'-0") because the toilet sits on floor framing that is supported by the log wall at a height of 96". Settling allowance is 6". This means the small diameter pipe must have at least 6" below it in the larger diameter pipe that is free and clear for settling. But the plumber had better allow for at least 8" here because of the nearby 90° bend in the big pipe and the fact that the two pipes must start out overlapping by at least 1". It does not matter that the settling space is in the basement not upstairs—this does not change the fact that the involved height is 8'-0" or that the settling allowance is 6".

"B" An interior partition wall is built on the first floor. The settling boards are attached to the ceiling above by a 2x4 top plate. The involved height of log wall is 96", or 8'-0". The settling allowance is 6". This means that the frame wall will be 7'-4½" tall when it is built (8'-0" minus 6" settling minus 1½" top plate thickness). This partition wall bears no weight (the second-floor joists don't touch it), and so it does not need to be supported in the basement.

"C" A screwjack is placed at the bottom of a long log post that holds up a log roof purlin. The log post is not attached to the second-floor framing. The involved wall height is 132" (not 192"). This is because the roof frame is supported by a log wall that is only 132" tall. The settling allowance is 8¼". The eave of the roof will be coming down 8¼", and every part of the roof, including the purlin, must come down 8¼". If it didn't then the roof would change pitch!

So, the distance between the nut on the screwjack and the plate of the screwjack must be 8¼"—that means 8¼" of exposed thread. Also note that the hole drilled in the bottom of the post must be extra deep to take up that 8¼" of threaded rod as the post is let down.

For those with keen eyes, you could allow a bit less than 8¼" of settling allowance at "C," because the green log purlin is 12" diameter and will become 11¼" after it dries and shrinks in diameter. This means the post will have to be let down ¾" less than 8¼", and so a settling allowance of 7½" would be enough. But 8¼" is okay—it's a little too much, but at least it's not too little!

Note that directly under the screwjack, in the basement, is a bearing wall that sits on a footing—you must transfer all point loads in a straight line to a concrete footing. Without the bearing wall below, this post would break through the first floor framing!

"D" Bathroom frame wall upstairs. Involved wall height is 36", or 3'-0". This is 132" minus 96". The second floor framing is sitting on a ledge in the log wall that is 96" off the first floor. The roof sits on a plate log that is 132" off the first floor. This makes the involved height for this wall 3'-0". The settling allowance is 2¼".

"E" Window settling space. The involved height is 48" (4'-0"), which is 84" minus 36", and the settling allowance is 3". See photo on the first page of this chapter—the space above the framing and below the log header cut, would need to be 3". But if you add a 1½" thick 2x4 in that space to attach the settling boards to, then you need to increase the settling space to 4½".

C-1 Lifting truss onto building. The tie beam will be under the plate logs—look at the notch in the plate log on right. Posts will support the truss (see C-2). Photo by Graeme D. Mould.

C-2 Log shell assembled and erected.

C-3 Common rafters and dormers framed on top of log purlins and ridge.

C-4 Purlins and ridge are supported by posts made of built-up 2x framing at one gable end of the house—interior view. Some windows installed. (The log truss shown in C-1 supports log purlins and ridge at the other gable end.)

C-5 Exterior view of gable end with conventional framing support for log purlins and ridge, instead of log posts. Common rafters have been installed.

(All photos on this page are of the same house.)

C-1

C-5

C-4

C-2

C-3

C-6

C-7

C-8

C-11

C-6 Setting log ridge onto mitered trusses. Ridge has been spliced over a kingpost. Note tenons on posts.

C-7 Log common rafters with sawn timber ridge and forked tree post, which will conveniently hide the ridge splice. Same roof as C-11.

C-8 Sawn timber valleys and dormer ridge with log jack- and common-rafters. A 2x frame wall will support the dormer ridge over large door opening below.

C-9 Sawn timber jack-purlins are scribed into the log hip-rafter of an octagonal roof.

C-10 Octagonal roof made of log rafters and sawn timber plates (bottom). Same roof as C-9.

C-11 Log common rafters, timber ridge; forked post with second half of ridge now installed, with splice hidden by forks—compare with C-7.

C-9

C-10

C-12 Door installed, settling space above, 1x3 wooden cleats nailed to log header will later hold the settling boards.

C-13 Log opening cut and ready for spline and buck. Sample buck shape shown. Keyway slot was cut for a 2x2x1/4 angle iron spline.

C-14 Window bucks and window installed. Settling space above window, and 1x3 cleats nailed to log header will hold settling boards. Also, just left of the window, note the drywall samples that have been trial-fitted into slots cut in the log wall. Compare with C-15.

C-15 Slots cut in log wall with chainsaw, for drywall. (The gang of 2x studs that are visible next to the log joist can also be seen in C-5.)

C-16 Buying trees "on the stump," meaning not yet cut down, and talking terms with the forest manager. We bought these trees.

C-17 The logs in this shell were peeled with high-pressure water, and no drawknife was used—this is not the normal process in my company. I usually prefer the drawknifed look.

C-18 A well-peeled log has graceful facets that show expert drawknife work with a sharp knife. The long, even knife strokes look much better than short, choppy ones. Consider this the "wallpaper" of your log home!

C-19 A sharp drawknife in skilled hands produces long, smooth shavings about 2" (50mm) wide. The outer bark was knocked off with a shovel or "spud" (in background on right), then the logs were peeled with razor-sharp drawknives. Freshly-peeled, green logs are almost white, and they become tan after a day in the sun (like C-18).

C-20 Purlins notched into a true log gable-end wall.

C-21 Detail of the purlin notch we used in C-20. The notch is both housed and also has a square tenon, or "block," inside, for strength. Note that there are no saddles—it is a round notch, not a saddle notch.

C-22 Saddle notch—the straightedge shows the proper concave shape at *top* of notch: thin where the square is resting. In contrast, the *sides* of the notch are strong and not very concave. You can see that there is no extra wood along the notch's perimeter—it was scored just barely inside its scribe line. "B5?"— I label logs by their corner (B) and their layer (5th from bottom—the half-sill log is Layer 1).

C-20

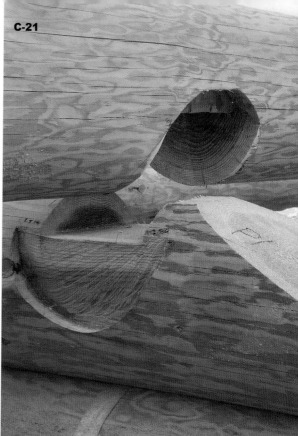

C-21

C-22

C-23

C-24

C-26

C-25

C-27

C-28

C-31

C-30

C-23 Long groove being ripped. Note my relaxed stance, my arms hang naturally, and my thumb is on the trigger so that I don't have to twist my hand.

C-24 Header log with 3 window headers already cut. The long groove is being ripped—the chainsaw is being held "wrong handed." Pro's will hold the saw either way, depending upon which gives the safer, or more accurate, or more comfortable, angle.

C-25 A 4-1/2" grinder sands to the groove's scribe line—a very fast and accurate method. The grain is always changing direction at the long-groove's edge, and a sander is unaffected by grain direction. Edge tools, like a chisel, scorp, gouge, and axe, are thrown off by grain direction.

C-26 Stapling gasket into a long groove and corner notch. (Note that this log has a keyway cut for a 2x4 window spline instead of an angle-iron spline—compare with C-13.)

C-27 Interior log wall that ends with exposed end grain. The long grooves have been scribed and cut to fit tightly and neatly, so they look good.

C-28 Log floor joist system. A horizontal slot has been cut in the wall logs that the floor material will be slipped into. Doing this supports the flooring's edge where it meets the log wall between joists, and means you don't have to scribe the flooring to the waney wall logs.

C-29 Another view of log floor joists, and the floor groove (slot) cut into wall logs. There is a door in the middle of this wall that leads to the balcony, and that is why the top log on the gable end also has a door-sill-flat cut into it.

C-30 A flyway, or log overhang, cut in the staggered-end style. The tips have not been trimmed to final length yet, the butts have been trimmed.

C-31 A flyway cut straight in the plumb style. There are many different flyway styles you can choose from—C-6 shows alternating bevel cuts, C-7 a double curve flyway, and C-32 a single curve.

C-29

C-32

C-33

C-36

C-35

C-32 Log shell assembled. Note the numerous interior and exterior gable 2x walls that support roof logs like ridges and purlins. I use very few log posts in exterior gable ends, though I do use log posts inside for supports. Logs posts in exterior gables are difficult to weatherseal, are not very stable, are time consuming, and are mostly covered up by 2x gable walls and not seen.

C-33 A forked tree is used to hold ridges in both directions and let's us express our imagination. This is the same house as C-32 and C-34. Heavy timber valleys at 45° are supported in custom-made steel brackets that are fastened to the log ridge. Photo by Graeme D. Mould.

C-34 Forked-tree post during construction. There is another log ridge that fits on the tops of the forks, and it is 90° to the log ridge seen here (the second ridge is on in photos C-33 & C-32). Steel brackets will hold valley rafters. Long 2x's and ratchet straps are used to temporarily brace log posts in position, and are later removed when no longer needed.

C-35 The art of scribe-fitted log walls—the goal is: this tight everywhere.

C-36 Same house as C-32, with the roof on (from a slightly different angle).

C-34

Roof Basics

Many people have difficulty visualizing how a roof goes on top of log walls. One problem seems to be the tops of the walls—they have round portions and do not look level. Where can you even hold a level on the plate log? Another problem is the layers that make up a log roof—there are log rafters, but then there are also 2x common rafters, and how do these relate? A 2x4 house, in comparison, seems simple.

First, we need to review the basic concepts of putting a roof on a building. Even if you are an experienced log builder or carpenter, you should read this chapter because some of the terms used, like *span*, and *level*, are defined in specific ways so they can be used in building a log roof.

Then we will move on to roofs with log structures: ridgepole, purlins, trusses, plate logs, posts, and rafters. In later chapters we will explore building roofs with advanced designs, like mitered trusses and valley logs.

Preparing for the Roof

The Basics

First of all, you must know the pitch (slope) and the span of your roof, and have someone qualified review the loads on the roof so it will be strong enough to do its job.

Roof systems can be very complex, so start studying now, instead of waiting until the log walls are done. Think about the roof, and plan for it, before you lay the sill logs. The roof starts with paper, not with trees: it starts with the blueprints.

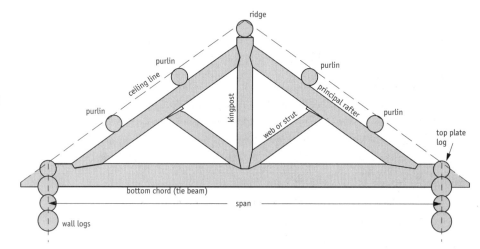

Figure 1: A log truss with roof parts labeled. A truss is supported only at each end of its tie beam—usually it rests on the eave log walls. It stretches freely in-between without needing mid-span supports like posts.

The span is the distance from the center of one eave log wall to the center of the other eave log wall.

A truss is only one way of the many ways to use logs in the roof.

We will keep things fairly simple in this chapter. You should learn to build a good *post and purlin* roof, or a log truss, before you try the tempting, but very challenging world, of towers, eyebrow dormers, and broken-back valleys.

I need to strongly caution you about trying to copy a log roof system clipped from a magazine. Many of these seem underbuilt, poorly designed, and some are even dangerous. Do not try to duplicate the roof you see in any photo—it may have been built improperly. It was certainly built for loads and spans that don't match your house.

Roof Pitch & Roof Span

Pitch is the steepness of the roof—some builders use rise:run (like 12:12); some use degrees, like 45°, to describe the slope of the roof surface. Both methods work.

The *span* of the truss (or of any beam) is the distance between the walls that support it—we will use span as the horizontal distance from the centerline of one eave log wall to the centerline of the other eave log wall. (*Figure 1*)

This is an important definition, worded carefully. The foundation width is not the same as span. The distance from the inside of one log wall to the inside of the other log wall is not span, either.

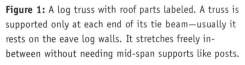

Figure 2: In the US, roof pitch is usually expressed as a ratio like 4:12, or 12:12.

Personally, I feel that 12:12 roofs are too steep for most log homes. The roof will feel more sheltering, and the room it encloses more cozy and natural, with a pitch between 5:12 and 8:12.

Steep roofs do provide somewhat more headroom upstairs, and can shed snow more quickly.

When we get to designing and preparing to compute the lengths of rafters, etc., you must know how far it is from the center of one wall to the center of the other.

Roof Load

Know the loads on your roof—this is essential to making the purlins, ridge, and trusses strong enough. Some parts of the world have very heavy snow loads (snow loads can go well over 150 pounds per square foot, psf). There are wind loads and dead loads to consider, too (*Figure 3*).

The local building inspector can tell you what the roof should be designed to carry. If you cannot get that information locally, then hire a structural engineer who works with timbers and logs (there is a list later in this book).

How many purlins do you need? What diameters should they be? Do your trusses need webs (braces) in order to support the loads? Do not guess that "it looks strong enough." How would you know that for certain?

The actual diameters of the roof and truss logs are determined by the loads that must be supported—and you will need an engineer to tell you that your ridge needs a mid-span diameter of 15" (for example), or that your principal rafters must be 16" (41cm).

FORMULAS AND DIAGRAMS FOR STATIC LOADS

Beam Overhanging One Support—Concentrated Load at Any Point Between Supports

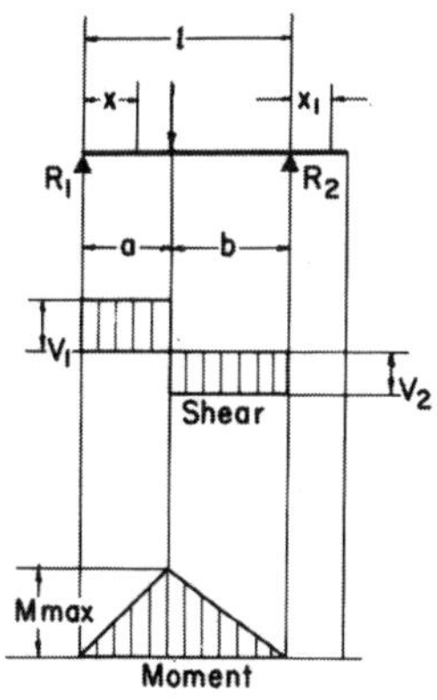

$$R_1 = V_1 \left(\text{max. when } a < b \right) \quad \ldots \ldots = \frac{Pb}{l}$$

$$R_2 = V_2 \left(\text{max. when } a > b \right) \quad \ldots \ldots = \frac{Pa}{l}$$

$$M \text{ max.} \left(\text{at point of load} \right) \quad \ldots \ldots = \frac{Pab}{l}$$

$$M_x \left(\text{when } x < a \right) \quad \ldots \ldots = \frac{Pbx}{l}$$

$$\Delta \text{max.} \left(\text{at } x = \sqrt{\frac{a(a+2b)}{3}} \text{ when } a > b \right) = \frac{Pab(a+2b)\sqrt{3a(a+2b)}}{27EI\,l}$$

$$\Delta_a \left(\text{at point of load} \right) \quad \ldots \ldots = \frac{Pa^2b^2}{3EI\,l}$$

$$\Delta_x \left(\text{when } x < a \right) \quad \ldots \ldots = \frac{Pbx}{6EI\,l}(l^2 - b^2 - x^2)$$

$$\Delta_x \left(\text{when } x > a \right) \quad \ldots \ldots = \frac{Pa(l-x)}{6EI\,l}(2lx - x^2 - a^2)$$

$$\Delta_{x_1} \quad \ldots \ldots = \frac{Pabx_1}{6EI\,l}(l + a)$$

Figure 3: Huh?? But this is the stuff someone must do to make a safe and strong roof.

The actual diameters you will use come from the loads on your roof, your local building codes, the species of wood you use (pine, spruce, Douglas fir, poplar), how the parts are joined (are you using only wood?, will there be metal fasteners?), and so on. These are complex questions and that is why you need an expert's help.

Roof Layout

Before you start building the roof system, the walls should be finished. The top logs on the walls, the plate logs, must be scribed in place, though they are still round on top and not yet cut at the roof pitch. If you are building trusses, the tie beams must be scribed and notched into the log walls.

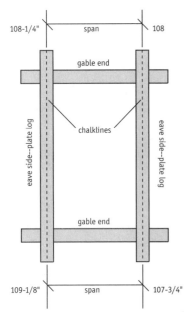

108-1/4"   span   108

gable end

eave side—plate log

chalklines

eave side—plate log

gable end

109-1/8"   span   107-3/4"

Figure 4: Example floor plan for roof layout. Note that the plate logs are eave logs and they are notched over (are higher than) the top logs on the gable ends. The elevations of the plate log ends are shown. Span is the distance from the center of one eave wall (chalkline) to the center of the other eave wall.

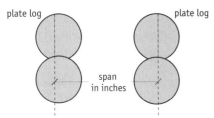

plate log   plate log

span
in inches

Figure 5: Here are the top two logs on the eave walls. Draw a plumb line down from the chalklines that are on the tops of the plate logs. Do this at both ends of each plate log. Check to make sure the span is the same at both ends of the building.

The plumb lines may not be in the "center" of the endgrain (like the left log above) because they come from the top-of-log chalklines and not from dividing plate log end diameters in half.

Figure 6: At both ends of both plate logs, measure from the sill logs (or floor) up to where the plumb line meets the top log surface. Note that the plumb line may not meet the plate log surface at the highest part of the plate log. Measure the height of the wall along the centerline of the log ends, even if it is not the highest part of the plate log.

The roof plate logs should be nearly level on top. And, of course, you have also been keeping parallel walls about the same height *(Figure 4)*.

The roof layout procedure described here applies to roofs that are supported with post-and-purlin systems, and to other roofs, and not just to roofs with log trusses.

Plate Log Centerlines

The endgrain of the plate logs need plumb centerlines drawn on them—use a 2-foot level, climb a ladder, and plumb down from the chalkline on the top surface of the plate logs. The distance from this plumb line to the plumb line on the parallel wall is your center-to-center wall distance (the *span*). The span must be the same distance at both gable ends of the building. Check to make sure that it is *(Figures 4 & 5)*.

If the spans at the gables are not the same, then your walls are tipping out or in a bit—you haven't been keeping the center of the logs in the center of the wall. You will need to redraw one or more of the plumb lines so that the spans at both ends of the building are within ⅛" (3mm) of being the same.

You can also plumb up from the sill log centerlines to draw the plumbline on the plate log ends. This is more accurate, but you still must check that the span is the same at both gable ends of the house.

Plate Log Heights

Next, measure up from the floor or sill log bottom to where the plumblines you just drew meet the top, round surface, of the plate log. Do this at the ends of all the plate logs.

Note that you are not measuring to the highest point of the plate logs, you are measuring up to the spot where the plate log plumb line meets the upper round surface of the plate log—this might not be the highest point *(Figure 6)*.

In our simple example, you would have four plate log elevations to measure, and they are each a little different *(Figure 4)*. This is typical—you cannot keep the walls exactly level, after all, and here is how you deal with this.

To lay out the plate log roof-pitch cuts you find the lowest corner and start there. If your plate log elevations are, say, 108", 107¾", 109⅛", and 108¼" then the lowest number, 107¾", is where you start *(see Figure 4)*.

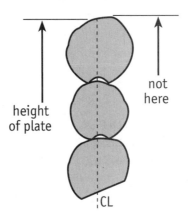

not
here

height
of plate

CL

Draw the Roof Pitch on the Plate Logs

Go to the corner that has the lowest plate log elevation, and draw the roof pitch on the endgrain. I use a digital level that displays roof angle in degrees or in pitch. You could also use a carpenter's rafter square.

The roof pitch line goes through the point where the plumb line and the elevation intersect (107¾" in our example). This is where the centerlines and the dotted horizontal level line meet *(Figures 7 & 8)*.

Go to each of the other corners in turn, and mark the same height (107¾" in our example) on each plate log centerline. Then use a digital level or a rafter square and draw the roof pitch at the correct angle and going through the height point.

The higher plate logs will have more wood removed than the lowest plate log—but this is necessary in order to get the plate logs level from end to end. There is no handy place to put a level on a plate log to see if it is level end-to-end—it is still round on top and a level will just rock around! But we know the plate logs are level because at their ends they are all 107¾" off the floor *(Figure 8)*.

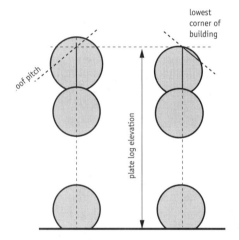

Figure 7: Once you know the elevation at your lowest corner, measure up that amount at all the plate log ends and mark that height onto the plumb centerline of every plate log. Draw the roof pitch at the correct angle, and going through the height point you just marked. This is where you will flatten the plate logs.

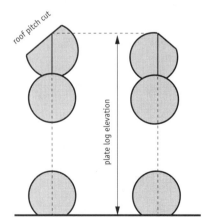

Figure 8: After the plate logs have their ends marked, bring them down to the ground and snap two chalklines from end to end for the roof-pitch lines. Rip and then brush to the chalklines with a chainsaw.

Put the plates back on the walls. Note that the two plate logs in this drawing do not appear to be the same height (the left plate looks taller than the right plate)—but they are exactly the same height because you measure plate log heights *only* on their centerlines.

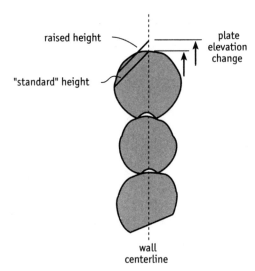

raised height

"standard" height

plate elevation change

wall centerline

Figure 9: Steep roofs make for very wide cuts on the plate logs. Here is the lowest plate log—if we cut it at the "standard" height we would be removing more wood than needed (the flat would be >4", 100mm). The other corners, being taller, would remove even more wood than this!

Instead of measuring plate elevation where the centerline meets the top surface, we can go a bit higher than this—up to where the "raised height" line meets the centerline. This also happens with large diameter plate logs that tend to have wide roof flats even at lower pitches.

If the pitch cut at the lowest corner looks wide (> 4"), then consider raising the plate elevation a bit and see if it looks better.

Figure 10: Photo of a plate log layout. You can see the centerline, the "standard" roof pitch elevation, and the "raised" height (because these are large diameter logs. The plate log has been sawn flat at the roof pitch and the common rafters are nailed on to it.

Steep Roofs, Big Logs, and Bowed Plate Logs Can be a Little Different

There are a few special situations to watch for when laying out and cutting the roof pitch on plate logs.

First, if the roof pitch is steep (usually 10:12, 40°, or steeper), then you might want to raise the pitch cut-line a bit to avoid removing so much wood from the plate log. The flat on the plate log needs to be only about 4" wide, or so. Figure 9 shows how to raise the roof pitch cut for steep slopes. As before, once you have the pitch drawn on the plate log at the lowest corner, then all the other plate logs get marked at that same height.

Second, with very large diameter plate logs (> 14" tips) you also may want to consider raising the plate elevation, even if the pitch is less than 10:12 (*see Figure 10*). Once you have the pitch drawn on the plate log's end at the lowest corner, then all the other plate log ends get marked and cut at that same height.

Finally, if a plate log bows *in* towards the inside of the building, then you might need to lower the plate elevation a bit to get a continuous flat for the roof. If you roll the plate log (before you rough notch it or scribe it) so that the bow is in the same plane as the roof pitch then you can avoid this situation.

Laying Out the Roof

Now that the plate logs are cut, and you know the span and pitch of the roof, you can find other important measurements and distances—the height of the ceiling peak, and the length of the rafters.

Here's how to draw a cross-section drawing of your roof. On a large piece of paper make the following lines:

1) span line, which is level;

2) centerline of roof, which is plumb, and half way between the centerlines of the eave walls.

For example, if your roof span is 26'-8", then starting where the span line and the centerline meet, measure to the left 13'-4" and make a mark, and right 13'-4" and make a mark. The distance between these two marks is equal to your roof span (26'-8" center-of-wall to center-of-wall), and the roof peak is in the center of the span and at right angles (90°) to the span line *(Figure 11)*.

Next, you need to find the peak of the ceiling. To find this height you must know your pitch. Then you can use a proportion to find your ceiling peak (example below). If the span is 26'-8" and the pitch is 10:12, then you will measure up 11'-1 ⁵⁄₁₆" from the span line to the peak.

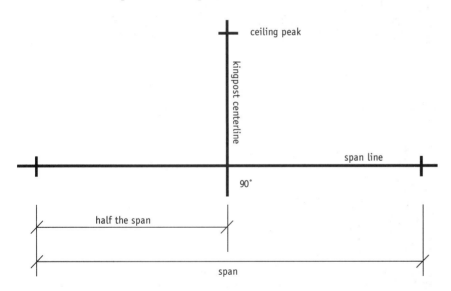

Figure 11:

Figuring Ceiling Height

$$\frac{\text{Your ceiling height}}{\text{Half your span}} = \frac{\text{Your rise}}{\text{Your run}}$$

$$\frac{\text{Your ceiling height}}{13'\text{-}4''\ (160'')} = \frac{10}{12}$$

Use a calculator and divide 10 by 12, then multiply that number times 160. The answer is 133.33", which means your ceiling peak is 133-⁵⁄₁₆" above the span line, or 11'-1⁵⁄₁₆" .

There are other ways to find the ceiling peak height: trigonometry, or a builder's calculator that works in feet and inches.

Now, draw a line that connects the ceiling peak to the span to the left and another line from the ceiling peak to the span to the right *(Figure 12)*. These are the *ceiling lines*. This is not where your 2x12 rafters sit, and these lines are not your shingles—the ceiling line is the bottom surface of your finished ceiling material. Think of it as the paper layer on the sheetrock, or the varnish on the tongue and groove ceiling.

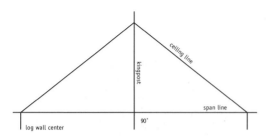

Figure 12: Roof design starts with the triangle of the ceiling and span. All support logs go inside this triangle, and all 2x rafters, insulation, and roofing go outside it.

Finally, use a square and draw lines 90° down from where the ceiling lines meet the span line—these are the centers of the log walls.

These lines: roof center, span, log wall centers, and left ceiling and right ceiling are the lines that form the basis of your roof support system. Whether you build your roof with log post and purlin, or with log trusses, or with heavy timber frame, or even with conventional 2x12's or other lumber, you must start with this simple drawing (*Figure 12*).

All logs or timbers that support your roof are *inside* this triangle, as we will see in detail when we lay out and build a log truss, in the next chapter.

Figure 13 shows the roof triangle we have drawn with one example of a log roof system we could use to support this roof.

Note that the flats sawn on the purlins and ridge touch the ceiling lines, and that the log rafters of the truss touch the underside of the purlins.

We plan for all the parts of the roof by starting at the ceiling lines and working layer by layer *in* (for the log supports); and layer by layer *out* (for the common rafters, insulation, and roofing materials).

Figure 13: Note that the span line is not necessarily the same as the second floor elevation.

Also, note that the top of the log rafters in the truss are not snug against the ceiling, they are below the ceiling by an amount about equal to the purlin diameter.

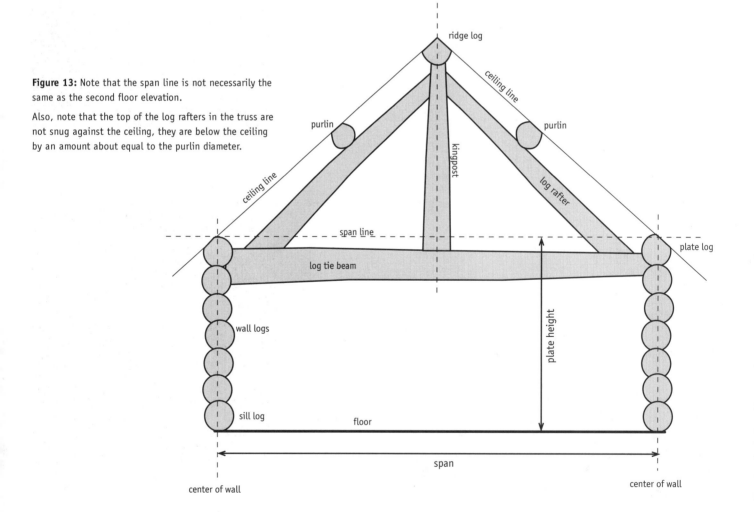

Supporting a Roof with Logs

There are several ways to use logs to support a roof: trusses, post and purlin, log gable ends, and log common rafters. Sometimes, as we will see, several methods are used in a building. Let's take a quick look at each of these and learn about their strengths and weaknesses.

Log Gable Ends

A popular log roof system has gable ends that are made of log walls. These have a distinctive, attractive look—triangle-shaped gable walls at the ends of the building *(Figure 14)*. These log gable ends, in turn, support purlins and a ridgepole.

A problem with log gable ends is that the roof changes pitch over time, as the logs shrink and settle. If log gable ends support purlins or ridge, then they should not be mixed with trusses. A truss never changes it pitch and the log gable ends are constantly changing pitch until settling is done—so something has to give.

A simple way to understand this problem is to look just at the gable end and measure how the roof is supported. The ridge is supported by, let's say, 17' of log wall, while the plate logs sit on 9' of log wall *(Figures 14 and Color C-20)*. The settling is allowance is 6%, which means that the ridge will have 5¾" (146mm) more settling than the plates (17' x 6% = 12¼" , while 9' x 6% = 6½"). Headroom upstairs will decrease by 5¾"—and the roof pitch will change, becoming less steep over time.

The changing roof pitch of log gable ends causes concerns for chimneys and roof framing—the common rafters have to be able to slide across the plate logs, and past chimneys. These concerns make log gable ends less popular than they once were. For small buildings, where purlins and ridge can free-span from gable end to gable end, log gable ends work well, and many people think that log gable ends are essential for the complete 'log look.'

Log gable ends can be mixed with post and purlin, but only if the posts are supported by adjustable screwjacks so the roof pitch can be changed to match the changing pitch of the log gable ends.

Post & Purlin

Very popular among professional log home builders, the post and purlin roof is good looking, and is probably the easiest log roof system to build. A purlin is a beam that runs parallel with the ridge—you can think of the ridgepole as the top purlin. The purlins and ridge are supported by posts that are supported by log walls or log beams *(Figures 15 & 16 and Color Figure C-3 & C-5)*.

Sometimes the posts are tilted and you might mistake them for the webs of a log truss. But there is a big difference: trusses can freespan long distances without sagging; simple log beams cannot.

Sometimes the purlins and ridge are supported by log posts, sometimes by framed walls made of 2x and plywood. Interior supports are almost always log posts. A disadvantage to using log posts in the exterior gable ends is that green logs shrink

Figure 14: Log gable end: the gable wall is built of scribe-fit whole logs stacked on top of each other. Here you see one purlin on each roof slope, and one ridgepole. The gable end is a solid wall, unlike the roof systems below that are mostly open spaces. See also Color Figure C-20.

Figure 15: Post and purlin roof: Log posts sit on a log wall or a log beam—one post under each purlin or ridge. Common rafters are layered on top of the purlins and ridge. See also Color Figures C-3 & C-5.

Figure 16: The posts are tilted, but that does not make them into a log truss — this is a post-and-purlin with pole common rafters.

Figure 17: A log common rafter roof in Latvia. In most of North America we would space the rafters closer together.

Figure 18: Pole rafters run from plate to ridge, and are spaced closely. This building combines pole rafters with log gable ends. Also see Color Figures C-8 & C-11.

Figure 19: Trusses are triangles made of logs—if there's no triangle, then it's not a truss. Here's a log truss being lifted into place on top of log walls. Also see Color Figures C-1 & C-6.

and can cause gaps that leak cold air and rainwater. Plus, you don't see much of the posts, because the gable ends get filled in with framing, insulation, and glass.

For these reasons, many professional log builders prefer to frame the gable-end walls out of 2x lumber and leave pockets to support the log purlins and ridge *(Color Figures C-2, C-4 & C-5)*. It's easy to nail thick log slabs onto the sheathing under the purlins, and they look like log posts—from the ground you cannot tell the difference.

Log Common Rafters

While purlins are parallel to the ridge, rafters are 90° to the ridgeline. *Common rafter* is a carpenter's term for the rafters that support the roof decking—it is different than the *principal rafters* found in trusses that support purlins, not sheathing. Because sheathing cannot span very far between supports, log common rafters are usually spaced close together.

Log common rafters look great but take a lot of time and care to get them installed right, and so they are not as *common* as they once were *(Figure 17 & 18)*. They need to be just 2' or 3' on center (depending upon the sheathing that is used), and this means that even small buildings need to have a lot of log rafters.

Once all the rafters are installed, the ceiling material is usually nailed to the flats sawn on each rafter's top surface—carpenters will need roof jacks for a place to stand while they do this. If it rains before the roof is weatherproof there is a good chance the insulation will get wet and the ceiling waterstained. This is a drawback to log common rafters.

There are numerous spaces between the log rafters and the plate log, each with a unique log shape—and they all must be closed off to prevent air infiltration. A lot of trim work is needed at the eaves where the rafter tails extend outside. Roofs with log common rafters have a tremendous, traditional appearance, but they are time consuming to build *(see Color Figures C-8 through C-11)*.

Log Trusses

Trusses are strong and good-looking—they really show off the skill of the builder and the style of log craftsmanship *(Figure 19 & Color Figures C-1 and C-6)*.

Log trusses are an excellent way to provide mid-span support for purlins and ridge logs. Log purlins can only span so far—in my company we never go more than 18' between supports, and 14' to 16' is much more common. (There is no rule-of-thumb for purlin span: you must do the engineering calculations for each roof load, wood species, and building design.)

Trusses are built of triangles with strong and stiff connections—and this is at the heart of their purpose: they can free-span long distances without losing strength or stiffness. You can support the mid-span of a log ridge and log purlins with posts or interior log walls, but the posts and walls might be in the way. A log truss provides support without any posts or extra walls.

Trusses are not well understood, even by some professional log builders, and are often used in places where they are not structurally needed, like the gable ends of places to put posts to hold up purlins and ridge. It is very rare to need a truss on a gable end. I prefer to use trusses where they look good and are necessary. Mixing trusses with post and purlin, or other support systems, also saves time and money.

Trusses take skill and time to build right. And a truss that is built wrong (weak joints, small diameter members, non-structural logs, no professional engineering) can be downright dangerous. We will cover log truss construction in step-by-step detail in the next chapter.

Timber Frame Trusses

Timber-frame trusses are used in the same sorts of places that log trusses are found—they are just made of different materials. Timber-framers use large, squared and planed timbers connected with mortise and tenon joinery.

Timber frame trusses are much easier to build than log trusses because all the pieces can be laid out and the joints cut just by measurement *(Figure 20)*. Round log trusses require more thinking and fitting and handling if the unique diameters and shapes of natural logs are to be respected.

There are many good books on timber frame joinery, and many hands-on workshops are available, so I won't be covering this topic here. I have used timber frame trusses in log buildings and they have a pleasant, refined appearance that contrasts well with the rugged organic shapes of the logs.

Hybrid Trusses

A frequently seen truss is made of round logs that have been squared off. Think of them as partly log, but completely timber frame at the joints. This is done to make it easier to build log trusses just by measurement, without lofting or scribing.

If a hybrid truss is large and has long pieces, then there is quite a bit of round log that remains visible *(Figure 21-left photo)*. But in smaller spans, or in trusses with webs, the round parts can get very short and look a bit bulged-out, and strange to me.

I have used this technique frequently, in the past. I have stopped building hybrid trusses—I just do not think they look as good as mitered joinery—they don't respect the natural shapes of the logs.

Figure 20: A timber-frame truss is lifted onto a log building. Another timber truss can be seen in the background, already in place.

Figure 21: Round logs are used for hybrid trusses, but at every joint the logs are shaped to be square timbers.
Photo courtesy: Beaver Creek Log Homes

Figure 22: 2x common trusses from the lumber yard make a strong and simple roof.

Figure 23: Twenty-two onion domes form the roof of the log Cathedral of the Transfiguration, 1714, Kizhi, Russia.

Figure 26: 2x common rafters are supported by log purlins, log ridge, and out of sight in this photo, the log plate. See also Color Figures C-2 through C-5.

Other Roofs

I have built three homes that used manufactured 2x trusses for the roof. They are inexpensive, span long distances, and are easy and fast to install *(Figure 22)*. By using scissors trusses you can even get a cathedral, or vaulted, appearance. It is also possible to hand-frame the roof with 2x or engineered lumber like TJI. Logs are not necessary, though, of course, many people like they way they look in their roof.

Framing the Roof

You will need to frame a roof structure on top of the roof support logs—on top of the log ridge, log purlins, and log plates.

This "second" roof frame provides space for insulation, a place to attach the finished ceiling materials, and a nail base (plywood or OSB) to attach the shingles. You might hear builders say that log roofs are "double" framed—and this is what they are talking about *(Figure 26 & Color Figures C-2, C-3, C-4 & C-5)*.

Figure 24 shows how the common rafters (2x12's, for example) sit on the ridge, purlin, and plates. There are usually no bird's mouths cut in these rafters. Instead, we cut the roof support logs (ridge, purlins, and plates) at the roof pitch, and the common rafters rest flat on this flat.

Figure 25 shows how easy it is to slip in the ceiling material (sheetrock, 1x4 tongue & groove, or etc.), and nail it to the bottom of the rafters—as long as we nail the 2x rafters to a spacer instead of directly to the roof support logs. The framing and finish detail at the wall plate logs is nearly the same as at purlins.

The spacer blocks are about the same thickness as the finished ceiling material (or slightly thicker), and are usually made of plywood or 2x lumber, or a combination of lumber and plywood.

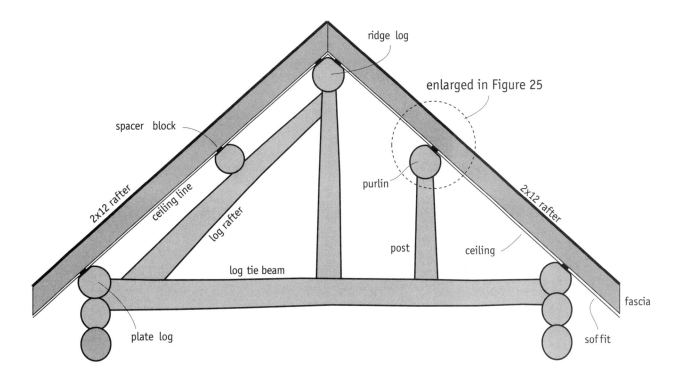

ridge log

enlarged in Figure 25

spacer block

2x12 rafter

ceiling line

log rafter

purlin

2x12 rafter

post

ceiling

log tie beam

fascia

plate log

soffit

Figure 24: Another roof sits on top of the roof support logs—this second roof could be 2x12 common rafters, 'TJI's', or stress-skin foam panels. This is true whether the roof support logs are a truss (left side) or post and purlin (right side)—here the roof is half truss and half post-and-purlin, something you would not do. The area circled above is detailed in the Figure 25 below.

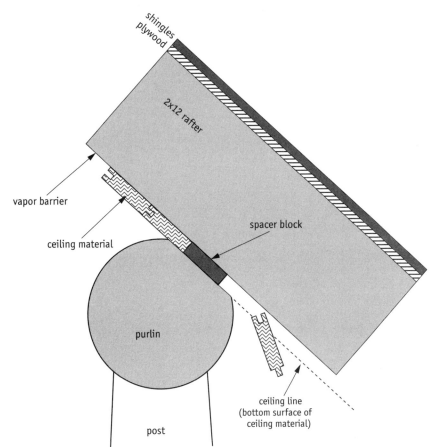

shingles
plywood

2x12 rafter

vapor barrier

ceiling material

spacer block

purlin

ceiling line
(bottom surface of
ceiling material)

post

Figure 25: Detail of one purlin from above. A spacer block that is slightly thicker than the finished ceiling material is nailed to the ridge, purlins and plates. 2x common rafters are then spiked through the spacer blocks into the log. This method allows the insulation and finished ceiling to be installed *after* the roof has been shingled and is weathertight.

The framing and finish detail at the ridge and at the wall plate logs is practically identical to this drawing.

Mitered Log Trusses

There are many ways to build log trusses. The technique I like best is the mitered log truss. Invented by Ed Shure and the crew at Timmerhus in Boulder, Colorado, just a few years ago, this is something entirely new to log building.

The mitered joinery where logs meet appears to be scribe-fit (*coped*, or round-on-round)—it looks great. But another advantage to this technique is its strength: inside the joints of a mitered log truss are large flat-on-flat surfaces that are much stronger, and easier to cut, than scribed joints.

Read this entire chapter before you start building a truss, because there are measurements you need to make on your log walls before you take them apart. For example, you must know the difference in elevation between the top plates and the top of the ends of the tie beam.

Let me caution you once again about those color photos in log home magazines showing "glorious" log roof systems. Many seem to me to be underbuilt and poorly designed, and could even be dangerous. Do not try to duplicate what you see in any photo.

My goal in this chapter is to give you instructions for how to build a simple and strong log truss. It is not difficult to do it right—but the roof system must be designed and built for your walls, with your log species, snow and wind loads, and your span.

Mitered Log Trusses

Where to Use a Truss

Trusses should be used where it is not possible to support the ridge and purlins with posts or walls. This is not well understood even among some professional builders. A truss resting on a gable-end log wall is not a truss because it is not spanning any distance between supports. A truss is supported only in two places: near the ends of the tie beam—and not anywhere else, it 'free spans' *(Figures 1 & 2)*.

Think of the 2x4 roof trusses in a garage: they sit on the eave walls and free-span without posts or other supports. Using trusses here makes sense because it eliminates posts or walls that would be in the way of parking a car.

Log trusses are beautiful and, when designed and built correctly, are very strong. But they are expensive, and are best used where they are the structural solution to a design problem *(Figure 3)*. In other places it is often better to support the purlins and ridge with posts and log walls.

Choosing Logs for Mitered Trusses

To make mitered trusses look good and the joinery really fit, there are important rules of thumb about the logs you choose to use for a truss, and their relative diameters.

When your logs first arrive, and before you start the walls, go through and choose important roof logs and mark them so they won't be used in the walls. Your best chance of getting the roof logs you need is to set them aside right from the start, when you still have plenty of goods logs to choose from.

And choose smooth logs and round logs for trusses—logs that have lots of bumps, or are oval in cross-section, are difficult to miter.

Purlins & Ridge

Purlin and ridge logs must be large enough to support the expected roof loads. The engineer or designer usually specifies their *mid-span* diameter—this is not the tip or butt end diameter, but the average. The number of purlins and the

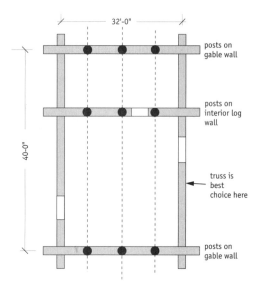

Figure 1: Floorplan of a simple log home. Does this log house need four trusses? No—it needs only one truss. The two log walls on the gable ends and the interior log wall can support the roof with log posts. In the large room a truss is a good choice—it means no posts will break up the room.

Figure 2: A log truss and other roof parts—purlins and ridge. A truss is supported only at each end of its tie beam—usually it rests on the eave log walls. It stretches freely in-between without additional support. The span is the distance from the center of one eave wall to the center of the other eave wall.

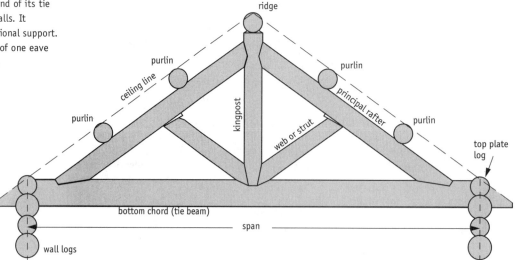

distance they span between supports (for example, the spacing between trusses) must also be determined by the engineer. The tree species you must use may be specified, too—lodgepole pine is half as strong in bending as Douglas fir.

Purlins can be bowed logs if they bow in one direction, and then are rotated so that their bow is mostly in the roof plane. All other orientations remove too much wood (weakening them) when they are sawn flat on one side to support the common rafters and roof deck.

When you are deciding where to locate the purlins remember to consider any intersecting dormer ridges and their heights, and the need for second-floor headroom. If you move a purlin a bit upslope could you walk under it easily? Make a scale drawing to see what's up.

Ridge logs are often the best long logs you have. They have to be sawn flat on two sides, unlike purlins that are sawn on just one, and so straighter is better.

Principal Rafters

In a truss, the principal rafters (also known as *top chords*) need to be large enough to support the loads, and in addition, the tip diameters of both rafters should be about two inches (50mm) smaller than the tip diameter of the kingpost (or 80% of the diameter of the kingpost tip). The rafter butts should be two inches or so smaller than the tie beam diameters where they join the tie beam.

It is best if both rafters have the same tip diameter, and one rafter tapers more than the other rafter. The rafter with the smaller butt joins the tip end of the tie beam; and the rafter with the bigger butt joins the butt end of the tie beam *(Figure 4)*.

Tie Beam

The tie beam (bottom chord of a truss) should be large enough that the notches for the principal rafter tails remove less than one-quarter of the tie beam diameter (more on this later).

For the tie beam, choose a log that has little taper—the butt and tip diameters nearly the same. This is because the butts of both principal rafters meet the tie beam *(Figure 4)*. Here's why: the butt of one rafter meets the butt of the tie beam (not much of a problem); the butt of the other rafter meets the tip of the tie beam (this can be a problem if the tie beam tip is too small). It is the tip diameter of the tie beam that often determines which log you will use.

It is probably best if the tie beam is straight, but if it is bowed, it should bow slightly in just one direction, not two. Orient the bow *up* when it is in the truss because if you put the bow down the tie beam will look like it is sagging.

The tie beam is almost always scribed and notched into the log walls at both ends, and this is completed before you start to build the truss.

Figure 3: To support the ridge there are two trusses in the middle of the building, and a post at each gable end. A truss is not needed at the gable ends. (There are no interior log walls in this cabin, so I needed two trusses for this span, unlike the building floorplan in Figure 1.) No purlins were needed, so the roof frame will be supported just on the wall plate logs and the ridge.

Hint   Nothing mentioned here about making the mitered joinery look nice overrules what your engineer tells you about what diameter logs to use to support the loads. Strength comes first.

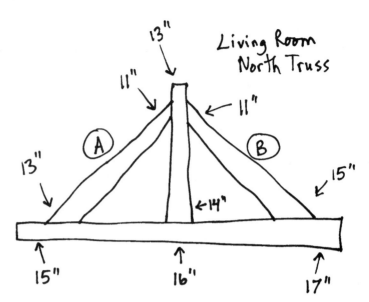

Figure 4: A sketch of a mitered truss showing relative diameters of the log parts. The tie beam tapers as little as possible (the tip is 2" smaller than the butt). Note that both rafters have the same tip diameter (11"), but rafter "B" (11-15) tapers more than rafter "A" (11-13). This makes it possible for the butt ends of both rafters to be about 2" smaller than the tie beam diameters.

Kingpost

The kingpost tip diameter needs to be two inches (50mm) or so larger than the rafter tip diameters. The kingpost butt diameter needs to be about two inches smaller than the tie beam's mid-span diameter. It joins the tie beam at the tie beam's center, so measure the tie beam there. There are times when I put the kingpost tip down and the butt up (at the peak of the truss)—this can sometimes make it easier to find appropriately-sized rafters.

Webs

If the truss has webs (also called *struts* or *braces*), then they should be about 70% of the mid-span diameter of the tie beam, and should be smaller than the rafter diameters where they join the rafters.

Summary of Relative Sizes

Try to obey the rules of relative diameter listed above. Joinery that looks mitered depends upon these relative sizes being followed closely. Choose the parts for each truss using the rules above, and make a sketch of each truss showing the actual diameters at the butts and tips of all the parts you will be using—this will let you catch any mistakes before you start building each truss.

Compare your sketches with the rules listed: is the kingpost diameter at its tip a couple inches bigger than the tip ends of both rafters? And so on.

If you are building more than one truss, then label all the logs so that you don't mix the rafters for one truss with the tie beam of another truss. Keep the parts for the 'north' truss separate from parts for the 'south' truss—they are not interchangeable unless your logs are identical in shape and diameter.

Hint The relative diameters of the truss parts is the key to mitered joinery.

Laying Out the Roof on the Deck

The plate logs are done, and you know your span and your pitch (we covered all this in detail in the chapter on roof basics and plate log layout), now we will make a *full size* drawing of each truss and then use the drawing to mark the log parts for cutting.

The drawing proceeds in 3 steps: first, draw the lines that the trusses share (centers of walls, kingpost center, span line, and ceiling lines) *(Figure 5)*. Second, draw the log parts that are used in each particular truss (the kingpost of truss 'A' is 14" at the butt, and the kingpost of truss 'B' is 16½", and so on). Finally, draw the joinery details where the log parts are mitered to each other.

Layout Steps:

1- draw roof lines
2- draw log parts
3- draw joints

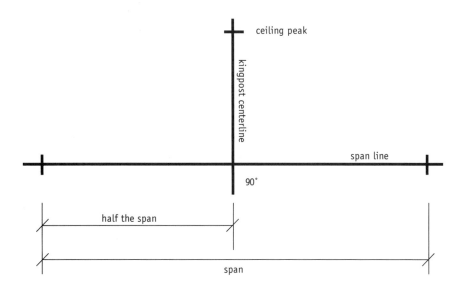

Figure 5: The roof lines.

For the drawing you need a large, flat, smooth, level area that is strong enough to support log truss parts, that you draw on with pencil and chalklines, and that you can screw the scribing jig to. A stiff plywood floor deck works well, though you'll probably want to use a 4x8 sheet of medium density fiberboard (MDF) at every place that gets joinery—on its dense surface it is easy to draw accurate lines.

Here's how to start drawing a full-size cross-section picture of your roof. Using a chalkline, snap the span line, and make it several feet longer than the span of your building *(Figure 5)*.

Snap the center of kingpost—make it several feet longer at both ends than the kingpost is tall. It is 90° to the span line, and crosses the span line near its center.

Starting at the intersection of the span line and kingpost center, measure half the span to left and half the span to the right, along the span line.

For example, if your roof span is 26'-8", then measure left 13'-4" and make a mark, and right 13'-4" and make a mark. The distance between these two marks is equal to your roof span (26'-8" center of wall to center of wall), and the kingpost center is in the center of the span.

Hint   Make a full size drawing of each truss and then use it to mark the log parts for cutting.

Figuring Ceiling Height

$$\frac{\text{Your ceiling height}}{\text{Half your span}} = \frac{\text{Your rise}}{\text{Your run}}$$

$$\frac{\text{Your ceiling height}}{13'\text{-}4''\ (160'')} = \frac{10}{12}$$

Use a calculator and divide 10 by 12, then multiply that number times 160. The answer is 133.33", which means your ceiling peak is 133 - 5/16" above the span line, or 11'-1⁵/₁₆".

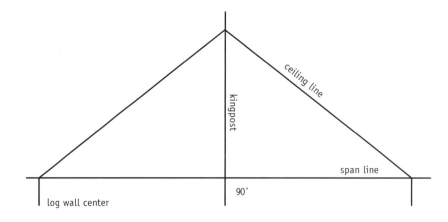

Figure 6: Here is the drawing, so far. These are the lines that will be used by all the trusses—they don't change, so use permanent chalk, or drive nails at the intersections. Remember that this is a full size drawing: if your span is 26'-8", then your span line will be 26'-8" long in the drawing.

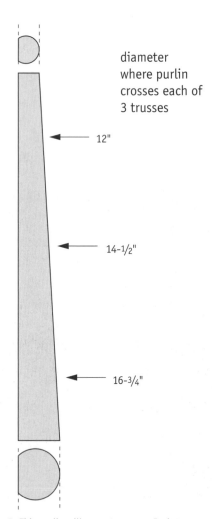

diameter where purlin crosses each of 3 trusses

12"

14-1/2"

16-3/4"

Figure 7: This purlin will cross 3 trusses. Each truss must be built to a different "size" in part because the purlin diameter changes. To lay out each truss we must know the diameter of each purlin at this location. We must also know which end of the building will get the butt of this purlin. Make this measurement for every purlin that crosses the truss you are working on.

Now measure up the kingpost center from the span line to mark the peak of your ceiling. To find the height of your ceiling you can use a proportion (example on the previous page). If your span is 26'-8" and your pitch is 10:12, then you will measure up 11'-1 5/16". There are other ways to find the ceiling peak: trigonometry, or a builder's calculator that works in feet and inches. For more information, see the chapter on Roof Basics.

Now, snap a line that connects your ceiling peak to the span to the left and another chalkline from the ceiling peak to the span to the right. These are the ceiling lines. The ceiling line is the bottom surface of your finished ceiling material *(Figure 6)*.

Use a rafter square and draw lines 90° down from where the ceiling lines meet the span line—these are the centers of the log walls (that is, the plumb lines you drew on the ends of your plate logs).

These lines: kingpost center, span line, left and right ceiling, and centers of log walls are the lines that all your trusses for this span and this building have in common. Draw them in ink, or use dark red (permanent) chalk because you will use these for *all* your trusses of this span and this pitch. Drive a small finish-nail at the four intersection points so you can find them later when you start the next truss.

All the other lines we add to the drawing are unique to each particular truss. If you are making more than one log truss for this span, then you will have to do a separate layout for each truss, even if they have the same span and same pitch. This is because the actual diameters of the log parts used in each truss are slightly different. So, for example, start and finish the 'middle' truss before drawing new layout lines for the 'great room' truss.

Laying Out Lines for this Truss

Now we will draw lines that continue the full-size 'picture' of a particular truss. To do this we need to have the actual logs that are going into this truss so we can measure their diameters at various places.

To go any further we need to measure the ridge log diameter where the ridge crosses this truss; the purlin diameter from flat to round where the purlins cross *this* truss *(Figure 7)*; and the actual diameters of the principal rafters, kingpost, and webs you will use in *this* truss.

Find the logs that are going into this truss and get them near the truss-building platform. Spread them out to look like the truss has been laid over on its side and gently pulled apart: every log is pointing the right way, and has been rotated to the orientation you will use it. For example, if the kingpost has a burl, then rotate it so it is pointing towards the great room.

Start this stage by drawing a line on the deck representing the top surface of the tie beam. At both ends of every tie beam you need to know the actual vertical difference in height between the top center of the plate logs and the top surface of the tie beam log. You can get this only by measuring them on the building. If the plate logs and tie beams are on the ground then you will have to put them back on the building. (I warned you about this back on the first page of this chapter!) You also need to mark the mid-span point on the tie beam—that is, the point that is halfway between the two log wall centerlines. This can only be done when the tie beam is still on the wall. You'll need this center-point later.

Figure 8 shows the height you need to measure at both ends of every tie beam. (Another way to get these numbers is to use a transit and shoot the elevation of the span line and the elevations of both ends of the tie beam.) On your full-size truss drawing, measure down these distances you just found, and snap a line that represents the top surface of the tie beam log.

If you use really bent or bowed logs in your trusses, then instead of snapping a straight chalkline you can put the tie beam log on the plywood deck and "map" its shape down to the deck, and then connect the dots. I use a framing square held on the deck and slide it until it touches the log—you are, sort of, lofting *down*.

Draw your purlins, ridge, and plate logs—using the actual diameters and locations they have—they must be sawn flat with a roof pitch already. The top (sawn) surfaces of each of these logs is matched up exactly to the ceiling lines. You must decide which purlin will be on the east side and which on the west; which near ridge and which near plate; and whether each purlin and ridge tip end points north or south.

Next, draw the top surface of the log principal rafters onto the truss-building deck. To lay out the top surface of each rafter the drawing must have the purlins and ridge drawn on first. Double check that you measured the purlins in the right places—that is, the diameters where each actual purlin will cross this truss.

I like to have the purlins overlap the log rafters by about two inches (50mm). This means that the purlins and rafters will lose very little of their strength. Every notch you cut weakens these roof logs, and shallow notches are stronger than deep notches *(Figure 10)*.

Note that the chalkline for the top surface of the log rafter may not be at exactly the same pitch as the ceiling, but that's okay—it's the sawn surfaces of the plate log, purlin and ridge that must be in a straight line and at the exact roof pitch.

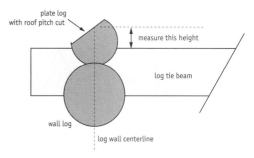

Figure 8: Some important measurements must be made while the tie beams and plate logs are still on the building. Also see photo below. (Did you notice that the horizontal dotted line in this drawing is the "span" line of your full-size truss drawing?)

Figure 9: In this photo you can see the relationship between the top of the truss tie beam and the roof pitch cut on the plate log. In fact, the plate log can be considered to be the bottom purlin of the roof.

The center-of-wall chalkline on the plate is sometimes on the roof-pitch flat, and sometimes on the round of the plate log.

Figure 10: The finished 'picture' of a truss. (I have shown all the wall logs and the floor—usually I just draw the top two wall logs, but I thought it might be more clear if you saw the whole thing.) We started with the lines common for all the trusses: span line, centers of walls, center of kingpost, and ceiling lines. Then we added the lines that are unique to this particular truss. To draw these lines, we must make measurements of log diameters and log positions. One after another we added: the purlins and ridge, the top and bottom of log principal rafter, top and bottom of log tie beam, and both sides of the log kingpost. Once the purlins are drawn at their actual size, then snap the top surface of each rafter so that it crosses the purlin bottoms by about 2"—just enough for a nice, shallow notch.

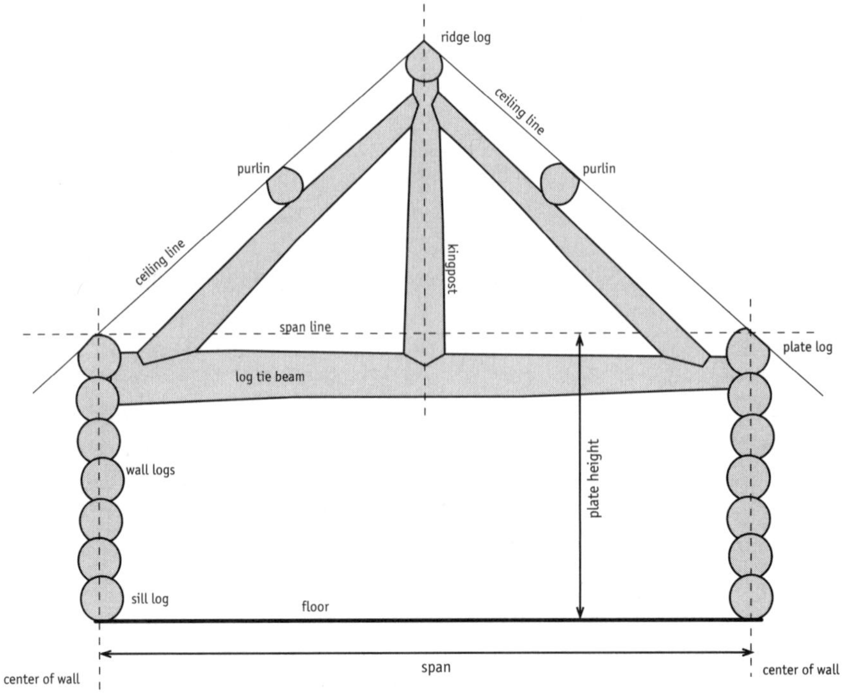

Hint   Note that the ceiling line is exactly at your roof pitch, but the top surface of the log rafters can be at a pitch that is slightly different than the roof pitch. This is because I like to snap the top-of-rafter lines based on the purlin diameters.

The log rafter can be at some other pitch, though usually it's not too far off. The bottom surface of each rafter is measured down from the top surface you already snapped. All you need to do this is the actual tip and butt diameters of each rafter.

Also note that if you have more than one truss supporting the purlins and ridge, then the truss that supports the butt ends of the purlins will actually be smaller than the truss that supports the tip ends. This looks better than making all the trusses the same size, and it is also stronger—the purlin notches at the butt ends would be too deep where they cross the principal rafters, if the trusses were all the same size.

Keep adding the lines that represent the surfaces of the actual logs that will be used in this particular truss. Snap centerlines on the log that is the kingpost, and measure the distance between the centerlines and the kingpost sides, and then transfer these lines to the drawing: the kingpost sides.

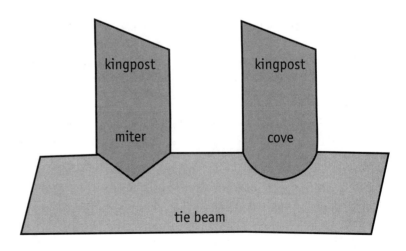

Figure 11: The miter is simply two flat surfaces that approximately match the shape of a coved (scribe-fit) joint. Compare with Figure 14.

Lay Out Each Joint

We have come to the third step in the process of making the full-size drawing. First we snapped lines common to all the trusses. Then we snapped lines representing the surfaces of the log parts we will use for one truss, projected down onto the deck. Third, we will lay out the joinery at each connection area.

The basic concept of a mitered joint is to find the place where two logs have the same diameters, and then lay out joinery from that point. (Note that it is actually the *chord lengths*, not diameters, but we'll get to that soon enough.) The miter is simply two flat surfaces that approximately match the shape of a coped, or scribe-fit joint *(Figure 11)*.

Position the Log Parts Like a Truss

Cut the logs for the truss to approximate lengths with smooth end-cuts that are square to the eye. Make sure they aren't too short, of course, but also no more than a few feet too long. You can see just how long they need to be by measuring your full-size drawing.

It helps to have all the log parts arranged as if the truss were lying down, and in the positions and orientations that each part will have when it is in the truss.

Have the tips of the rafters pointing towards the peak. Have the tie beam pointing the right way and spun right. You don't need to use a level to get them in position, just have them close. The kingpost should be rolled and have chalklines on all four faces.

Arrange the log parts so they look like a truss that is gently pulled apart. This positioning makes it far easier to imagine the truss, and keep from getting confused when you make the measurements that you need to lay out each joint.

Kingpost Bottom

Let's start with the joinery where the kingpost meets the tie beam—in many ways, this is the easiest joint to understand *(Figure 14)*. Once you see what's happening here the other joints will make more sense.

Measure the diameter of the kingpost bottom, and find how deep into the tie beam you will need to notch it to come to that same measurement on the tie beam (*"X" in Figure 15*).

For example, if the kingpost bottom has a diameter of 14¼", then how deep into the 16" tie beam do you have to go to find a spot that is 14¼"? I set an outside caliper to 14¼" and then walk over to the middle of the actual tie beam log and slip the caliper over the tie beam from the top surface. Where it stops is point "A". Now measure depth "X" *(Figures 12 & 15)*.

Next, working on the full-size layout drawing, measure depth "X" along the kingpost centerline down from the chalkline that is the top of the tie beam, and mark point "A" *(Figure 15)*.

Figure 12: Set an outside caliper to the diameter of the kingpost bottom and test to see how deep it goes into the tie beam. This is how you find point "A" (see next page).

Figure 13: The miter has been cut into the top of the tie beam and is being sanded flat, and to the lines.

Figure 14: The miter at the kingpost bottom— it approximately matches the shape of a coved (scribe-fit) joint. Compare with Figure 11.

Photo courtesy: Terry Davis

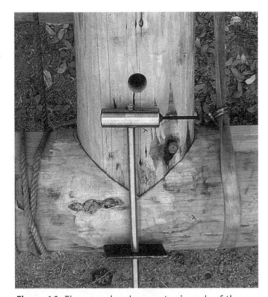

Figure 16: The cross-dowel connector is made of three parts: a long bolt, a large plate washer, and a piece of steel rod or dowel tapped and threaded for the bolt. The small bolt makes it easy to hold and maneuver the dowel in its hole while you get the bolt threaded into it.

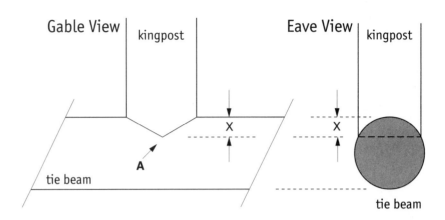

Figure 15: The concept behind mitered log joinery is: find the place where two logs have the same chord lengths, and then lay out joinery from that point.

Caliper the smaller part and compare it to the larger part.

Actually, the fit of the miter can be improved by going a little bit deeper than "X" into the tie beam, but only about ⅛". For example, if you found that point "X" is 5⅛" into the tie beam, then mark 5¼" on the full-size drawing.

Now draw the miter: connect point "A" up to the left and up to the right where the tie-beam-top chalkline and kingpost-side chalklines intersect.

Joinery at the Peak

Next we will draw the joinery where the rafters meet the kingpost top—the peak of the truss. Start by measuring the diameter of one of the rafter tops, set an outside caliper to that amount, and slide the caliper onto the tip end of the kingpost, just as you did with the tie beam. If you measured the left rafter, then slide the caliper onto the kingpost from the left side. The caliper stops at Point "B" at depth "Y." Measure the depth "Y" on the kingpost *(Figure 17)*.

On the full-scale drawing, find where the rafter centerline meets the kingpost side, and use a rafter square to draw a line at 90° to the kingpost side. From the side of kingpost measure in the depth "Y" along this line (go in an extra ⅛" to make the miter fit better). This is point "B." Now, draw the miter by connecting "B" to "A," and "B" to "C."

Points "A" and "C" are easy—this is where the surfaces of the log pieces cross each other on your drawing (the top surface of the rafter meets the left side of the kingpost at point "C").

This joint looks best if the points "B" for the rafters are level across the kingpost. So, fudge one up, or one down, or both, to make "B" come out level across the kingpost. I usually just split the difference *(Figure 18)*.

Rafter tails

Next we will draw the joint where rafters meet the tie beam *(Figure 19)*. Structurally, this joint requires a shear plane and bearing surfaces that are large

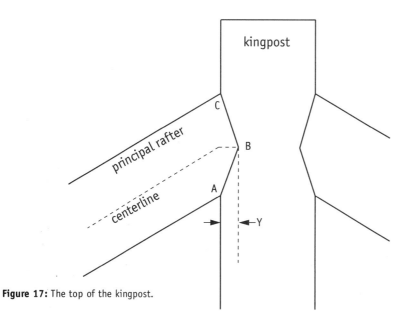

Figure 17: The top of the kingpost.

Figure 18: The top of the kingpost and principal rafters. A drawknife is used to clean up the joinery, and make the logs fit seamlessly.

enough to withstand the loads, and we lay it out with these in mind. This joint often requires some hidden steel to make it strong enough—a shear plate and bolt are typical.

This joint is prepared a bit differently than the others because miter "A" must be deep enough into the tie beam to make shear plane "B" big enough to resist the loads.

Your engineer will tell you that area of shear plane "B" needs to be 345 square inches, for example, and then you go to your full-scale drawing and to your logs to find how deep you must go into the tie beam to get a cross-sectional plane of that size (it is just length times width).

The miters fit best if the rafter and tie beam chords are the same length at point "X." You can test this, and go a bit deeper if it improves the fit. But "A" should probably not go deeper than one-quarter of the diameter of the tie beam, or shear could be a problem in the tie beam.

The slope of bearing plane "A" is usually one-half the outside pitch of the rafter. If your roof is 34°, then the slope of "A" is ½ (180° - 34°) = 73°. Draw "A" on your full-size layout drawing. Measure down "A," the depth determined for the area of the shear plane, and then draw the long miter surface *(Figure 20)*.

Figure 20: Where principal rafter meets tie beam. The angles shown here are an example only.

Figure 19

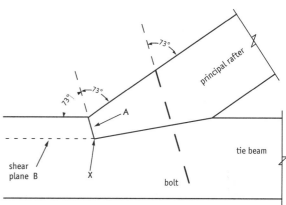

Webs

In some trusses, the principal rafters need extra support to keep them from bending under the roof load of the purlins. Usually this means adding short logs called *webs* or *struts*. Your engineer can tell you if you need webs. Customers often want them, whether they are needed, or not—because they look nice *(Figure 21)*.

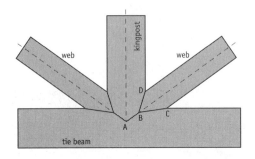

Figure 21: This truss has webs, also called *struts*, that help support the principal rafters.

Figure 22: The top end of the web can get a double tenon (above), or a simple mitered joint (below).

If you use webs then you need to decide what angle they'll have, and where they will join the tie beam. The engineer may specify their angle from horizontal, but if not, then a good choice is to have webs slope the same as the principal rafters. If the rafters are 8:12, then the webs will look good at 8:12, too *(Figures 2 & 21)*.

I have seen many trusses with webs that are 90° to the rafters. They do not look as good, and they often support the rafter too high—not near its mid-span.

You also need to decide where the web will meet at its bottom. The web can join only the kingpost, or only the tie beam, or can be joined to both the kingpost and tie beam *(this last choice is shown in Figure 21)*. Usually, if the web touches the tie beam, then a cross-dowel connector is needed to join the kingpost to tie beam.

Snap the centerline of the web, and then measure out to its top and bottom surfaces at both ends, and snap them on the deck drawings. Point "A" is where the kingpost and tie beam have the same chord lengths—part of the kingpost layout. Point "B" is where the web and tie beam have the same chord length (it is also where the web and kingpost have the same chord length). Points "C" and "D" are where the surfaces of each set of parts meet, just as before.

One popular joint where the web meets the rafter is a double tenon and mortise. Hardwood wedges are knocked into the space between the tenons to keep the bottom web joint fitting tightly as the kingpost shrinks in diameter *(Figure 22)*.

Transferring Layout Lines from the Deck to the Logs

Now that layout is complete, the next step is to mark each log part for cutting. We will do this by placing each log above its picture on the drawing and transferring the miter joint lines with a jig. In boat building this is called *lofting*.

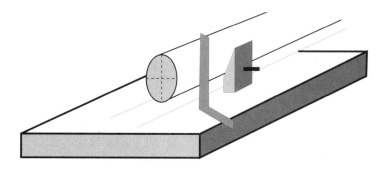

Figure 23: To prepare a log part for marking, level each end above the truss layout deck, and then slide it around until it is plumb above the lines that represent it. Note the *cross-hairs* on the end grain.

Set a log on two leveling blocks and adjust the blocks to level the truss part. For example, bring the center (the *cross-hairs*) of both ends of a log truss part to, say, 8" off the platform, or whatever is convenient for you, the C-jig you build, and the diameter logs you are using.

Move the truss piece so that it is plumb above the chalklines that represent its edges, or surfaces, by using a rafter square to help you position the log *(Figure 23)*.

For the tie beam only, position it plumb above the drawing with its end elevations plumb above the elevations on the drawing, and also have the tie beam's mid-span point plumb above the center-of-kingpost line on the deck drawing.

Figure 24: The C-jig: this one is made of wood: vertical grain 2x4's covered on both sides with furniture-grade plywood—the jig must be stiff and flat. Glued to the bottom is 1/2" baltic birch plywood that is wider than the "C" itself. The straight edge (not shown here) must be the same thickness as the offset overhang on the base so that when the C-jig edge is on a line on the deck the cut-line is lofted in that same plane. (12" ruler is for scale.)

Figure 25: A tie beam is in leveling blocks, the C-jigs are screwed to the decks in the proper places, are plumbed, and steadied with opposing chains and turnbuckles that have been snugged up. A straightedge is being used to transfer the cut lines onto the log.

It can useful to close-off the open end of the C-jig once the jig is in place. Here a removable vertical piece of straight wood has been screwed on. This turns the C-jig into, well, an "O-jig" and gives the straightedge more places to rest while it is being used.

Figure 26: The joinery lines have been transferred from the deck onto the log. This process is called lofting. Can you tell which truss part this is? (The answer is on the bottom of this page.)

Figure 27: Each joint is scored to prevent splintering and then cut close to the line with a chainsaw. Trim the surface flat with a sander, handplane, or electric plane. This is a rafter tail.

Answer: It's the top of a principal rafter where it meets the kingpost top. It easiest if you look at the deck.

Place the "C-shaped" jig on the line you want to transfer to the log, screw the baseplate of the jig to the platform, plumb the face of the jig, and adjust the arm and turnbuckle to hold it steady *(Figure 25)*.

Transfer the joinery lines from the jig to the log with a straightedge that is exactly the same thickness as the offset amount in the baseplate *(see Figure 24)*. Hold the straightedge on the plumb, flat face of the jig and slide it until it touches the log. Rock the straightedge on the log surface, and mark the log everywhere it touches.

Move the C-jig to all other joints you need to draw on the log, and continue until all the joinery is completely transferred (lofted) to this log. Each joint will require two positions for the C-jig—one for each of the two mitered surfaces. Do not move or bump the log until all its joints are marked *(Figure 26)*.

Use a level to draw horizontal and vertical centerlines on the ends of this log truss part—you may need these for free-tenon layout or other uses.

Remove the log you have completed, and place the next truss part on the leveling blocks, in position over its layout lines, and transfer the joinery to the log with the C-jig and straightedge. Continue until all the joints on the parts of this truss are lofted.

On each truss part, snap chalklines joining the centerlines (cross-hairs) that are on the log ends, if they are needed for tenon layout, or other purposes (joining purlins to the principal rafters, for example).

Plywood *free-tenons* can help hold the log parts in place while assembling the truss. Placed inside the joint, on the flat surfaces of the miter, they cannot be seen when the truss is assembled. I use ½" baltic birch plywood, and cut the slots using a router with ½" plunge bit. These free-tenons have the added benefit of preventing light from getting through hairline gaps in the miters.

Cutting the Parts

Score the joinery lines with a utility knife to prevent splintering, and then cut the truss joints with a chainsaw, and plane the joints flat with a hand or electric planer, or a grinder-sander *(Figures 13 & 27)*. Test often with a straightedge. All the surfaces at joints are flat planes, so cutting them and getting them trued up is quite easy—unlike cutting a coped truss notch with its concave hollows and fragile edges.

Truss-cutting Jig

As I said, mitered joinery is quite new, and builders continue to experiment and innovate with the techniques. The first tool for marking joints was a special bubble-scriber that could handle the very wide settings that are needed. Then came the C-shaped jig with a straightedge and pencil. This was improved by using a straightedge fitted with a piece of metal-cutting bandsaw blade—it marked and scored in one step. I have tried a $50 hand-held laser-pen that projects a dot, and I have also used a laser that projects a line—but the dot is a bit large in diameter.

Brooks Minde, a log builder from Minnesota, has developed a jig that holds a chainsaw bar like a drop-saw, so there's no marking at all—he just puts the log in place above the deck and slides the jig into place to cut the joints. Obviously this is very fast (*Figure 28*).

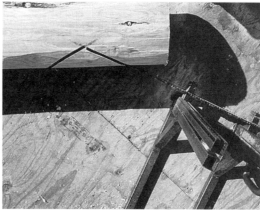

Figure 28: Using Minde's jig to cut the male truss parts; and on the right, cutting the female side of a joint.

Photos courtesy: Brooks Minde

Pads are bolted to the chainsaw bar that slide in slots in the vertical steel tubes of the jig—this guides the cut. The tubes can be plumbed easily with thumbscrews. You can find out more about Minde's truss-cutting jig in *Log Building News* #28, or purchase the jig from Brooks at 218-525-1070 (central time) or by e-mail: mindelog@cpinternet.com

Fitting & Bolting

Test fit the truss, with the truss lying down flat, but high enough off the ground that you can see how well the joints fit on both the top and bottom. Support the truss pieces a few feet away from the joints or you won't be able to see the fits (*Figures 31 & 33*).

The joints usually fit nicely the first time. If not, there are several possible causes. If the truss deck was not flat or stiff, the C-jig would not have lofted accurately— could you feel a bounce when you walked on the deck? If a truss part moved after one joint was lofted but before another was done, then the part will not fit.

As needed, dress the surface of the logs at the joints with a drawknife or spokeshave to make the surfaces match together perfectly. Remember that we have been trying to match a curved line with flat planes, so there will always be a few places that need a little work.

When the fits are good, drill holes for the through-bolts. Long auger bits really help—these holes can be two or three feet long (*Figure 31*).

For the cross-dowel connector, two holes must line up closely: the bolt hole up from the bottom of the tie beam must meet the dowel hole that comes in from the side of the kingpost. A simple tool (*Figures 29 & 30*) makes this easier.

Figure 29: The kingpost is usually bolted to the tie beam. To drill these holes so they intersect, a simple jig guides the auger bits.

Figure 30: Cross-dowel drilling jig—this one was made by Timmerhus. First drill through the tie beam into the end grain of the kingpost (for the bolt). Then insert the solid rod of the jig (top in photo) into this hole. Next, drill the hole for the cross-dowel connector by feeding the drill bit through the short piece of pipe (bottom left).

Figure 31: Drilling a truss through-bolt where rafter tail meets tie beam. Threaded rod can be seen on top of the truss logs and is being used to help guide the drilling angle.

Note the plumb and level lines on the end grain of the tie beam.

Figure 32: The housed square notch. The tie beam is positioned on the wall (usually rough-notched only very shallowly, if at all) and then the notch is scribed. Using the scribers, mark spots on the scribe line on the tie beam and on the log wall so that you'll know where to lay out and cut the square block inside the notch. You will need a scriber that holds a pen in both ends.

See also Color Figures C-21 and C-22.

Once the truss is bolted together it is ready to be lifted onto the building—I use two heavy nylon straps around the top of the kingpost, below the joint with the principal rafters—if you lift by the rafters, there is a danger of stressing those joints too much. Hook a strap on each side of the kingpost so the truss will hang plumb.

Where Tie Beam Joins the Wall

There is a great deal of weight on most structural trusses. All the weight from the roof that a truss is carrying is transferred to the log walls at just two points: one at each end of the tie beam.

To make sure this joint is strong enough, consult with your engineer. Figures 31 & 32 and Color Figure C-21, show a joint that I often use at this location—a housed square notch. It is scribed just like a round notch (there are no saddles, so it is not a saddle notch). During scribing, you use the scribers to mark where the block of wood inside the notch will be.

I like to use large, thin, sheet-metal shims between the flat surfaces of the square notch to provide a large bearing area to transfer the roof weight safely onto the log wall. Once this joint is assembled it looks just like a round notch—all its strengths are hidden, but working.

Mixed Roof Systems

Log roofs are often made of a mix of support systems: a log truss for places that need to free-span; log posts on top of interior log walls; and 2x frame walls or log posts on gable ends (*Figure 34 & Color Figures C-2 & C-5*).

If you made a mitered truss, then you can use the full-size drawings on the deck to measure the lengths of log posts in a post-and-purlin system, or the dimensions of 2x framed gable walls. Use the actual diameters of the purlins and ridge for each location, draw them in, and measure down to the span line on the deck drawing.

Here's a useful trick for 2x framed gable walls that support log roof beams: frame the pockets ½" lower than the actual height needed, and then when you are ready to place the logs into these pockets, shim up each framed pocket with a piece of ½" plywood. This way, if anything goes awry, you can always take a shim out and replace it with a thicker shim to raise the log beam, or a thinner shim to lower the log beam. If the pocket in the gable 2x wall is framed at *actual* height, and you need to lower a purlin for any reason, you would have to cut and re-frame the gable wall—which is not much fun while the crane is sitting there waiting!

Make sure you note that posts on log walls, and framed gable ends, usually are several inches below the span line—so make sure you measure this height difference on your building and add that onto the length you'll need. This is because the flats that you cut on the top of gable log wall and interior log walls are not at the same elevation as the span line, they are usually somewhat below the elevation of the span line (see Figure 10, page 176).

Figure 33: Set the truss parts on three spare logs, as shown. Test fit each joint, and then the entire truss. Come-alongs and straps hold the truss together while you drill holes.

Figure 34: Log roof systems often combine trusses (this building has 3, though it is hard to see all of them); posts and purlins (all 3 gable ends, a mid-wall, and the covered entry). Sometimes the gable ends are framed with 2x walls to support the purlins and ridge.

From this perspective you can see how the flats sawn on the plate log, purlins, and ridge all line up in one surface—they "plane-out."

Engineering & Designing Log Trusses

By Tom Hahney, MS, BSCE

If you design or build structural log trusses, that is, trusses that carry more than their own weight, dig into this chapter with both hands and start appreciating how important it is to have professional engineering advice. You don't want to be fooling yourself or endangering the people who live in the homes you build.

In this chapter you will learn something about wooden truss design and structural analysis in general. You will not learn everything you need to know to design a truss, but you will come away with a clear sense of what must be known before you can safely begin. You will know what you have yet to learn, and the questions you need to ask.

Note from Robert W. Chambers:
This chapter first appeared as an article in **Log Building News #18.** *It has been modified for this book. Thanks to Tom Hahney, I have been able to include it here as an example of how a structural truss made of logs is designed; and to show you why you should seek professional engineering advice.*

Engineers can find rare log building design data at my website www.LogBuilding.org — then click on the "free information" link. PDF downloads are free. Topics include seismic design, fire resistance of log walls, noise transmission, design values (MOR, MOE, shear), energy efficiency of log walls, R-value calculations, and more.

If you choose not to read all of the specific details, there are some comments at the end that you will find interesting.

Log Building News 17 has a wonderful article by Russ Detwiler and Ed Shure entitled "A New Way of Building Log Trusses," that came out of a truss workshop at the CLBA-ILBA Conference at Banff, Alberta in 1994. The Timmerhus crew and I worked together on that presentation—I did the engineering, and they built the truss. The truss for that workshop is shown in Figure 1.

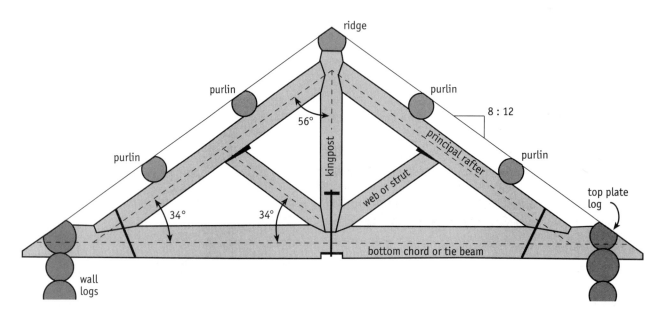

Figure 1: The Banff truss—a typical truss made of round logs and mitered joinery.

Determine the Loads

Let's take it from the top, which is often the place to begin in the structural analysis of a building. Start at the roof, find out what loads are anticipated on the building, and find ways to get the loads safely into the foundation and thus to the earth. I want to emphasize the final three words, "to the earth."

While this chapter focuses on the roof, remember that all loads have to follow some path to the earth, and from a structural point of view, the shorter and more direct the path the better.

The first step is to determine the loads imposed on a structure by snow, wind, earthquake, building materials, rain, ice, etc., and this is where many design errors occur. The design of the truss for the Banff workshop entailed a snow load of 60 pounds per square foot (psf) on a cruciform-shaped building, which meant that the trusses that were most critically loaded were supporting purlins which were, in turn, supporting the load from snow that accumulates in valleys *(Figure 2)*. While it is not always true that a truss supporting valley snow loads will be the most heavily loaded truss in a building, it is a good place to start.

The whole issue of determining roof loads is an article unto itself, and valley snow-load situations are especially interesting. It is beyond the scope of this chapter to address this critical issue. I would, however, suggest Appendix Chapter 16 in the 1997 *Uniform Building Code* (UBC) for a discussion of snow loads, and how roof geometry and wind drifting influence the design of roof systems.

I am assuming that the proper analysis has been done and that we know the critical loads on the truss, and where these loads are located. I am also assuming that the general geometry of the truss is decided: we know its span and the roof pitch, and the type of truss—a kingpost truss with purlin braces. I am also assuming that the species of wood is known, and in this case it is Western larch.

It is worth noting that for the Banff truss a total of ten pages of calculations, diagrams, and computer-generated shear and bending moment diagrams were needed to reach the point where the loads on the truss were known—which is where we are starting now. This was about one-third of the total work done on the design.

Joinery Drives Diameters

Now we will look at the rest of the story: how big are the log members of the truss? How are we going to join them together so that the loads from the roof are carried safely to the earth? In general, the sizes of the parts of a truss are driven mostly by the joinery used in connecting the pieces together. This means that decisions about the type of joinery need to be made early in the design process. What is the aesthetic effect to be achieved by the joinery? What size, species, and quality of material is available? How large are the loads? The interrelation of these three questions strongly govern joinery design.

For example, will steel be used in the joint, and, if so, is it visible or hidden? Must all joinery use a traditional oak-pegged mortise and tenons? Is the species and size of the wood sufficient to allow mortise and tenon joinery? Are the loads so large and/or the material so slight, that steel fasteners are necessary?

When designing a truss for a timber frame building I am often asked to use oak-pegged mortise and tenon joinery, using steel pins where the oak just won't cut it. For most log trusses, however, I tend to rely more on bolts and bearing surfaces. By *bearing surface* I mean the area of contact where one truss member touches (i.e. is joined) to another.

Let's say that we are working with a relatively low-slope roof, significant snow load, a material like pine (which is not all that strong) and an owner who doesn't care if he sees steel or not. This is a familiar project: there will be lots of bolts and steel plates on both sides of all truss members. I know you have seen pictures of this type of truss in log home magazines. If your wood is a little bigger in diameter, or the loads are smaller, you might use a single steel plate located in a kerf cut on the centerline of each truss member, and that has bolts that pass through this wood-steel-wood *sandwich*. This approach hides the steel plate, and you end up seeing only the nuts and bolts.

Have you noticed the trend here yet? When you remove wood from a truss member you weaken it. Yes, even a bolt hole weakens it. And, if you have enough of those holes spaced in a certain way, you can be in real trouble. The joinery with

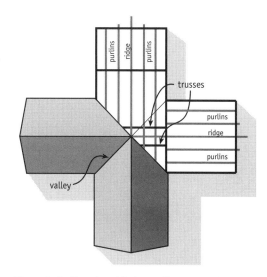

Figure 2: The floorplan with the Banff trusses. It is a *cruciform* (cross-shaped) building, with four valleys. This is a bird's eye view, cut-away to show the two trusses, purlins and ridge logs.

visible steel plates on both surfaces doesn't remove as much wood, and so weakens the joints less. But, for joinery that uses an embedded steel plate, we must use larger diameter truss members, if all other issues are equal, because the kerf for the steel plate removes wood from the truss members.

As we move from joinery relying on bolts, plates, and bearing surfaces, to full mortise and tenon pegged joinery we need even larger and stronger material, if everything else remains constant, because a mortise and tenon removes even more wood from the truss members.

For the Banff truss we used a combination of bearing surfaces, hidden bolts, and some mortise and tenons. The bearing surfaces come from using mitered truss joinery, which gives wonderfully large and consistent bearing areas, when compared to a similar joint that is only scribe-fit or coped.

A scribed joint will typically bear only along the edges of the notch, and some of these edge fibers will crush when load is applied. A mitered joint, however, provides bearing for the total area covered by the joint. This means that less wood can be removed to create the required bearing areas.

I chose a 12-inch (300 mm) mid-span diameter principal rafter as a starting place in sizing the members of the truss. Using the guidelines found in the previous chapter in this book, the size of the other truss members can follow from rafter size. I chose to start with the rafter because they are often structurally critical, though each and every truss-member is checked to see if its size is adequate as we go through the analysis.

Beams

A truss carries load from one support to another (such as from one wall to another wall) in a manner entirely different from the way in which a beam carries that same load. A beam carries a load to its supports by what is appropriately called *beam action*.

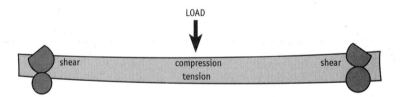

Figure 3: A simple beam with a point load at the center of its span. The load causes the beam to bend, and this causes compression, tension, and shear.

Two types of stresses (forces that tend to tear the fibers of wood apart in one of several ways) are induced in a beam as it resists the loads imposed upon it. One of these stresses is called *bending stress*, and the other is *shear stress (Figure 3)*.

Bending stress causes the wood fibers in approximately one-half of a beam's cross section to compress against each other, end to end, and it causes the wood fibers in the other half of a beam's cross section to be in tension to each other, end to end. For example, in a single-span beam, supported at each end, with a single point-load at its center, the fibers in the upper part of the beam are in compression, and those in the lower portion are in tension. In this beam the maximum bending stress occurs at the center of the span.

The other part of beam action is shear. For the beam just mentioned, shear is greatest at the support at each end of the beam. As the support (often it is a wall) pushes up and the beam pushes down, there is a tendency for the beam to be *sheared* off.

What is interesting about this is that wood is much, much weaker along the grain than across the gain. It is obviously much easier to split a piece of wood along the grain than it is to shear or tear it across the grain (as anyone who has split firewood can attest). And, due to the *steering wheel phenomenon*, the vertical shearing tendency becomes a horizontal shearing tendency along the weak splitting-path parallel to the grain.

Trusses

In contrast to simple beams, many trusses carry their loads by stressing the truss-members only in tension or in compression (instead of in bending, like beams). Because wood is stronger both in tension and in compression parallel to the grain than it is in shear—about 6 or 7 times stronger—the log members of a purely

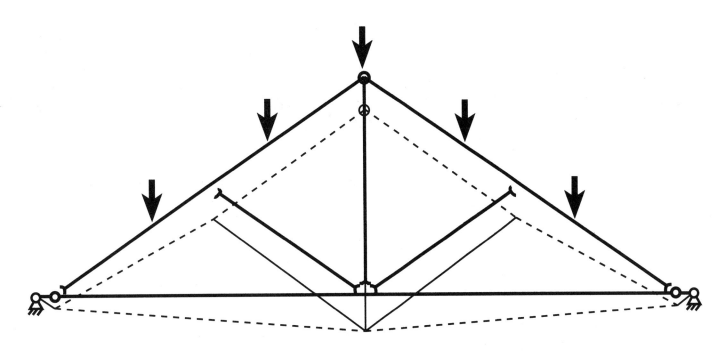

Figure 4: Truss schematic from computer program. Dashed lines indicate deflection in the truss. Arrows show locations of loads at ridge and purlins.

tension- and compression-member truss can be much lighter than a beam that does the same job. (This is also why stick builders can use trusses made out of 2x4's, even though 2x4's are too small if used as beams to do the same job).

Some of the members of the Banff truss have only tension or compression stresses. In general, the kingpost is in tension only, and the purlin braces are in compression. The bottom chord (tie beam) is primarily in tension. But, because of the geometry of the truss and the location of the imposed loads, the principal rafters are in compression but also in bending, that is, they have beam action, too *(Figure 4)*. We will see how this influences joinery design.

Designing Joinery

I will be describing the design of the joints at just two locations on a truss: where the principal rafter connects to the tie beam, and where the kingpost connects to the tie beam. I will then say something about the connection between the purlin brace (also known as a truss web) and principal rafter.

Rafter to Tie Beam

The first joint we'll consider is where the principal rafter meets the tie beam. To begin, we need to know the force that this joint must withstand. I use a computer program to build a mathematical model of the truss's geometry, assign sizes to the members, and place loads where they occur. This program has a graphic interface so that it actually looks like a truss on the screen with loads on it *(Figure 4)*. I wouldn't attempt to design a truss without one of these programs, now. It's way too much work and time the old way.

For the Banff truss, my computer indicated that each principal rafter is thrusting against the tie beam with a force of 39,700 pounds (18,000 kg). So, how do we attach the rafter to the tie beam to resist this load? I usually begin with a bolt. For this truss I chose a 1-inch (25 mm) diameter bolt with conventional washers. The washer and nut on the tie beam can be recessed and the hole fitted with a wooden plug. This will take care of some of this thrusting load, but I begin here because the truss must also hold together as it is lifted onto the roof, and a bolt also deals with the roof uplift caused by wind or seismic activity. The weight of the truss materials themselves is often not enough to keep things in place during windstorms and earthquakes.

The remainder of the thrust must be handled by the wood-to-wood joinery. So the question is: how much of the rafter thrust does the 1" (25mm) bolt handle? Notice several things about this bolt *(Figure 5)*. First, it is in "single shear." That means that the bolt passes through one plane where the two truss members join, and it is resisting the rafter thrust occurring at this plane by developing interior shear stresses (similar to the shear in the beam that we discussed earlier). Also notice that the bolt is at an angle to the grain of both the tie beam and the rafter, and that it is the same angle in each member. If the length of bolt in the two members is the same, then the forces passing from one member into the bolt, and thus to the other member, are balanced.

If the angle to the grain is different, or if the length of bolt in one member is different than the length in the other (and both of these situations are common), then the weaker of the two bolt-member resisting forces is used in the design. In our case, the angle to the grain is the same, and the length of bolt in both members scaled out to be about 9.5" (240 mm)—nuts and washers are not included in this length.

Also note that the allowable load that we assign to this bolt assumes that the bolt is in a hole that is no more than ¹⁄₁₆" oversized, that the shaft of the bolt is smooth (it is not all-thread), and that the steel is A307, or better.

Next, I look in an engineering manual for the allowable value of this bolt. This truss was designed using the *Timber Construction Manual*, 3rd Edition, though recently I tend to use *National Design Specification, Design Values for Wood Construction*. In any event, the values from the table need to be adjusted to account for the bolt being at a 17° angle to the grain. This is done using the Hankinson formula, which combines the angle to the grain (a), the allowable load parallel to the grain (P), and the allowable load perpendicular to the grain (Q) to give us a value that fits our situation (N). **Here is the equation:**

$$N = \frac{P \times Q}{P \sin^2 a + Q \cos^2 a}$$

A bolt used in this manner transfers forces between itself and the wood at 90° to the length the bolt. When I completed this calculation I found that the bolt will transfer 2,750 pounds (1,250 kg) perpendicular to its long axis, which is in a direction 17° to the grain of the wood in both members. At this point we come back to the question of how much of the 39,700-pound rafter thrust is being taken care of by the bolt.

To do this we need to know how much of the thrust coming down the axis of the rafter (39,700 pounds) is going in the same direction as the bolt resistance is acting (that is at 17° to the grain). I use trigonometry to discover this and looking at Figure 5 you can see that it is 37,965-pounds (14,650 kg). Subtract the bolt's resisting capabilities from this value and this leaves 35,215 pounds yet to go—more than 90% of the thrust is not handled by the bolt. Is anyone still with me? Hang in there!

Bearing Planes

This remaining 35.2 kips ("k," a kip is equal to 1000-pounds) must be taken care of by the wood-to-wood joinery. This joint must prevent the rafter from thrusting outward.

In Figure 5, notice that the rafter bears on the tie beam on two flat planes, 'A' and 'B'. I refer to these as bearing planes because one truss member 'bears' upon the

Note In this truss, the 1" bolt handles only 7% of the total rafter thrust, and we must learn some lessons from this fact. First, bolts cannot do it all, we'll need something else to handle the remaining 93% of the thrust.

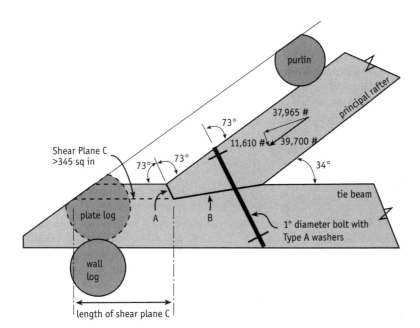

Figure 5: Rafter tail to tie beam joint. Bearing Planes A and B, and rafter thrust shown. Bearing Plane A must be at least 30 square inches (194 cm²) in area, for this truss. The bearing planes, and not the bolt, do the vast majority of the work in this truss.

other and transfers force to the other by compression. The size of the surface area of these bearing planes is important in this type of truss design because they usually do the majority of the work transferring loads in the truss.

In this truss, Plane 'A' is at 73 degrees to the grain of the tie beam, and Plane 'B' begins at the bottom of Plane A and slopes up to the top face of the tie beam. Plane B has more surface area than Plane A. Plane B is also taking less load from the principal rafter because the pitch of the roof is relatively flat (8:12, or 34°). We can see in Figure 5 that the amount of the rafter thrust transferred across Plane B is about 11,610-pounds.

Both of these factors more than balance out the one weakness of Plane B, which is that the load imposed on it by the rafter is almost perpendicular to the grain, and wood is about 60% weaker in compression perpendicular to the grain than it is in compression parallel to the grain. This means that Plane A is the critical one and the one whose area we will need to determine first.

Earlier we chose a bolt angle of 17° to the grain of both the rafter and the tie beam in order to help balance the load-transferring abilities of the bolt. By choosing the angle of Bearing Plane A in this same manner we once again transfer the load from one member to the other in a balanced fashion. We do this by dividing the angle between the top of the tie beam and the top of the rafter in half, and using this bisection as the bearing plane between these two members.

In Figure 5 notice that the total angle between tie beam and rafter is 146°; and dividing this in half leaves 73° on each side. Notice also that, similar to the bolt, force is transferred across this bearing plane in a direction perpendicular to the plane, which means that the force transferred to each member by the other is at the same angle to the grain (17° again) in both members. Once again, this balances the transfer of forces.

In practical terms this means that the required area of the bearing planes can be the same in the tie beam and the principal rafter. And this means that the depth of notch cut into the tie beam that creates Bearing Plane A in that member, is as small as possible, thereby weakening the tie beam as little as necessary. If, for example, an angle other than 73° was used to create Bearing Plane A, then we would have to increase the area of this bearing plane and therefore increase the depth of the notch cut into the tie beam, which we'd just as soon not do.

(I also worked out a design for this joint which employed a 90° angle between the top of the rafter and the bearing plane, which changed the other angle from 73° to 56°. This resulted in a 21% increase in the required size of the Bearing Plane A—36 in^2 instead of 30 in^2. The convenience of cutting this one angle on the foot of the rafter at 90° in a jig may offset the additional weakening of the tie beam. My goal here is to make you aware of what happens when Bearing Plane A is defined in a manner other than by bisecting the angle between the top of the tie beam and the top of the rafter. The required bearing area increases because of the different capabilities of wood in carrying loads parallel to the grain versus perpendicular to the grain.)

Back to our story. You will remember that the thrust load that must be handled by the bearing plane is 35.2 kips, and is acting at a 17° angle to the grain in both members. We use the Hankinson formula once again, this time using the allowable value for larch in resisting loads parallel to the grain and perpendicular to the grain, and we discover that the allowable load for larch is 1190 psi (pounds per square inch). Therefore, the required bearing plane area must be at least 29.6 square inches (35.2 k ÷ 1190 psi = 29.6 in²).

Shear Plane

As the thrust is transferred from the foot of the rafter to the tie beam some of this thrust is trying to push, or split off, the top of the tie beam. Figure 5 shows this shear plane, and it needs to be large enough in area to withstand the rafter thrust. The portion of the 35.2 k that is trying to split off the top of the tie beam is equal to 35.2 k x cosine 17° = 33.68k. Dividing this load by the allowable shear strength of the tie beam tells us that the area of this shear plane must be at least 345 square inches (2226 cm²).

Dividing this area by the length of the chord at the bottom of the notch (ie, where Planes A and B meet) cut into the tie beam will yield the length of tie beam that needs to protrude beyond the bottom of this notch (*Figure 6*). I typically expect the builder to calculate this length based on the actual diameter of his tie beam at this joint and the actual depth of the notch he is cutting into the top of the tie beam to produce Bearing Plane A.

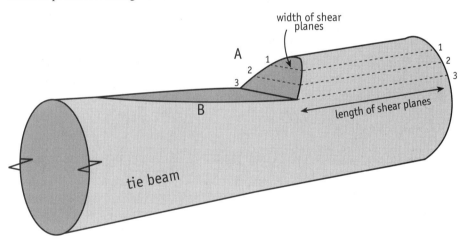

For example, if the notch into the tie beam makes a shear plane that is 9" (23cm) wide, then you'll need 38.3" (97cm) from the notch to the end of the tie beam in order to have a shear plane that is at least 345 in² (2226cm²) in size.

This completes the design of this joint. To communicate the necessary information to those crafting the truss, I typically provide a drawing like Figure 5 showing the size, location, and orientation of the bolt, the angle and size of the required bearing area for Bearing Plane A, and the required area of the Shear Plane C.

There are some general guidelines in the previous chapter about choosing the relative diameters of truss members. In choosing the size of the tie beam I like to have one that is a least four-times larger than the depth of my rafter foot notch. The rafter itself is often the most critically loaded member of a truss because as I mentioned previously it may be carrying the loads imposed upon it via beam

Figure 6: Cutting the Bearing Planes A & B into the top of the tie beam creates a shear plane that must be large enough to resist being split (sheared) off by the outward thrust of the rafter tail.

The area of the shear plane is found by multiplying its width by its length. The deeper Bearing Plane A is cut, the wider the shear plane becomes—option 3 is wider than option 1. The length of the shear plane is most easily increased by making the tie beam longer, outside this joint.

The plate log often crosses over this area, above the shear plane. Do not notch the tie beam in a way (saddles slabbed off, or square notch cut, for example) that reduces the size of the shear plane.

action as well as by compression parallel to the grain. I will talk more about this when we discuss the purlin brace to rafter connection.

Kingpost-to-Tie Beam Joint

Let's go next to the connection between the kingpost and the tie beam. You will recall that the kingpost is in tension. In essence, it is holding up the center of the tie beam, which is being forced downward by the loads from the purlin braces (webs).

In our truss, this joint *(see Figure 7)* is subjected to a load of 18.35 k (8340 kg). I like to use steel to handle tension loads of this magnitude, and I often specify a bolt with a large, rectangular plate washer. You can tuck this custom-made washer up into a mortise in the bottom of the tie beam, and disguise it with a rectangular wooden plug that looks like a through-tenon. This looks good, and can be removed to tighten the joint as the wood shrinks.

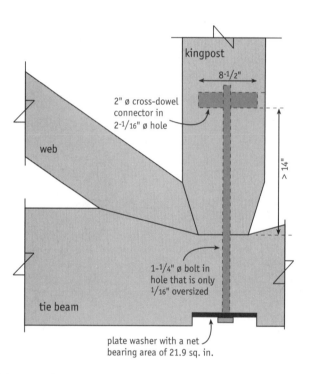

Figure 7: Kingpost to tie beam joint. There can be a lot of tension at this joint, and the cross-dowel connector and large bolt (all hidden) work well here. The washer is often a thick steel plate that is very large (here it could be 3" x 8", 75mm x 200mm) to prevent the wood of the tie beam from crushing.

In the kingpost, the connection is a cross-dowel joint connector made of smooth round steel stock that is threaded to receive the bolt. All of the steel used must be structurally designed. The washer must be large enough to support the 18.35 k load perpendicular to the grain of the tie beam. In our case this means that it must be at least 21.9 square inches plus the area of the hole for the bolt (18.35 k ÷ 839 psi = 21.9 sq. in). The plate washer must be thick enough or it will flex and not transfer the load as it should. I will not cover the engineering calculations needed to determine its thickness here.

The bolt itself is often sized based on the amount of tension it needs to carry—in our case this is 18.35 k. The *Manual of Steel Construction* says that an A307 bolt of 1J" (30mm) diameter will hold 19.9 k. I specified a bolt 1¼" (32mm) in diameter, which gives me a bit of a safety factor.

The bolt threads into a 2" (50mm) diameter cross-dowel that must be long enough to provide the required bearing area between it and the kingpost. We find this length by dividing the load by the allowable unit stress for our wood loaded parallel to the grain, and dividing again by the diameter of the cross-dowel (18.35 k ÷ 1265 psi ÷ 2" = 7.25")

To this I add the room taken up by the hole for the bolt itself (1.25") to give a total length of the cross-dowel of 8½". We also need to consider the distance from the cross-dowel to the end of the kingpost. In softwoods, this should be at least 7 times the diameter of the cross-dowel, or 14 inches (7 x 2" = 14") (7 x 50mm = 350 mm).

Purlin Brace Joint

Let's move on to the purlin brace (or *web*) and its joint with the rafter. My computer tells me that the compression force in the web is 12.9 k. Before we design the joint we need to find the diameter of the log for the web, which acts like a *short* column. (In a larger truss, a web member may be long enough to be an *intermediate* column, which uses a different equation for calculating its minimum diameter.)

For a *short* column, the minimum cross-sectional area is equal to the compression in the member divided by the allowable compression parallel to the grain. In our case this area = 12.9 k ÷ 1265 psi = 10.2in² (66cm²). This translates into a web member whose diameter is 3.6" (9cm). A web member this small would look puny in this truss, so I chose a web of 10" (25cm) in diameter, and the builder could make it larger, if desired.

For this particular truss, the builders at Timmerhus Inc. wanted to use a wedged, double-tenon joint *(Figure 8)*. This joint has the advantage that it can be tightened as the rafter, kingpost and tie beam shrink in diameter as they season. If we use a wedge that is of harder material than the truss members, such as oak or maple, then the critical bearing area is between the wedge and the rafter. The geometry of the truss tells us that the web pushes on the bottom of the rafter at an angle that is 68° to the grain of the rafter. We use the Hankinson Formula again to find that

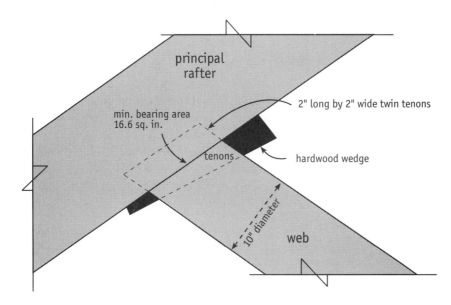

Figure 8: Joint where web meets rafter—a double tenon, with hardwood wedge between the two tenons.

the required bearing area between the wedge and the rafter needs to be at least 12.9k ÷ 776 psi = 16.6 square inches (so the wedge must be about 1¾" wide— 16.6 in² ÷ 10" = 1.66") (42 mm wide).

The twin tenons fit into twin mortises in the rafter, and the wedge fits between the tenons *(see Figure 9)*. As the material in the truss shrinks, and the wedge is driven in to tighten things up, the tenons withdraw slightly from their mortises. Make certain that an inch or so of tenon remains in the mortise at all times.

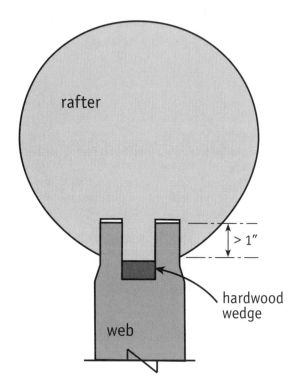

Figure 9: Cross section of rafter at web. The mortises in the rafter remove wood, and weaken the rafter, which is why we must know the section modulus and cross-sectional area.

Rafter Diameter

Figure 9 is a cross-section of the rafter at the web joint, and shows the material removed to make the mortises and the bearing surface for the wedge. If we estimate that the total mortise depth needs to be 2" (50mm), and that the mortise is 2" wide, then we have some of the information we need to check the size of the rafter. Does the remaining cross-section have enough strength to safely carry the load on the rafter?

I went back to my structural-analysis computer program and found that the rafter is both in compression and in bending at this joint. That is, it is acting like a truss member by being in compression or tension, and additionally it is acting like a beam, because it has shear and bending stresses. Here are the magnitudes of the internal forces acting on the rafter at this joint: compression = 39.7k; shear = 6.34 k; bending moment = 99.1 k-in. In the analysis of this joint, the shear is checked with one equation, and the compression and bending is checked with a different equation.

Looking at the shear first, how large does our rafter cross-section need to be to withstand the 6.34 k shear load? We find this by dividing the shear by the effective allowable unit shear stress. The effective allowable unit shear stress is ⅔ of the allowable unit shear stress, or ⅔ x 97.75 psi. Therefore the cross sectional area must be equal to or greater than 6,340-pounds ÷ ⅔ of 97.75 psi = 97.3 square inches (628 cm²).

Let's check on the combined axial and bending situation. For this we use the interactive equation shown here:

$$f_c \div F_c + f_b \div (F'_b - Jf_c) \leq 1.0$$

where:

$$J = [(le/d)-11] / (K-11)] \ and: \ 0 \leq J \leq 1.0$$

and where

$$K = 0.671 \sqrt{(E/F'_c)}$$

Is this fun, or what!?!?

Years ago I developed a computer spreadsheet to figure out this sort of mumbo-jumbo; and it means I don't have to go through these formulas time and again. The bottom line is that if we have a cross-sectional area of 97.3 square inches (628 cm²), then the section modulus at this location must be at least 87 cubic inches. (The section modulus is a measure of how much bending stress a beam can withstand.)

Referring back to Figure 9, how do you figure out the area and section modulus? One way is to use calculus and develop equations that will give you these values. However, I use an engineering geometry computer program that tells me the area and section modulus of any shape that I import from my CAD program. When I did that for a 12" diameter rafter with the shape in Figure 9, I came up with an area = 101 square inches, and a section modulus = 130 cubic inches. This tells us that our original choice of a 12-inch diameter rafter for this truss is okay both for the loads imposed and for the joinery used at this location.

If either the area or the section modulus had been too small (less than 97.3 square inches for the area, or less than 87 cubic inches for the section modulus) then we would have had to use a larger rafter, or change the joinery. Sometimes I guess the right size the first time (remember that one of the first things I did was guess that a 12" diameter rafter is a good place to start), and sometimes it takes me several tries to find the best combination of log-member size and joint design. Every situation is different.

The following variables are constantly changing from building to building: species and quality and size of wood available; snow, live, dead, wind, and seismic loads; geometry of the truss; pitch of the roof; span; aesthetic considerations; the slickness of the roofing material; and the geometry of the roof such as valleys, hips.

Review

Before I move on to some other comments, let's review what we've done so far. I began by talking about discovering the loads which come from the roof, and about making choices regarding the type of structural system to be used to carry these loads safely to the ground. Whether stick-framed, post and purlin, trusses, or combination systems, determining the correct loads is where we start, and this often involves considerable work.

Beams and trusses both can carry loads from the roof on to the walls, but because most truss members are loaded primarily in compression or tension you can carry a greater load for a longer span if you use a truss. Aside from the loads and the span, the most critical element in sizing truss members are the choices about joinery. Steel plates and bolts remove little wood, and so members can be smaller. Mortise and tenon joinery removes much more wood, and members need to be larger. The Banff truss relies on bearing planes and hidden steel fasteners; a system that is in-between these two. Each situation is different as the number and value of the variables is always changing.

Other Comments

Now I want to leave this specific truss and consider some general issues about trusses. I have given you a taste of what is involved in designing a truss. If the actual roof load is carried in some other fashion, such as a post and purlin roof, then decorative trusses can be concocted in a wide variety of shapes and styles. But when a truss is really carrying the roof load, then structural design, detailing, and construction are serious matters.

It is important to make a distinction between a *working* truss and a *decorative* truss. Too often a builder or designer will see a beautiful truss in another building and will then adapt that truss to their own project. When done properly this is a fine. *Properly* means that the new truss must go through structural analysis to size members, and design joinery for the loads imposed, the geometry of the roof and truss, the quality and species of wood, and so on.

But, maybe the beautiful truss that was copied was never intended to be a working truss in the first place. Was it even designed and engineered properly— and how do you know that? Unless it has been there at least 100 years, then it is foolish to use it as a model for a working truss. And even then the quality and species of wood, the joinery, the loads, etc. need to be evaluated again because they are very likely different.

One of the interesting things about wood as a structural material is that its failure is very time-dependent. As a result, a truss that is okay today may not be okay tomorrow, or next year, or in 20 years—it can take years to find out that a log truss was not designed or built right.

'Collar Tie' Truss

One type of truss that often falls into the *decorative* category is the so-called *collar tie* truss (*Figure 10*). This does not act as a truss at all until after it has already failed. In fact, the collar actually increases one of the problems that trusses were designed to solve: the rafters thrusting out on the walls. Yes, it is convenient to be able to walk under the collar and be able to use this space as living area. But beware the real problems.

The collar was called a collar *strut* in old carpentry and timber framing books. Which, in fact, is what it really is—a compression member, a strut. It is not a tension member—it is not tying things together. As load is applied to the principal rafters of a collar truss, they want to bow inward. The collar strut keeps them from doing that, and it goes into compression. When it does this it is pushing back on the rafter, and this pushing-back force is transferred down the rafter and actually increases the out-thrust at the rafter foot.

If the wall can't resist this additional thrust, then it is pushed out, causing the rafters to slump. The ridge drops, and eventually the collar goes into tension to try and keep things from getting any worse. We are in trouble. Was the collar-to-rafter joinery designed and built to withstand the tension load that now exists? Does that joint remove so much wood from the rafters that they are not strong enough? Will the rafters be strong enough without the support of the strut, which has now turned into a tie?

Collar tie trusses may have some usefulness in a 2x lumber roof where the load per rafter is smaller, but they have little or no place in a heavy timber roof as a structural truss.

I once asked the best timber engineer I know how high above the floor one could raise this collar and still expect it to act correctly as a tying member. He said, "It must be low enough to step over." Is everyone scared now?

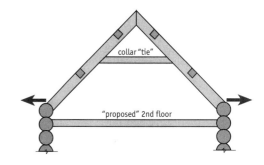

Figure 10: A collar "tie" truss. You must be very careful and seek professional engineering when building these.

It seems simple: "I can get a lot of extra headroom just by moving the tie beam up high." But the consequences can bring a catastrophe. The arrows indicate thrust outwards of the rafter tails and wall plate logs.

Summary

For those of you who made it through this whole chapter, my congratulations! For those of you who read the introduction and ending, I know you are curious about what all the stuff in the middle was about—ask your engineer. And for those of you who didn't read any of this, *good luck* to you and your clients.

Purlins & Ridge

If a home has logs in the roof, the chances are good that it has log purlins and a log ridge. This was not always true, and as recently as the 1940's, log common rafters were typical. Why the change? Mostly because 2x rafters became so popular: 2x rafters have spaces that are easy to insulate; installing a fire barrier and finished ceiling material to the bottom of 2x's is simple; and 2x material is inexpensive. Since there is no structural reason to have log common rafters underneath 2x common rafters, roof systems based on log purlins (instead of on log rafters) became popular with log builders.

In this chapter you'll learn some helpful tricks about choosing logs for the purlins and ridge, laying them out, and cutting them.

I am often asked how many log purlins are needed, how close they should be to each other, how large in diameter, and how far they can span between supports. The answers to these questions can be determined only by engineering calculations. There are no *rules of thumb* that apply to all log homes. Someone must do the numbers.

If the upstairs will be used as living space, then the size, number, location, and the height of purlins off the upstairs floor must be worked out on paper while the home is being designed. Too many times I have been given plans that wasted usable loft- and second-floor space because purlins were positioned too low in the roof.

Purlins are not just for holding up 2x rafters. For example, a log purlin can support a dormer roof, valleys,or an adjacent log ridge, and it should be put at a height that makes this easy and practical.

Some home-design computer applications have 3-D walk-through, or fly-around, tools that can help, but they help only when you are sitting at your desk. So, sometimes I build a scale model of the roof system using ½" dowels and a hot-glue gun—it helps me visualize how things will go together in three dimensions, and I can take it outside.

Purlins and Ridges

Choose the Logs

For purlins and ridges, the best logs are straight, long, and don't have much taper. Ridge logs are the straightest logs that you have—log builders call them *gunbarrels* or *candlesticks*. While purlins can have bow in one plane, they should not bow in two planes *(Figure 1)*.

Avoid spiral grain when choosing structural roof logs. Left-hand spiral is not allowed because it is unreliably weak in bending (engineers call it "F_b"); and right-hand spiral is allowed, but only if it is low or moderate (see *Log Building Standards*, Section 2.A.4).

Your engineer or architect will specify the mid-span diameter and species of these important logs. If you are experienced at doing engineering calculations, then you could use the *Log Span Tables* to size simple roof logs like purlins and ridge. I use a structural engineer for all roof systems that I build, however, and I advise you to do the same.

The diameter and species of roof logs affects both the strength and stiffness (the amount they sag under a load, engineers call this "E," the *modulus of elasticity*). Purlins can be spliced at, or between, supports, but stiffness and integrity improve if you use long logs that are supported in more than two places.

Once you have selected which logs to use for purlins and ridge, then you need to decide where in the roof each will go—north slope, west slope; at the top, middle, or bottom of the slope; and which way it will point. Label each log with its location and orientation, and draw a map of where every log goes in the roof. The map will help you cut flats in the right places, rotate the bow so it is upslope, and have all the butt-ends of purlins and ridge point the same direction.

Lay Out the Roof-Pitch Flat

Purlins, plates, and ridges almost always support 2x rafters or structural insulated panels—and this means that they need to be flattened at the roof pitch, and all these flats must be in the roof plane. Some builders put the purlins and ridge logs up into the roof, and then stretch strings to mark them for cutting the roof pitch. But it is easier, faster, and more accurate to cut the roof-pitch flats *first*, not last.

In fact, the method described in this chapter lets you lay out and cut an entire post-and-purlin roof system on the ground, one piece at a time, without having to put any of the logs up onto the walls until the roof is assembled. This saves time and is safer, since all the cutting and fitting is done with your feet on the ground.

Log Building Standards specify that roof-pitch flats should be 1½" (about 40mm) wide, or more, if they support ceiling materials or framing lumber. They are usually wider than this, but remember that the wider the flat, the weaker the log will be—simply because of the extra wood that was removed. For large diameter purlins (12" tip and larger, 300mm) I usually cut roof-pitch flats that are about 4" to 7" (100mm to 175mm) wide, though they can get wider than this if the butt end of the log flares out (like the purlin on the left in *Figure 1*).

Figure 1: Log purlins and ridges are structural roof beams. The purlin on the left has some bow, and the bow has been rotated so it is at the roof pitch, and pointing towards the ridge. Support-flats have been cut on the bottom of the ridge and purlins where they meet the posts.

Also note that the roof beams are supported in 3 places: a post on the gable wall, a truss at midspan, and another post.

Figure 2: A level helps snap the chalkline plumb.

Orient the Bow

A purlin with a bow can be oriented so the bow points either up-slope or down-slope—that is, the bow should be approximately parallel to the roof pitch. Another way to visualize this is that a purlin's bow should point either towards the ridge log (up-slope) or towards the plate log of the wall (down-slope).

If the purlin bows straight down it will look like it is sagging; and if the purlin bows straight up through the ceiling, then you'll have to cut a lot of wood off of it for the roof-pitch flat. So, do not rotate the bow so that it is horizontal or vertical, instead put the bow at 8:12, or whatever your roof pitch will be.

Ridge logs should be as straight as possible. If the ridge has a slight bow, then I usually rotate it so the bow is down (*sagging*) to avoid sawing a lot of wood off the middle of the span when the roof-pitch flats are cut.

Lay Out the Lines

With the purlin bucked to length, and in V-blocks on the ground, rotate the log so the flattest surface faces up. To keep the purlin from sagging, have the V-blocks in about 20% of the length of the log (for example, 8-feet in from each end on a 40-foot log; 2.4m in on a 12.1m log) (*Figure 3*).

Using a 2-foot level, draw a horizontal line on both ends of the purlin at a depth that will produce a flat that is about 5" (125mm) wide, but never less than 1½" (40mm) wide, for the entire length.

It is crucial to have enough mid-span diameter remaining after the flat has been cut—you need that strength, and the engineer often specifies what cross-sectional area must remain.

Pin a chalkline to the level lines on the ends of the purlin. Stretch the string horizontally, and snap a line. The chalkline should be taut, so that it really does snap, when you let go. Re-chalk the line and snap the other edge of the roof-pitch flat. When purlins are longer than about 25-feet (7.6m), the chalkline will sag and doesn't leave a straight line. The solution is easy: rotate the log so the roof-pitch flat is plumb, and then snap the chalkline—you'll be pulling the string up plumb and then letting it go plumb: no sag.

Chalkline accuracy improves if you have a second person to spot your work. One person goes to the middle of the log, pinches the string between thumb and finger, pulls it straight up 2-feet (600mm), and holds it there until the spotter (who is at one end of the log) says it's plumb. When you open your fingers the line snaps straight down.

You must stretch the chalkline in the same plane as the flat you are going to cut. You pull horizontally if the log is rolled so the roof-pitch cut is level. Don't stretch the chalkline plumb unless you have rolled the log so that the roof-pitch flat is plumb. And never stretch the chalkline so that it snaps towards the center of the log (unless, of course, you're cutting the log in half). The accuracy of the roof-pitch flat depends greatly upon pulling the chalkline back so it is in the plane of the cut.

Figure 3: To keep a purlin or ridge from sagging under its own weight, place the V-blocks, or other temporary supports, about 20% in from each end. This improves the accuracy of the chalklines.

Figure 4: A ridge log, with both roof-pitch flats sawn. Note that the two flats do not always meet to form a sharp wooden peak on the ridge. At the end closest to us, there are two waney portions, but this causes no problems. Many ridges have wane between the roof-pitch flats.

The endgrain plumb line appears slightly off-center, but this is because this line is drawn after a chalkline is snapped on the visual center of the ridge, not by dividing the endgrain exactly in half. The plumb line is correct.

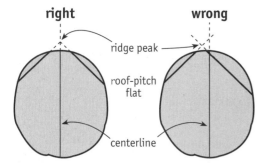

right   wrong

ridge peak

roof-pitch flat

centerline

Figure 5: Ridge layout. Looking at one end of the ridge log, with plumb centerline. If there is wane in between the roof-pitch flats, then the actual peak of the ridge will be above the top surface of the log, and not on the log itself. In that case, make certain that the roof pitch flats intersect plumb above the centerline, not off-center.

Figure 6: Ripping a roof-pitch flat. With the log at a comfortable height, hold the saw so your left hand is on the trigger and the gas fill points up, not down. By cutting with the top of the bar, the sawdust sprays away from you. The saw feels like it is pushing itself towards you, but your left thigh can push the saw back into the cut. If you keep your forearm parallel to the ground, then your right hand will pull the saw straight instead of rotating. I prefer the log to be a little higher than is shown here.

Now that both chalklines are snapped you can easily see how wide the roof-pitch flat will be. If the flat is too narrow or too wide, then you should change the depth of one or both of the lines on the ends of the logs and re-snap the chalklines. Use a different color chalk to make it easy to distinguish the new lines.

Ridge Layout

Layout for ridge logs is slightly different than for purlins. With the ridge in V-blocks on the ground, rotate it until the desired peak is on top (if there is a slight bow, then it usually is best bow down). Snap a top-center chalkline that divides the log in half, by eye. Draw plumb lines on the end grain at both ends of the chalkline.

Next, draw the roof pitch on the end grain. A digital *smart level* is just the trick for this job because it has an LCD display that read in degrees or in pitch (like 7:12). Or, you can use a rafter framing square held at the appropriate pitch.

I like the roof-pitch flats to be about 4" to 6" wide. This can mean that the actual peak of the ridge is not actually on the ridge log, it is above it *(Figure 5)*. This often happens at the butt end of a ridge, and when roof slopes are steeper than 8:12 (33°). The roof-pitch flats do not have to meet and create a continuous wooden-edged peak for the entire length of the ridge log. Having wane between the two flats is okay *(Figure 4)*.

But if there is wane between the roof-pitch flats, then you must be certain that the two flats intersect at a place that is plumb over the centerline of the ridge.

It is important to note that the two roof-pitch flats do not have to meet—that is, the very top of the ridge log can be round on top it's entire length, or just part way (it will depend upon the shape of the log). If the roof slope is very low, or if the ridge log is small in diameter, then the flats will probably meet to form a sharp ridge peak. But usually, the ridge comes back to round here and there. With large ridgepoles, or when the roof slope is steep, you would have to remove too much wood if you made the ridge peak continuous and without any wane.

Snap both the chalklines for one of the roof-pitch flats. Remember to stretch the string back in the plane of the cut. Then, only after you cut and finish the first flat, snap the two chalklines for the other roof-pitch flat.

Rip the Flat

A narrow flat is easily sawn with the log rolled so that the flat is level—this gives you a good view of the chalklines on both sides of the cut *(Figure 6)*. Ripping with a chainsaw is comfortable when the saw is about hip-height, so raise or lower the log to make the roof-pitch flat a height that is comfortable for you.

I like to use the top of the chainsaw bar, not the bottom, to do the ripping because it is easier to pull the saw than it is to push it, and because there is no handle on the bottom of the saw. An added benefit to ripping with the top of the bar is that the sawdust flies away from you. I hold the chainsaw (with the gas cap pointing up) *wrong-handed*—that is, my left hand is on the trigger. Keep your right hand on the top of the wraparound handle (your forearm in line with the sparkplug), so you are pulling straight down the cut. If your right hand is near the recoil (pull rope), then the saw will twist as you try to cut straight.

Do not rip right to the chalklines. Leave ¼" (~5 mm) of wood above the lines—until you become accurate at ripping. Remove the high spots left after ripping by using the saw with an overlapping brushing, or sweeping, motion *(Figure 7)*. A chisel chain (sometimes called *full chisel*) makes brushing difficult—the sharp edges of the teeth tend to catch and dig in—semi-chisel is better.

When you think you're done, sight down the length of the flat from one end of the log—the high spots or twists will show quite plainly. If the log has been rolled so that the level lines on the log ends are actually level, then at any place along the roof-pitch flat you can use a 2-foot level to check for twists. Finish the job with an electric planer, but it is surprising how smooth, flat, and true you can get the surface with just a chainsaw, a chalkline and level, and some practice.

If your chalklines were accurate, and you ripped and brushed to the lines, then the flat will be quite good. If all the chalklines are still on the log, but the flat is not straight, then the chalklines were snapped poorly. Figure out what happened so you don't make that mistake again.

Centerline

With the purlin in V-blocks, or on horses, and the roof-pitch flat finished, rotate the log so that the flat you just completed is at the actual roof pitch, but upside down. The roof-pitch flat should be pointing mostly downward—rotated 180° from its final orientation *(like the purlin in Figure 12)*. Hold a digital level on the roof-pitch flat, rotate the log as needed, and chock it to keep it in place. Note that the purlin need not be level from one end to the other, but it must be rotated to the correct roof pitch.

Snap a chalkline that will be the bottom center of the purlin. Just as with wall logs, this line should divide the purlin in half, by eye. If posts are attached to the purlin with tenons, then you'll want to move the chalkline so its is nearly centered on the purlin where the posts will be. You start with the bottom-center chalkline because the roof-pitch flat makes it difficult to decide where the top-center of the purlin is.

Use a 2-foot level to draw plumb lines on both log ends that go through the bottom-center chalkline. Then, rotate the log 180° and snap a top-center chalkline that joins the two plumblines on the end grain. In some places the top centerline may be on round wood, and in other places (where the roof-pitch flat is wide) it will be on the sawn flat.

Now the purlin has a top centerline and a bottom centerline—we need these lines for laying out purlin support-flats, for cutting the support posts to the right length, and for cutting mortises (if any) to join the purlin to the tenon on top of posts.

Figure 7: Brushing a roof-pitch flat. The technique is like brushing a saddle: medium-high chain speed with the bar nearly 90° to the surface, and a sweeping, arc-like motion towards you. Use the chainsaw like a planer (a precision power tool for flattening), not a sander (which only makes things smooth, not flat).

Make one or two sweeps, then take a step down the log. Overlapping the arcs keeps you from making *moguls* (washboard bumps) on the flat surface. For added stability, keep the back handle of the saw held against your thigh, as shown.

Use only the flat bottom of the bar for brushing—and brush all the way from one chalkline to the other in one pass. If you use the end or *nose* of the bar then the flat comes out dished. If you tip the saw up or down to touch just one of the chalklines, instead of *both* of them, then the flat will have a hump in the middle.

Build a Purlinizer

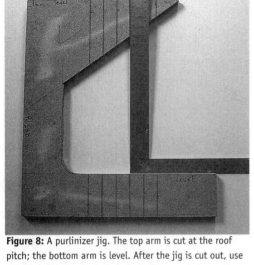

Figure 8: A purlinizer jig. The top arm is cut at the roof pitch; the bottom arm is level. After the jig is cut out, use a rafter square to find whole-inch distances between the top arm and the bottom arm, and mark them on the jig. Connect these with a series of straight lines, from top to bottom. Each purlinizer only works with one specific roof pitch (this one is for any roof that is 12:12, or 45°).

Purlins and ridges are often supported by log posts, a truss kingpost, timber posts, or 2x frame walls, and that's why the roof logs need flat spots cut on their bottoms. These *support flats* are level in both directions—along the length of the purlin or ridge, and from side to side. That's because the tops of posts are cut level.

The challenge is to determine what is level on a log purlin or ridge. A log's round surfaces and natural shape prevent the usual roof layout tools (a rafter square, or a level, for example) from working very well. I invented a simple and effective jig to lay out the support-flats on the bottom of purlins and ridges—I call it a *purlinizer (Figure 8)*.

The purlinizer jig is made from a piece of good quality, ¾" thick, particle board. Make it big enough to fit over the largest diameter log you are using. And each roof pitch needs its own purlinizer: if your house has two roof slopes (say 8:12 and 6:12), then you'll need two purlinizers—one for each pitch. The concept behind the purlinizer is simple: the diagonal portion of the jig is cut at the roof slope, and every other edge is either plumb or level.

The flat on the purlin bottom should be deep enough to make the flat a sufficient size for bearing, but not so deep that it unreasonably weakens the purlin by removing too much wood.

Once the jig is cut out, use a rafter framing square to draw lines that indicate the depth of the purlin *(Figure 8)*. To cut a post to length, you need to know how deep the purlin is at that point. The lines you are about to draw on the purlinizer indicate purlin depth.

Figure 9: Purlinizer in use. The jig's angled cut is set on the purlin's roof-pitch flat, and here the 14" line is matched with the purlin top center line. From the bottom arm, the flat is being measured up 1¹/₂". The depth of the purlin for this flat is 12¹/₂" (14 - 1¹/₂ = 12¹/₂), as can be seen marked on the end-grain.

Note that the purlin is rolled to the right (the roof pitch flat is almost level)—the purlinizer works with the purlin at any angle.

(The channel, or dado, marked on the roof-pitch flat is for a piece of heavy threaded rod that will splice this purlin to another purlin, end-for-end.)

I like to make the purlin depth an easy number, to reduce errors when I subtract fractions, and this is the purpose behind having a series of lines on the jig that show whole numbers of purlin depth. It is easier to subtract numbers like 88⅜" - 11" than it is solving 88⅜" - 11⁹⁄₁₆".

To draw the depth lines on your purlinizer, place a rafter square as shown in Figure 8, and slide it left and right until the distance from inside of the bottom (level) arm to the inside of the top roof-pitch cut is a whole number—14" for example—and then draw a vertical line and mark it "14". Slide the square until it reads 13" and mark it. Continue until you have marked all the whole-number distances that will fit on your jig. Connect the points with a long straightedge— the 14"line on the top leg of the purlinizer is one straight line with the 14" line on the bottom leg of the purlinizer. (Metric users: make a depth line for every 20mm increase in distance, or so.)

Lay Out the Support Flats

Working on the bottom centerline of the purlin or ridge, carefully lay out where support-flats are needed: gable ends, mid-span posts, that sort of thing. Make certain you measure from the correct end of the purlin—remember that you made a map showing tip-butt orientation and placement for every roof log. It is all too easy to lay out the support flats starting from the wrong end of a purlin or ridge.

You do not need to level the purlin or ridge from end to end, or from side to side, to use the purlinizer. The jig works no matter how the purlin or ridge is sitting—that's one of its beauties. Further, the log does not have to be rotated to the actual roof pitch, in fact, it is often better if it isn't.

Figure 10: Looking at the bottom of a purlin, and a support flat. In the roof, the support flat is level side-to-side and end-for-end. The post will have a tenon, in this case, so the support flat has a matching mortise.

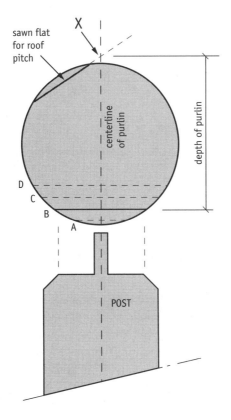

Figure 11: Cross-section of a purlin and post. In this example, there are four depths for the flat (A, B, C, and D)—each is a whole-inch distance from the top center of the purlin. It looks like "C" and "D" would be too wide, and "A" too narrow. "B" would work well.

Also note the top of the purlin ("X") is often in hanging in space, and not actually on the surface of the log. This is because Point "X" is where the center of the purlin and post intersects the ceiling line.

Purlin depth *must* be measured to Point X, and not to the top round surface of the purlin or anywhere else.

Figure 12: The flat has been cut (the purlin is upside down). This photo shows that the flat is parallel to the level arm of the jig—so the flat will be level.

Hold the purlinizer jig on the roof-pitch flat at one of the support locations, and slide it up the roof-pitch flat until one of the depth-lines representing whole-numbers is touching, or points at, the top-center chalkline of the purlin or ridge *(Figure 9)*. This makes the level leg on the bottom of the purlinizer a whole number distance from the ceiling line (12", 13", 14", and so on).

Decide how deep into the purlin or ridge to make the support-flat. For example, if the top of the post is 8" (200mm) in diameter, then you want to go deep enough into the bottom of the log to make a flat that is about 6" (150mm) wide *(Figure 11)*. Working from the bottom leg of the purlinizer, measure up a whole-inch amount and see how wide the flat would be. Try other amounts until you get a flat that would be about 6" wide. If necessary, measure up in half-inch increments (like 1½") instead of whole-inch, but avoid quarters, eighths, and sixteenths just to keep things simple. (Metric users won't have this problem because they don't use fractions.)

Once you have found a good depth, use the purlinizer to mark the four corners of the flat. Mark two points at one end of the flat, then move the purlinizer 8" (20cm) down the log, put the whole-inch line on the top centerline, and mark the other two points on the bottom of the log. Connect all four points with a flexible ruler—bend it only *in the plane of the cut* (just as a chalkline is snapped only in the same plane as the cut). And then cut and smooth the flat *(Figure 12)*.

If there will be a tenon on the post, then lay out the mortise on the purlin or ridge after the flat is finished, not before, because this gives you a good flat surface to work on.

Label the depth of the flat on the purlin or ridge, and on the support flat itself. For example, if the purlinizer was held with its 14" line on the top chalkline, and you measured up 1½" from the bottom leg to mark the four points, then the flat is 12½" below the ceiling (14 - 1½ = 12½). This is how you know that the length of the post that goes on this flat will be 12½" shorter than the ceiling height at this location.

Wane

In some places along a purlin or ridge, the top-center chalkline will be on a round, waney, surface of the log, instead of on the roof-pitch flat. Where this happens, it won't be possible to hold any of the purlinizer's depth lines on the chalkline. (In fact, the depth lines can touch the centerline only when the centerline is actually on the roof-pitch flat.) When this happens, just slide the purlinizer up the roof-pitch flat until one of the whole-inch depth lines points at the centerline. You can either sight down the depth line, or for more accuracy, hold a ruler on the line and slide it down until it touches the log.

On a log ridge, if the flats meet and create a peak, then that's the top center of the ridge. If the roof-pitch flats don't meet, and there's a waney round part between the two ridge flats, then snap a top-center chalkline there *(Figure 4)*.

Scribing Purlins

Where purlins are supported by posts or by 2x walls, a simple flat on the bottom is sufficient, as we have seen. But when a purlin crosses a truss or a valley rafter, then it is often scribed and notched over the rafter *(Figure 13)*. To do this, you'll have to lift the log onto the roof and hold it steady while you scribe the notches.

(Another option is to make a flat at a known depth in the bottom of the purlin, and a flat at a known height on the top of the truss rafter. But the math is a bit more difficult—you'll have to use trigonometry to get the flat at the correct depth and place on the rafter.)

Once again, you do not want to notch the purlin so deeply that it is overly weakened by the wood removed. To avoid this, you must have built the truss or valley rafter so it is below the ceiling line by a distance that is slightly less than the diameter of the purlin (after the roof-pitch flat is cut) that crosses it *(Figure 14)*. You can see the necessity of planning ahead!

To scribe a purlin, it's handy to use a simple device to steady it on the roof. The jig is made of steel angle iron, and is bolted to the rafter to keep the purlin from sliding down hill *(Figure 13)*. Bolt the device lower than needed so you can add shims to get the purlin into the right place—that's a lot easier than having to unbolt the jig to move it downhill when it's too high up the rafter. Use a digital level to rotate the purlin's roof-pitch flat to the correct slope.

When the purlin is supported in just two places *(as in Figure 13)*, I don't bother trying to level it end-for-end before scribing. Instead, just use scriber settings that will bring the purlin roof-pitch flat down to the correct height at each end. When a purlin is supported by scribed notches at three or more places, however, you'll probably have to level it end-for-end before scribing its notches.

Figure 13: A purlin is lifted onto the roof, and temporarily held in place so it can be scribed down onto the rafter of the truss (background), and, in this case, also onto a valley rafter (foreground). The devices are bolted to the rafters several inches lower than necessary so the purlin can be shimmed up to the right location (2x6 and plywood shims shown here). The crane is still holding the purlin—note straps. Also note that the purlin has a slight bow, and it has been rotated into the roof slope, and is pointing towards the ridge.

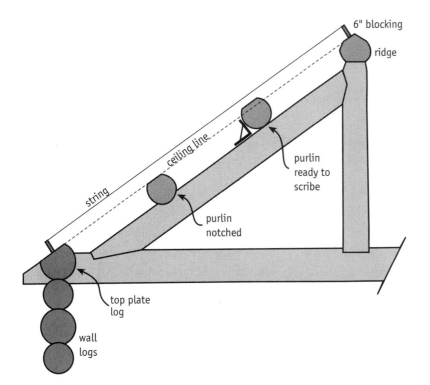

Figure 14: When scribing a purlin down onto a support log like a roof truss or a valley, you can't use the ceiling line because the purlin itself is in the way. Instead, block up a string 6" (150mm) at both the plate and the ridge, and then stretch a string above, and parallel to, the ceiling line.

The scribe setting must be measured vertically, and NOT at right angles to the roof-pitch flat on the purlin. You must figure out what 6" at your particular roof pitch will be. (If the roof is 12:12, then the plumb distance related to 6" is about 8½".) (In metric: if the roof is 45°, then the plumb distance related to 150mm is 212mm.)

Figure 15: An outrigger purlin (or plate) can help add headroom in the loft. As you near the top of the walls, the flyways are extended to support the outrigger. In this house, there are actually two outrigger logs—one scribed to the other—for extra strength. Note that the 2x common rafters are supported by the outrigger, not by the wall plate log.

In these cases, it can be useful, or even essential, to have the wall plate logs and ridge in place before you try to scribe a purlin down to the correct height. Stretch a string from the roof-pitch flat on the ridge to the roof-pitch flat on the plate log and you now have a reference for exactly where the ceiling line is *(Figure 14)*. In practice, however, when the purlin is in position to be scribed it will be in the way of a ceiling-line string, so prop the string up by exactly 6" (150mm), or some other amount, at both the ridge and the plate.

Keep the crane attached to the purlin and safely holding its weight. The devices are used only to stabilize the purlin for a few minutes while you scribe the notches —they are not adequate for holding the weight of the purlin.

Outrigger Plates

Outrigger plates can be used to provide a little more headroom in the second story or loft. They were more popular a decade ago than they are now, but they are still used from time to time *(Figure 15)*. As the walls get taller, flyways are extended to support the outrigger logs.

Outriggers get laid out and cut at the roof pitch, while the top log of the log wall (the plate log) is cut flat and does not support the roof. This moves the ceiling line from the wall plate log out to the outrigger plate log. I recommend that a structural engineer review any plans using outriggers.

Log Valley Rafters

When two roof planes meet you can get a hip, or you can get a valley. Which one you get depends upon the floor plan: where the gables are, which direction the ridges run, and the shape of the building in plan *(Figure 1)*.

In this chapter, you will learn about valleys (and much of this also applies to hips). The emphasis will be on how to lay out and cut the *backing angles* of valley logs—the wedge cut out of the top surface of a valley log to provide the proper roof slopes.

I feel strongly that structural log valley rafters are not for beginners, they are for experts. The information in this chapter will help bring you above the beginner level, but it will not make you an expert. I always consult a structural engineer on every project that has structural valleys or hips made of logs.

First we'll tackle the easiest valley: when both roofs have the same slope, or pitch. Then we'll go on to the more challenging situation when the two roofs have different slopes.

Valley and hip layout requires that we use trigonometry—things like tangent, sine, and cosine, and we will use a hand-held calculator with *trig* functions. (There are other ways of figuring backing angles, but this method is quite simple.)

But even if you do not know trigonometry, I have organized this information so that you can just choose the correct formula and plug in the right numbers to find the backing angles that you need in order to lay out and cut a log valley rafter.

Getting Started on Valleys

Convert All Pitches to Degrees

The formulas we will be using work only with slopes in degrees, so, if you are using roof pitches like 8:12, or 6½:12, then you first must convert every pitch to degrees of slope.

This is easy to do. Using a calculator with *trig* functions, enter your roof pitch as follows: enter the rise, push divided by, and enter the run, and then push equals. So for a roof that is 8:12, enter "8" then "÷" then "12" then "=". Your display should say "0.6666667". This is 8 divided by 12.

Now find the inverse tangent button (or buttons). On some calculators you'll have to use "INV" and then "TAN". Other calculators have one button that is labeled "TAN $^{-1}$" or "ARC TAN"—but these all mean the same thing. Refer to your calculator instruction book, if necessary.

With "0.6666667" still in the display, push the buttons for the inverse tangent, and now your display says "33.6900" (and it may go on for several more decimal places to the right). This means that an 8:12 roof is 33.69° slope from horizontal. Try some others. 8½:12 is 35.31°. A 12:12 roof is 45°. A low-sloping roof like 4:12 is 18.43°.

Figure 1: When two adjacent roofs meet they can form a valley (top) or a hip (bottom).

Figure 2: This roof has a valley and several hips. (The log building style is called piece-en-piece and features vertical posts with panels of scribed logs in between.)

Figure 3: This is the same roof, showing most of the logs in place that support that roof, including a log valley and several log hips.

Start in "Plan"

Find your *roof plan*—it's the bird's-eye view. Figure 5 is an example—it shows the ridges, the roof slopes, valleys, and indicates which ends are gables.

We will start with valleys when the roof slopes are the same—let's say all roofs in the building are 12:12, or 45°.

We'll be using several letters for important roof angles: D, S, SS, R1, DD, and so on. I suggest you use these letters because engineers and designers all use the same letters to mean the same thing. "SS" is always the slope (in degrees) of the main roof. "R1" is always the slope of the valley.

Figure 4: Looking straight up a log valley rafter. At the top, it starts at the ridge (in this building the ridge is a glu-lam), and at the bottom, it ends at the plate logs. You can clearly see the backing cuts—the wedge of wood removed from the top of the valley to match the slopes of the two adjacent roofs and make a flat bearing area for rafters or stress skin panels.

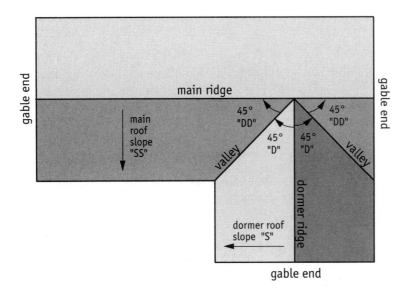

Figure 5: I have added shading here to make it easy for you to "see" the roof slopes, but the angles shown are measured level, in plan, *not* in the planes of the roof pitch.

Deck Angle

First, figure out the angle of the valley to its ridge, in plan—"D" in this drawing. The formula for the adjacent deck angle is:

D = inverse tangent (tan SS ÷ tan S)

So, if both the main roof and the adjacent roof slope are 12:12, which is 45°, then
D = inverse tangent (tan 45° ÷ tan 45°)
D = inverse tangent (1)
D = 45°

In fact, whenever the two roof pitches are the same, then the deck angle is 45°. If both roofs are 7:12 then D is 45°. If both roofs are 21.7°, then D is 45°, and so on.

The main deck angle formula is:

DD = inverse tangent (tan S ÷ tan SS)

SS	is the slope of the main roof
S	is the slope of the adjacent roof
D	is the adjacent deck angle
DD	is the main deck angle
R1	is the slope of the valley (or hip)
C5m	is the main backing angle
C5a	is the adjacent backing angle

Hint The formulas are always done in the same order:
1) deck angles
2) slope of valley
3) backing angles

Slope of the Valley (or Hip)

Next, figure the slope of the valley, which is called "R1." The formula for this is:

R1 = inverse tangent (tan SS x sin DD)

Using our example where both roofs are 12:12, or 45°.
R1 = inverse tangent (tan 45° x sin 45°)
R1 = inverse tangent (1 x 0.707106)
R1 = inverse tangent (0.707106)
R1 = 35.3°

The trick to doing these quickly and without errors is to not re-enter any numbers into your calculator. (I even avoid using the calculator memory.) The method will vary from brand to brand of calculator. On mine, I push: 45, tan, x, 45, sin, =, inv, tan — and out comes 35.264°, which I rounded to 35.3°. Note that I didn't use any calculator memory, and I never had to re-enter one of those long numbers.

Note that while both roofs slope at 45°, the valley has a lower slope: about 35°. Valleys always have a lower slope than the roofs, and this is why there is a "headroom problem" with log valleys. Make scale drawings to discover if you have a headroom problem.

Backing Angles

Next, figure the backing angles for this valley. We need these so we know what shape wedge of wood to remove from the top of the valley log. The formula for the main backing angle is:

C5m = inverse tangent (sin R1 ÷ tan DD)

Using our example of both roofs at 12:12, or 45°:
C5m = inverse tangent (sin 35.3° ÷ tan 45°)
C5m = inverse tangent (.57735 ÷ 1)
C5m = inverse tangent (.57735)
C5m = 30.0° and
C5a is also 30.0° because both roofs have the same slope

Hint When both roofs have the same slope, then the backing angles (C5m and C5a) are also the same.

Layout on Valley Logs

Snap a chalkline for the top center of the valley log—a straight log works best, but if it has a slight bow, make it bow down. Cut the log square at both ends, and draw plumb centerlines down from the chalkline.

Next, decide how deeply into the valley you want to cut the backing angles. The deeper you cut, the less strength the valley has; the shallower you go, the smaller the backing flats become—meaning less bearing area for rafters or stress skin panels.

Draw a level line through the depth at both ends. Where the plumb line and the level line meet, lay out the backing angle—in this example it's 30° from horizontal *(Figure 6)*. Do this at both ends of the valley log. Now snap chalklines for the cuts—make sure you pull the line back at the same angle as the backing cuts! Or, roll the log until one backing cut is plumb and snap the chalkline plumb. Roll the log until the other backing angle is plumb and snap plumb. Layout is complete.

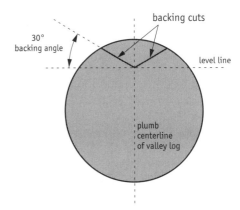

Figure 6: Layout of backing angles on the ends of a valley rafter. The ends must be cut square to the log.

Hint   Lay out the valley with both ends of the log cut square. Then, if the valley has an end that is cut plumb, do that last.

Figure 7: The log valley rafter is cut and in place. Note the two backing angles (the wedge cut out of its top surface). One backing cut is part of the roof on the left, the other is part of the roof on the right. Getting these angles correct is essential.

In this photo we are placing a log jack purlin. It is a "jack" because it ends in a valley instead of going to a gable end. It is a "purlin" because it is parallel to the ridge.

The building is cross-shaped, and so has four valleys.

Because of the very long spans, the engineer specified that the ridges had to be glue-lam timbers instead of logs.

When the Two Roof Slopes are Different

Sometimes the main roof and the adjacent roof have the same slope. But often they are different: the main roof might be 45° and the dormer roof 33° (*Figure 8*). If the two roofs are different, then we end up with two different backing angles instead of one backing angle. Here's how we figure these.

Deck Angle

The formula for the adjacent deck angle is:

D = inverse tangent (tan SS ÷ tan S)

So, if the main roof slopes at 45° and the adjacent roof slopes at 33°, then

D = inverse tangent (tan 45° ÷ tan 33°)
D = inverse tangent (1 ÷ 0.649407)
D = inverse tangent (1.53986)
D = 57°

The main deck angle formula is:

DD = inverse tangent (tan S ÷ tan SS)

DD = inverse tangent (tan 33° ÷ tan 45°)
DD = inverse tangent (0.649407 ÷ 1)
DD = inverse tangent (0.649407)
DD = 33°

SS	is the slope of the main roof
S	is the slope of the adjacent roof
D	is the adjacent deck angle
DD	is the main deck angle
R1	is the slope of the valley (or hip)
C5m	is the main backing angle
C5a	is the adjacent backing angle

Hint   D + DD = 90˚

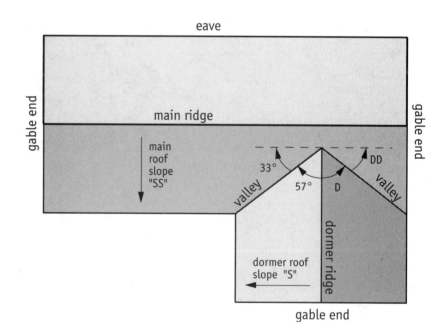

Figure 8: I have added shading here to make it easy for you to "see" the roof slopes, but the angles shown are measured level, in plan, not on the roof shape.

Slope of the Valley (or Hip)

Next, figure the slope of the valley, R1.

The formula for this is:

R1 = inverse tangent (tan SS x sin DD)

R1 = inverse tangent (tan 45° x sin 33°)
R1 = inverse tangent (1 x 0.544639)
R1 = inverse tangent (0.544639)
R1 = 28.6°

On my calculator I pushed: 45, tan, x, 33, sin, =, inv, tan — and out comes 28.5744°, which I rounded to 28.6°.

Backing Angles

Next, figure the backing angles for this valley. Because the two roofs have different slopes, we will have to figure two backing cuts—one for the main roof and another for the dormer roof.

The formula for the main backing angle is:

C5m = inverse tangent (sin R1 ÷ tan DD)

C5m = inverse tangent (sin 28.6° ÷ tan 33°)
C5m = inverse tangent (0.478300 ÷ 0.649407)
C5m = inverse tangent (0.736517)
C5m = 36.4°

The formula for the adjacent backing angle is:

C5a = inverse tangent (sin R1 ÷ tan D)

C5a = inverse tangent (sin 28.6° ÷ tan 57°)
C5a = inverse tangent (0.478300 ÷ 1.5398649)
C5a = inverse tangent (0.3106117)
C5a = 17.3°

When you need two different backing angles for a valley rafter, make certain you have been careful in layout. It makes a VERY big difference which side of the valley has which angle. Use the main backing angle (C5m) on the side of the valley that meets the main roof. Use the adjacent backing angle (C5a) on the side of the valley that meets the dormer (*Figure 9*).

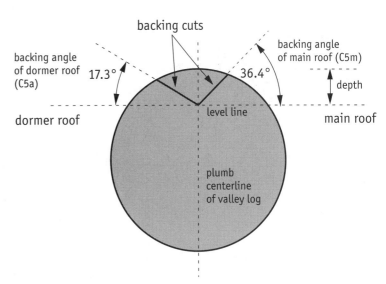

Figure 9: In this example, the main roof slopes at 45° and its backing angle is 36.4°. The dormer roof slopes at 33° and its backing angle is 17.3°.

Cutting the Valleys

Cutting backing angles on logs is tricky. Here's why: for each cut you have only one chalkline to follow, and you don't know exactly how deep to go. This is a *blind* cut.

Figure 10: When working with logs we use centerlines for the layout. Here is a simple roof plan—note how the centerlines all meet at one point.

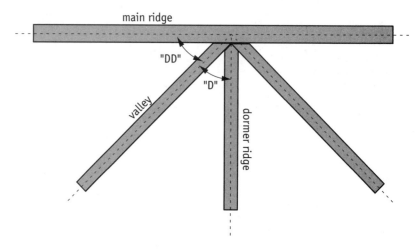

main ridge

"DD"

"D"

valley

dormer ridge

Here's some tricks that makes cutting backing angles more accurate. First, roll the valley log so that the cut you are about to make (start with either one) is plumb. Hold the chainsaw bar so it is plumb, and cut to a depth that is shy of the depth of the bottom of the *other* backing cut.

You are guessing here. And it is best to guess too shallow rather than too deep. If you cut too deeply, then the valley will be weakened—and you definitely want to avoid that. It is best if the two backing cuts *just* meet each other at the center of the valley *(Figure 4)*.

Next, roll the log so the other backing cut is plumb, and cut it—again keeping a bit too shallow. Use a pry bar to break out the wedge of wood. Then, if you need to, cut a bit deeper on one side or the other, and pry the remaining wood out.

Now, with most of both backing cuts exposed, it is much easier to go over each side again carefully with the chainsaw to make the two cuts meet cleanly at the bottom center of the valley. Sight down the valley from both ends to correct any high spots you see. If you cut too deeply, however, there is no way to fix it—don't even try, or the valley will become crooked.

Figure 11: A complex roof. The floorplan is cross-shaped and the two main ridges are glue-lams. This makes for four valleys and eight jack-purlins, and most of them can be seen in this photo.

Take a look at Color-Figures C-8 & C-32 for valleys, jack purlins, and jack rafters. A "jack" is a shortened piece found in roofs that have hips or valleys, and a "common" is a full-length rafter or full-length purlin.

For More Information

Complex roof framing is very well known by timber framers. It is not always the same thing—since they are working with timbers that have 90° edges, and we work with round logs.

On the Internet, go to: ***www.tfguild.org*** From the Guild's home page, search for "hawkindale." You will find a ready-made Hawkindale spreadsheet that you can download onto your computer for free.

There are many angles that timber framers use that have no application in log building. Just use: D, DD, S, SS, R1, C5a, and C5m—and ignore the many other angles on the spreadsheet that you won't need.

Figure 12: Complex roofs can be difficult—but they sure look great. Old Faithful Inn, Yellowstone National Park.

Log Building Tools

Log building requires some tools that are easy to find (like chainsaws), and also some tools that are highly specialized—like scribers. In this chapter I'll help you decide what tools you need, and offer suggestions on where to buy them.

Safety: All of the power tools that are used for building log homes are dangerous, and many of the hand tools can be risky as well. Wear safety glasses, hearing protection, and steel-toe boots. When using a chainsaw I also wear kevlar chaps and a helmet with face screen. Read and understand the manuals and warnings that come with your power tools. Avoid dangerous chainsaw kickback by never cutting with the top of the nose of the bar. Develop safe working habits. But, reading about tool use is not a substitute for hands-on training by a professional—get hands-on training.

Chainsaws

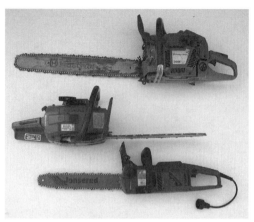

Figure 1:
Top: Husqvarna 362XP with 24" bar.
Middle: Husqvarna 254XP with 18" bar.
Bottom: Jonsered 2016 electric with 16" bar.

Husqvarna   800-GET-SAWS   www.husqvarna.com
Stihl   800-467-8445   www.stihl.com
Jonsered   503-653-7791   www.jonsered.se

Hint   With practice, you can easily start Huskys when they are hot. They flood easily, so use a half-choke or even a quarter-choke, not a whole choke when re-starting the saw. It also helps to stop a Husky with its choke, not the off-button.

A top-quality chainsaw of the right size and in top working order makes all the difference. Without question, the way you will feel about log building (and to a large degree, your skill and speed at notching) hinges on your saw and chain.

I have taught more than 600 people to cut notches, and I have seen chainsaws of almost every model, age, and condition. I'll be blunt—an average student with a great saw does better than an outstanding student with a poor saw.

Husqvarna and Stihl are the saws that I recommend. Most other chainsaws are not suited for log building. Buy a professional model saw, not one designed for homeowners *(Figure 1)*.

My favorite notching saw—the Husqvarna 254XP—is no longer available, and the Husky 357XP is the next best saw I've used. In fact, in some ways it's even better: it has slightly more power, and less vibration.

Each saw has its own feel and character. These differences are not easy for beginners to recognize, but they are real and important. Stihls are easy to start—when cold or hot. They have a *softer* suspension than Huskys—the handles have a more flexible attachment to the motor, and the bar also has a softer connection to the motor. Stihls drive like a Cadillac. My choice for heavy ripping is a Stihl: the big Stihls (bigger than 80cc) have power, are easy to start, and have soft suspension.

Husqvarna saws are more difficult to start than Stihls, especially when they are hot. Huskys also have a harder feel to their suspension, which I really like. I have more control over the bar and chain—it's like there is a more direct link between what my hands are trying to do, and what happens. Huskys drive like a Ferrari. My choice for notching is definitely a Husky: great power-to-weight ratio, high chain speed, finesse, and superb control.

Size of the Motor

44cc to 66cc is the correct range of size for a notching saw. Saws that are smaller than 44cc are probably too small for building with logs. Saws larger than 66cc are heavy and too difficult for beginners to control with accuracy. Many professional log builders cut notches with saws that are 66cc and larger—but it is just not needed because almost all of the cutting done while notching is only a few inches deep.

A saw that you love for cutting firewood is a very poor choice for log home construction if it falls outside the size and chain recommendations made here. We need speed, finesse, and light weight, not brute horsepower. And keep in mind that when you are notching you can be running a saw for several hours a day—it's easier not having to carry that extra weight.

Bar Length and Bar Shape

A bar that is 16" (40cm) long is about right for notching. Huge 26" (66cm) diameter logs can be notched with a 16" bar because most of the cutting is done less than half-way through the log. Bars 20" (50cm) long or longer are difficult to control while notching, and will make it difficult and slow for you to learn notching techniques: cut, carve, brush.

If you own just one saw and one bar, then cutting saddles often determines what length bar you will use—those are some of the widest cuts you'll make. For cutting the long groove, 14" of bar length is plenty, and some professional log builders use a saw with a 13" bar for grooves.

If you use .325 chain, then get a *narrow profile* bar—it is not as wide as the standard bar—only about 2⅜". Unfortunately, it can be difficult to find narrow profile bars for ⅜" pitch chain. Carving-type bars are not recommended for cutting notches. Use professional bars only (they can be flipped over, unlike homeowner-type bars that cannot be used with the lettering upside-down).

For ripping, cutting door and window openings, and trimming log-ends you'll want a long bar—one that is 6" to 8" longer than the diameter of your biggest logs. So, for example, if your largest butts are 24", then a 30" long bar is about right.

Chain Oil

Use chainsaw bar oil to lubricate the bar and chain—do not use waste oil, motor oil, or hydraulic oil. Bar oil should be a clear brown color (not red). It is *tacky* or *stringy* which keeps it from slinging off the chain (and getting all over your logs). Waste oil is full of worn metal filings and carbon and will permanently stain the logs.

If the oiler on your notching saw can be adjusted, then turn it all the way down. This keeps you from slinging excess bar oil all around your corner notches. Because your logs are clean, free of grit, and the bark is removed you do not need much lubrication. (Felling trees and cutting dirty firewood requires more bar oil.)

Saw Chain for Log Building

There are thousands of styles and sizes of chain for chainsaws. A special chain was even designed just for cutting the trees knocked down on Mount Saint Helens that were covered with volcanic grit. There are chains made just for waterlogged trees.

There are no chains made especially for log building, but there are a few chains that serve quite well. Keep in mind, however, that there are thousands of chain

Figure 2: Parts of a chain. To the left is a side-view of two cutters. Below are two cutters in cross-section. The rounded one is a *chipper*, and the other has a sharp corner and is a *chisel*. Chipper is better for log building.

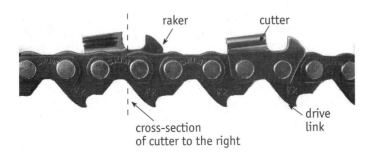

raker cutter

cross-section of cutter to the right

drive link

chipper chisel

models that are *poor* for log building. Unfortunately, your saw shop does not know which chains work best for log building—and this means that you have to know what to buy, and not rely on their advice. You are the expert, not them.

Here's how to avoid the wrong chains. First, do not use *chisel* or *full-chisel* chains—in cross-section, chisel teeth have a sharp angle. All new chainsaws come with chisel chain, which is not good for people who are learning to build log homes. Some professional log builders use chisel chain (though I do not). With chisel chain it is much more difficult to learn the chainsaw techniques of carving and brushing. Chisel chain is fast cutting, but it is not as suitable for the cutting we will be doing—I don't like a tooth that aggressively cross-cuts into the wood. You should start with *chipper* chain *(Figure 2)*.

The best chains for log building are chipper chain or semi-chisel (also called *micro-chisel* or *rapid-micro* by some manufacturers). In cross-section, the teeth are rounded. Figure 3 is a chart of some of the chains that are suitable for log

pitch	Oregon	Carlton	Bailey's	Totals	Stihl
.325	20VB, 21VB, 22VB	K1C, K2C, K3C	325RO	Z520PS, Z5288PS, Z523PS	23RM, 25RM, or 26RM
3/8	72DP, 73DP, 75DP	A1C, A2C, A3C	38RO	Z880PS, Z888PS, Z883PS	33RM, 35RM, or 36RM

Figure 3: Notes about this chart:

(1) Where there are 3 numbers listed in the table above, they are for the three gauges: 050, 058, and 063—your bar will use only one of these gauges;

(2) Total brand chains might be listed as "TT" instead of "Z"—they are the same chain;

(3) The last portion of the chain part number is usually the number of drive links, and it has been left off here because I don't know how long your bar is;

(4) The Carlton chains for 3/8 pitch might be listed in their catalogue as A1EP, A2EP, or A3EP instead of "C."

building. To use the chart, first you need to know whether your saw and bar are .325 pitch, or ⅜" pitch—it will be stamped on the bar. Then just use that row of the chart *(Figure 3)*.

The best all-around saw chain I have ever used for notching, cutting saddles, brushing, and ripping are the Carlton chains. Many chainsaw dealers do not know the difference between chisel and chipper style chains, and some stock only chisel chain. They will not want to order a 100-foot spool of the correct chipper chain for you unless you pay for the whole spool (enough to make about 30 chains).

To buy chain you will also need to know your chain *gauge* (the width of the groove in your bar, it will be .050, .058, or .063); and the number of drive links in your loop of chain (or you can use the length of your bar).

If your local store is unsure, or tells you "there is no such chain available," or "this one is almost the same," or says that Carlton (or Oregon, or Stihl) chain is junk, then take your business somewhere else—call Twin City Saw (651-645-3531 or 888-673-0451, they'll ship it to you) or Bailey's (800-322-4539). Total and Carlton chains are distributed by Tilton, Inc.— to find a local dealer call 800-447-1152.

Lately I have been using a *low profile* Oregon chain (91-VS) that works great for long grooves. (I still prefer the Carlton chain for corner notches). This style of chain works like the *thin kerf* circle-saw blades you can buy for table saws: they remove less wood, which means that they don't have to work as hard: they cut faster, and with less effort.

SAFETY WARNING

Note that ANSI B175.1, UL-listed safety chains are designed to reduce dangerous chainsaw kickback. They have special rakers or guard links or both built into the chain. Some of the chains listed in the chart are safety chains and some are not. Ask your dealer for advice.

If you are a professional log home builder then these differences are significant and useful. To be efficient you'll need one saw for cutting long grooves (~65cc with 14" bar and Oregon 91-VS chain), one saw for cutting notches (~55cc with a 16" bar and Carlton K2C chain), and a third saw with a long bar for saddles, ripping, and door and window openings (>70cc with a long bar). But those three saws will cost perhaps $2000, which is reasonable for a professional.

To begin, though, I recommend you start with just one saw (about 55-62cc) with a narrow profile 16" bar and Carlton A2C chain. Then buy one long bar for your saw to trim log ends (flyways), rip, and cut door and window openings. This saw is not ideal for the tough work of ripping logs in half—but it can definitely do it, and there really isn't all that much heavy ripping when you build. I have built several log homes using just one saw and two bars.

Chainsaw Gas

Premium unleaded gas (the highest octane) is best for saws. In some areas, high octane gas has alcohol/methanol mixed in (these are sometimes called *oxygenated* fuels). It is best to not use oxygenated gas. Follow the manufacturer's label for mixing fuel mix-oil with gas—I always use either Stihl or Husqvarna brand mix-oils—not some off-brand that might be okay for lower-rpm lawn mowers but not for high performance chainsaws.

Electric Handtools

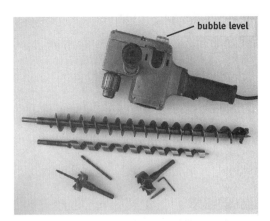

Figure 4: Above: Milwaukee Hole Hawg drill—note the bubble glued to the drill to help you drill plumb; 2" x 24" Northwest auger; $^7/_8$" x 16" Lenox auger; and two Self-Feed bits. On the left, the short feed screw has been replaced with a pilot bit from a hole saw. On the right is the short feed screw that comes with Self-Feed bits, and the hex wrench to change it.

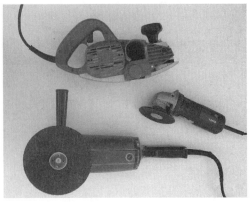

Figure 5: Makita curved-base planer (top); Bosch 4$^1/_2$" sander; and Makita 7" grinder (bottom).

Drills

You'll need a large and powerful drill to auger holes for through-bolts and electrical runs. Milwaukee Hole-Hawg ($300) is one of the favorites *(Figure 4)*, but Porter Cable and DeWalt make ½" models, too. Be careful with these tools because they have enough torque to throw you off a building if the bit catches.

Auger Bits

Northwest Manufacturing makes good long auger bits—up to 54" long and 2" diameter (available from Schroeder and others)—24" of twist is a good length for average-size logs. Try to simplify your needs—a 1¼" auger can be used for through-bolts and for electrical cable. The B&A carbide-tipped augers sold at Workhorse Supply are good bits. Milwaukee Self-Feed bits are good for drilling shallow countersinks and mortises (available in diameters from ½" to 4⅝", $30 to $140). The Self-Feed screw can be replaced with a hole-saw pilot bit so that you can start a Self-Feed countersink at an oblique angle to the log—useful when building trusses and stairs.

Sanders and Grinders

For cleaning up knots, scars, and saddles you'll need either a powerful sander-grinder or a curved-base planer, or both *(Figure 5)*. For grinders, get one that is 7" or 9", and has at least a 13-amp motor (about $160+)—you need a machine you can really push, and the grinders with less than 10-amps just won't last. I like a 4½" sander for trimming to the scribe lines of long-grooves. Many models are made, my current favorite is the Bosch ($90) because it has a smaller diameter barrel that is easy to hold, and a slide switch that you can lock "on." I use it with a hard-rubber backing pad and 24-grit discs.

Curved Planer

Makita makes a curved-sole planer that makes saddles (scarfs) and knots look great, once you get the hang of it. It is not sold by US Makita dealers, but is available in Canada and some other countries (Model #1002BA).

Hand Tools

Scribers

If you become a professional builder, then there are two models I recommend: the Timmerhus Ultra-Scribe ($425) and Jim Grieb's Gearhead scriber ($300) *(Figure 6)*. Both have professional features—both legs write, the scribers are always plumb, the tips can swivel without changing the scribe-setting. Grieb's scribers use Fisher "Space Pen" ballpoints in each leg—the best writing tips I have ever used—they write on wet wood and don't wear down or break off. Ultra-Scribes can hold either indelible pencils or felt pens. For amateurs, Lee Valley's Veritas scribers ($65) are sturdy, affordable, and easy to use. They must be re-set for plumb each time you sharpen a pencil, or change the scribe-setting distance.

Chalkline/Plumb bob

Use chalk colors blue, orange, or green. Dark red chalk is permanent, so use it only where it won't be seen. The cotton string that comes with chalklines can be replaced with 100-pound-test braided dacron fishing line. Rough-up the new string with 150-grit sandpaper and it will hold more chalk.

Some chalklines can double as a plumb bob—hang it on the log end to help you decide when the log is in the center of the wall *(Figure 7)*. But if there's a breeze, chalklines just aren't heavy enough to hang plumb. Buy several 2-pound (1 kg) lead *downrigger* weights, tie a strong string to each, and leave one in each corner of the building (so you don't have to search for one each time you set a new log on the wall).

There are some great Japanese chalklines available in catalogues—they all feature a pin (instead of a bent clip) for holding the *dumb* end of the line. Nice.

Rules

I always have a 12" flexible steel ruler in my back pocket—I use it for testing fits and finding hang-ups, laying-out saddles, drawing straight lines over the curved surfaces of logs, measuring shoulder heights, swatting bugs, and more. The General 1201ME ruler is perfect—try one. When you need a flex rule longer than 12", try a graphic-arts or art supply store, or cut a 3-foot piece of steel banding (used to hold heavy items onto pallets) *(also see Figure 10)*.

Pencils

For scribing and writing on wet logs you'll need *indelible* pencils. Several brands are available, my favorite is the extra-hard Veritas from Lee Valley.

Chisel

¾" or 1" wide carpenter's chisel with a guard (this chisel should be 9"-11" long) used for scoring notches *(Figure 7)*. This is not a mortise chisel, or a butt-mortise chisel. It's more like a paring chisel or general carpenter's chisel. An excellent 1" bevel-edge chisel is the Stanley UK Sheffield-Steel "Best Contractor Grade"—but it's tough to find. A good chisel for curved lines is a Japanese "scribing" chisel (Japan Woodworker #01.309.24, $55)—unlike a gouge, this chisel can be easily sharpened on a flat stone.

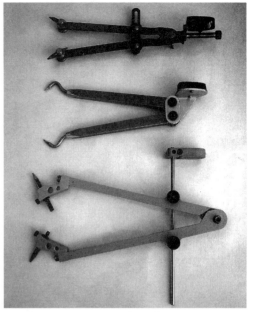

Figure 6: Scribers, from the top: Lee Valley Veritas with indelible pencils; Grieb Gearhead scribes with ballpoint space pens; Shure-Ultrascribes with indelible pencils (also can hold Staedler Lumocolor 318 felt pens).

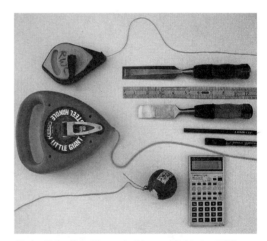

Figure 7: Small and large chalkboxes; 1" chisels; 12" flex ruler; Construction Master calculator that works in feet, inches, and fractions of an inch; indelible pencil and Lumocolor 318 pen; and a simple plumb bob—a lead ball on a string.

Figure 8: A Stanley lead-shot dead-blow sledge hammer. A Garland #3 rawhide mallet; and an inexpensive all-plastic & lead dead-blow. All-wood mallets just don't have the heavy thwack or easy control of a dead-blow mallet.

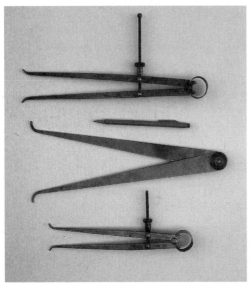

Figure 9: Top and bottom are spring-nut inside calipers; middle is a Starrett firm-joint 12" caliper.

Figure 11: Micro-Mill attachment clamps to the chainsaw-bar, is guided by a 2x6, and helps you make a straight and square cut in a log wall.

Mallet

I like dead-blow types with lead-shot, or rawhide faces with cast iron—my favorite is a 2½ pound Garland (Travers #99-005-003), it feels great and will last a lifetime (or more)—replaceable water buffalo rawhide inserts! Stanley makes a wonderful lead-shot dead-blow sledge hammer—an incredible tool, it is almost non-marking on logs *(Figure 8)*.

Calipers

You'll need an inside calipers with 8" to 12" legs—the brand doesn't really matter. The best inside calipers have a quick-nut for fast adjustment (Starrett #73B) or a tension-joint (Starrett #27-12)—Travers Tool stocks these *(Figure 9)*.

Drawknife

Barr Quarton makes the best drawknives ($135, and worth it) *(Figure 10)*. And no wonder, Ed Shure of Timmerhus tells me that he helped perfect the design (available from Schroeder or Barr). I also like the Künz Cooper's spokeshaves, which I modify by opening the throat wider and rounding the sole—I use them to clean up difficult grain around knots.

Chainsaw Attachments

For years people have been inventing attachments for chainsaws: drills, winches, peelers, planers, and more. I have been, for the most part, unimpressed with how these work and how long they last. But Accutech's Micro-Mill is the exception—I really like it. It can be used as a *lumbermaker*, but it excels at making the vertical cuts for doors and windows *(Figure 11)*. Screw a 2x6 guide to the log wall, clamp the Micro-Mill to your chainsaw bar, and, with care, you'll have a straight and square cut.

Figure 10: From the top, Künz Cooper's spokeshaves (Highland Hardware and others); large Barr drawknife; 12" flex rule; flexible Japanese square; 36" flex rule; Travers digital level.

Axes

I don't use axes frequently, but I own several. My favorite broadaxes (for hewing logs flat) are antique American axes, but Gränsfors Bruks makes very good modern axes (Schroeder & others).

Log Frogs

Also known as log *cleats*—in the old days builders used *log dogs* that were pounded into logs, now they use frogs. I like Eric Latocki's "Latocki Froggies" available from his website www.ptlogcrafters.com and from Schroeder *(Figure 12)*.

Figure 12: Top: log cleats; bottom, Latocki froggies.

Peavies & Cant Hooks

You'll want several of these. Get a few 4-foot cant hooks and a couple 5-foot peavies. My favorite peavey is the Dixie brand *rafting* or *Bangor-style* with a one-piece socket that is easy to push down between a pile of logs to pry them apart (Schroeder and others) *(Figure 13)*.

Crow Bar

I like a good crow bar/pry bar, and use them a lot for shifting logs into final position. The best, by far, are Gränsfors Bruks 36" remodeling bars (Lee Valley) *(Figure 14)*. Unlike cheap prybars, these are tempered their full length, and they have a thin, splayed end that can really reach into tight spaces. These bars are fabulous.

Boots

Get steel toes and non-marking soles (most black soles leave skids marks on logs). I do not recommend *loggers* boots (the ones with studs) for log building. Red Wing makes great steel-toe work boots: comfortable and absolutely non-marking.

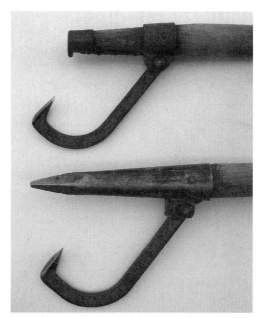

Figure 13: On the top a Dixie cant hook; on bottom a Dixie "Bangor" peavey.

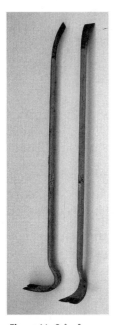

Figure 14: Gränsfors Bruks prybars.

Figure 15: Some log building tools cannot be bought at a store. These are "iron butterflies" or "helicopters" and are used for lifting logs slightly before scribing. You'll need to have them made for you by a local welder.

Where to Buy Tools

New Tools for Log Builders

Schroeder Log Home Supply, Grand Rapids, MN 55744, 800-359-6614. www.loghelp.com Full-service log home supplies: tools, sealants, gaskets, finishes, furnishings, scribers, peavies, auger bits. Get their catalogue. Gary and Kathy Schroeder, owners. Very highly recommended.

The Log Home Store, 800-827-1688, 503-843-3608, www.aLogHomeStore.com Tracy Johnston has assembled a full line of tools, books, and supplies for do-it-yourselfers and log home professionals. Excellent.

Lee Valley Tools, 613-596-0350 or 800-871-8158. I really like this big catalogue. Much, much more than just log tools. www.LeeValley.com Veritas tools (scriber, pencils), Gransfors Bruks axes, the Tove Renovator's prybars I like, and more.

Magard Ventures Ltd, 8365 Domagala Road, Prince George, BC V2K 5R1, 250-962-9057. Lots of log building tools, axes, Starrett and Mackie scribes. No website.

Timberwolf Tools, PO Box 258, Freeport, ME 04032, 800-869-4169. Stocks many specialty power tools like big beam planers, huge circular saws, drill guides, auger bits. www.TimberwolfTools.com

Garrett Wade, 161 Avenue of the Americas, NY, NY 10013, 800-221-2942. Fancy, great, expensive woodworking tools. www.GarrettWade.com

Woodcraft Supply Corp., www.woodcraft.com 800-225-1153. Good, general purpose carpentry and woodworking, plus a few log home tools.

Highland Hardware, 800-241-6748, www.HighlandHardware.com Not a lot of log building tools, but one of the few that has the Kunz Cooper's 18" spokeshaves in stock (#033915 $33)—these are good for fine peeling near knots.

The Japan Woodworker, 1731 Clement Ave, Alameda, California 94501, 800-537-7820 www.japanwoodworker.com I use their "scribing" chisel (gouge).

The Wood House, PO Box 801, Ashton, ID 83420. Basic log building tools.

Accutech Innovations Inc., Micro-Mill chainsaw attachment and accessories, 23 Gladiola Drive, Carlise, ON L0R 1H1 Canada, 905-690-6069 or 866-202-2345 dz@accutechinnovations.com   www.AccutechInnovations.com

Northwest Mfg, PO Box 299, Kelso, WA 98626, 360-636-0390. Makes specialty tools like long auger bits, log calipers, etc. Their scribers are a bit small and lightweight for my work—I prefer the Timmerhus and Grieb scribers below. No website.

Grieb "Gearhead" Scribes, Made by Jim Grieb, they use ballpoint "space pens" that can write underwater, upsidedown, so a wet log is not a problem. These scribes don't need to be plumbed, and feature rotating ends to change the "angle of attack." Wonderful to use. Call Jim, US eastern time zone, 231-529-6974.

Timmerhus, Ed Shure "Ultrascribes" PO Box 18748, Boulder, CO 80308, 303-449-1336. "The first of its kind." Not always in stock.

Barr Tools, Barr Quarton is a blacksmith and tool maker. His large drawknives are the best anywhere. Period. 800-235-4452 or 208-634-3641 (The Barr drawknife look-alikes sold by others are not recommended.) www.BarrTools.com

KMS Tools sells Makita log building tools, and more. Curved-bottom electric planer 1002BA $350US (in 2006, and expect to pay some US customs duty, too). 604-522-5599 or 800-567-8979 (Vancouver, BC, Canada). www.KMSTools.com

Spiral-grain "scribe" (grain indicator, or scratcher) — Gateway Graphics 416-674-2552, sales@gateway-graphics.com www.Gateway-Graphics.com

Antique Tools

Try junk shops, flea markets, antique stores. You might be lucky enough to find an old broadaxe, slick, or framing chisel. Here are a few dealers who specialize in antique tools:

The Fine Tool Journal; 27 Fickett Rd, Pownal, ME 04069. Includes articles in every issue about some type of old tool. Sells some, too www.FineToolJ.com 207-688-4962

Bob Kaune Tools; 360-452-2292, www.antique-used-tools.com

Philip Whitby's The New Boston Tool Room; 603-642-4054   www.newbtr.com

Bud Steere Antique Tools; 110 Glenwood Drive; North Kingstown, RI 02852.

Two Chiselers; Bob Finch 1864 GlenMoor Drive, Lakewood, CO 80215 303-232-1932

Martin J. Donnelly Antique Tools, 800-869-0695   Antique edge tools sold online at www.MJDTools.com   I often shop here.

eBay, www.ebay.com can be a good place to bid on (and maybe buy) old tools

Other Tools, Fasteners, etc:

Bailey's, PO Box 550, Hwy 101, Laytonville, CA 95454, 800-322-4539. Good prices on chainsaw and logging supplies: saw chain, safety gear, bars, chainsaw attachments. website is www.baileys-online.com

Seven Corners Ace Hardware, St. Paul, MN 55102, 800-328-0457. Fantastic 550-page catalogue, great prices, especially on electrical power tools--they have it all, 15,000 items in stock. www.7Corners.com

Cumberland General Store, www.CumberlandGeneralStore.com   Olde Time stuff.

Jamestown Distributors, 28 Narragansett Ave, PO Box 348, Jamestown, RI 02835, 800-423-0030. This is where ship builders shop, but also many other tools, long auger bits, huge galvanized bolts, & good prices. www.JamestownDistributors.com

Travers Tool Company, 800-221-0270 have deadblow and rawhide mallets at low prices; also calipers and flex rules. And they have a digital level that you can zero at any angle (model PRO-360, catalogue #57-020-505) www.travers.com   I have found the digital level for as little as $190. The price and the name on this tool varies a lot (TTC, Mitutoyo, Macklanburg Duncan, etc), but they all are identical.

Portland Bolt & Manufacturing has the biggest lag-screws and timber bolts I've found (3/4" to 1-1/2" diameter! and longer than 18"), also nice Ogee cast-iron washers, and more www.PortlandBolt.com   800-547-6758 or 503-227-5488

Log Construction Resources

In this chapter you'll find some of more resources that will help you with your log home project. There are a number of helpful books about log home design and how to be the general contractor for your own home. (I have not reviewed any videotapes or DVDs here because I have not yet seen any that I can endorse wholeheartedly.)

I strongly recommend you get hands-on training from a professional teacher (and this is different than a professional *builder*), and you'll find my recommendations in this chapter. There are many log building schools, but I have listed only those teachers and workshops that get my first-hand approval. You won't go wrong with the ones on my list.

There are engineers and architects that specialize in handcrafted log homes, and you'll find several pages of names and contact information. I recommend you hire a designer who is skilled and experienced with handcrafted log homes. It's strange that we'd never hire a surgeon who was performing their first operation, but many people eagerly hire an architect who will be designing their first log home—will this novice design be yours? I hope not.

Hire someone with experience because there are a lot of peculiarities to designing and engineering wooden structures. It will make a big difference to the success of your project to hire a person who knows what they're doing.

Finally, I'll finish up with a hodge-podge of other resources: chinking material, gaskets, log grading services, and a few Internet web sites. (Tool Resources are listed on pages 230-231.)

The Internet is a very deep source of information. But you must be *careful*. Some of the log home builders and log home schools that look great on the web are bad choices. Some log home "discussion groups" are dominated by log-home bullies—I guess they think that yelling their strong opinions is better than seeking true understanding. Is assaulting the best way to assist? I don't think so.

And there is even a large "log builders association" website that has no professional log home builders! Triple-check everything you find on the web. The websites that I have listed in this chapter are reliable.

CHAPTER TOPICS

Books & Magazines

- How to build log homes
- Log building in the past, and in other countries
- Magazines and newsletters
- Engineering, preservation, and maintenance
- Timber framing
- Carpentry, rustic furniture, and more
- Being an owner-builder, house design, tools, the business of building, and the kitchen sink
- Where to buy books

Log building schools & workshops

Architectural design for log buildings

Engineering for log buildings

Other resources

Hint: Start your looking at   www.LogBuilding.org

There is a LOT of log building information gathered there that is hard to find anywhere else: R-value of logs, Gin Pole construction, Engineering Design Values, Seismic Research on log wall stability, Wood Drying, Mold, Fire-Resistance of log walls, how-to Video clips, Window Installation tips, log home finishes . . .

Books & Magazines
Reviewed by Robert W. Chambers

1. How to Build Log Homes

Handbook of Canadian Log Building
By Dan Milne. It remains one of the better beginner's books. Many good color photos and simple descriptions of techniques and useful drawings. Some of the material is way out of date (it was published in 1984)—there is no mention of kerfing to control checking of logs, the notch shown is the round notch, which is not used much (the saddle notch is the notch of choice), bare fiberglass used in the grooves (don't do this!), and there is nothing on underscribing. The Log Home Guide Information Center, and the Log Home Store have it for $20.

Building With Logs and The Owner-Built Log Home
By Allan Mackie, Firefly Press. *Building With Logs* is the book that started it all. Much on Allan's philosophy on log homes and life; the newest edition has some more up-to-date technical information that the earlier ones lacked. It is a classic of the frontier, do-it-yourself style. About $25.   I like Mackie's *The Owner-Built Log House* (2001, $25) more than his *Building With Logs*. It is still not "modern" log building. Look to Mackie for *why* you build more than *how* to build.

Notches of All Kinds
By Allan Mackie, 1977. A how-to, step-by-step book on many log joints (dovetail, round-log, splices, trusses)—contains some material that his *Building with Logs* lacked. Some notches (the square notch, or lock notch) are much simpler than this description makes them seem. There is better info on log truss construction elsewhere, but this is an interesting book. $22

The Craft of Modular Post and Beam
By James Mitchell, Hartley and Marks Publishers, 1998, $25. An interesting book on how to build using short logs (logs that are shorter than the walls). Well written, clear, good drawings, detailed instructions. Stuff on Piece-en-Piece, expanded-style. I do not like every technique, but it is the best and only book on the topic. Originally published in 1983 as *The Short Log and Timber Building Book*.

Your Log House
By Vic Janzen, 1981, republished in 1999. Some good ideas, some not good ones, and a lot has changed in the past 25 years. Not the best how-to book, $27. Try Amazon to find this book.

Complete Guide to Building Log Homes
By Monte Burch, Published by Outdoor Life. $20. Save your money. The cover photo makes you think the book is worth buying. I learned nothing. It contains bad ideas, and some are even dangerous. Good appendix on sources for tools, etc. One of the biggest selling log home books of all time—oh well.

Log House Plans
By Allan Mackie, 1979, Log House Publishing. Page after page of blueprints and photos. Some of them look dated now, but still a useful book for ideas. $25

A Log Builders Handbook

By Drew Langsner, Rodale Press 1982. $10. This is a book about how Drew built his own dovetail, hewn, chinked home. Some good material on hewing, broadaxes, splitting shingles. Good photos and drawings. Not a text book.

Building a Log House in Alaska

By Alex Carlson, Cooperative Extension Service, Univ. of Alaska, Fairbanks, Publication # P-50A, 1977. Good little book for the homesteader, or goldminer. Not bad, really, though dated now. Still, if you have no electricity, and no generator, there is lots of stuff in here about hand tools. I like it.

Log Bridge Construction Handbook

By M. M. Nagy and others, Forest Engineering Research Institute of Canada, 2601 East Mall, Vancouver, BC V6T 1Z4 Canada, 1980. A one of a kind book—how to build bridges using only trees, and that will support heavy logging trucks. Okay, so not everyone needs to know this. But, if you do . . . . www.feric.ca

Log Home Guide Magazine and Books

Allan and Doris Muir used to publish a magazine and books on log building, like Dan Milne's *Handbook of Canadian Log Building*, one of the better early textbooks. Their publishing company stopped printing, and some past stocks of their books are available from Schroeder Log Home Supply, Grand Rapids, Minnesota 800-359-6614 or 218-326-4434, and from The Log Home Store 800-827-1688 or 503-324-0922. A "Millennium Issue" of the magazine was published. The Log Home Guide Information Center, including Doris's huge library, is now at 1107 NW Fourth St, Grand Rapids, MN 55744.

Alaska Northwest Publishing Company

They publish many books on Alaska. One is the *Alaska Log Home*—which has the best photos and drawings of any how-to book I've seen, but is somewhat lacking in instructions. Sales and Circulation Office, 130 2nd Ave South, Edmonds, Washington 98020 USA.

American Log Homes

By Art Theide and Cindy Teipner, 1994. A re-make of a coffee-table book printed earlier. Good photos. Helpful chapter on resources—architects and builders. The newer book, *The Log Home Book*, is probably better overall.

Log Home Appraisal Training Guide

By the Staff of Marshall Swift, 1998. ISBN 1568420706. I haven't seen a copy, but it is surely the only book on the topic, so if you have an appraisal question or problem, maybe this book can help.

Alaska Log Building Construction Guide

By Mike Musick, Alaska Housing Finance Corp, no date (about 1998). A good source of information on energy efficient log home construction. Much of the book deals with how to meet Alaska government requirements, so it won't apply to the exact code in other places, but if it's energy efficient in Alaska, chances are your climate is a lot warmer! Good drawings and photos, well produced on glossy paper, and spiral bound. It has information on preventing air leakage in log homes that is not in print anywhere else.

Log Houses: Canadian Classics

By Peter Christopher and Richard Skinulis, $40. Excellent color pictures of old and new log construction—probably my current favorite picture book. Good place to browse for ideas.

Picture Book of Log Homes

By Allan Mackie. Log homes photographed—great place to see the standards by which to measure your work. You won't believe how terrible the interior design is in these homes (some must have been done by Elvis Presley's decorator), but the logwork is good. A book to dream by. I prefer Cindy Theide's picture book or Christopher's *Canadian Classics*. Out of print. Try The Log Home Store, or Amazon.com.

The Log Home Book

By Cindy Teipner-Thiede and Arthur Thiede, Gibbs-Smith Publisher, Salt Lake City, 1993. Excellent color pictures of old and new log construction, chinked and scribe-fit. Short description of log home construction in the back, with some useful drawings. Good lists of log home architects, builders, and suppliers. $30

Small Log Homes

By Robbin Obomsawin, Gibbs-Smith Publisher, Salt Lake City, 2001. Excellent pictures of about 15 small log homes. In the back are floorplans. I love small places. Some of these are not very small, though—some of them are more than 2000 square feet. But one is only 24 square feet! $22.

Best Log Home Plans, and other books

By Robbin Obomsawin, Gibbs-Smith Publisher 2002. There are some very good designs in this book—I have built several of them for customers. You can buy these stock plans from Robbin, or have her make custom changes for you. $25

Robbin has also written *Log Home Classics* ($22) and the *Not So Log Home* ($40), which is about "log detailing" style construction. My own tastes run to horizontal, handcrafted log cabins, so big log McMansions and non-log homes don't appeal to me as much. But Robbin is a good designer, and her numerous books are useful resources for all things log.

A Boy's Big Book of Jigs

By John Boys, ILBA 2005. One-of-a-kind book that collects many useful, interesting, and way-out jigs, tools, and gizmos made by log builders for log building. This book is aimed straight at the heart of professional log builders. Chapters include tools for scribing and layout, cutting and drilling, holding and lifting, building log stairs, and lots more. 3-ring binder $95 (Canadian) available only from the International Log Builders Association 800-532-2900. "Work smarter, not harder!" is the motto here.

2. Log Building in the Past and in Other Countries

Japanese Joinery

By Yasuo Nakahara, Hartley and Marks, 1990 $30. Japanese timber framing in detail drawings. If you want to get fancy--or just dream--get this book. It will make you wonder how they do it, and then tell you how. Eye-opening. Almost 400 pages.

Testaments in Wood—Finnish Log Structures at Embarrass, Minnesota
By Wayne Gudmundson, 1991, Minnesota Historical Society Press, St. Paul, MN. A photo book. Finns populated northern Minnesota and built some very nice log buildings there around the turn of the century. Until recently, the local volunteer fire department had been practicing their techniques by burning these buildings down!

Building With Logs
By Clyde Fickes and Ellis Groben, 1945, Facsimile reprint in 1981 by Shorey Books, 110 Union Street, Seattle, Washington 98101 ($5). Did you think that log building was dead until recently? Wrong! In the 1930's and 1940's there was plenty of great log building happening, as this booklet proves. Details how to use bubble scribes with indelible pencils (in 1945), covers all the basics including settling, notches, scribing. Not at all up-to-date, but a very interesting historical pamphlet.

Park Structures and Facilities
U.S. Dept. of Interior, National Park Service, by Albert Good, 1935. Reprinted several times, and easily available in 2006. It is an encyclopedic survey of the structures built in US National and state parks during the 1930's by public works programs (WPA, CCC, etc). Many of these are log buildings, & many are still standing. Hundreds of pages of drawings, blueprints and photos of picnic shelters, bridges, cabins, outhouses, museums, etc. Great designs, great looks. One of my favorites. Deluxe copy $55 (6 pounds!), or as a budget reprint (3 pounds) re-titled as "Patterns From the Golden Age of Rustic Design" for $20.

The Craft of Log Building
Written in German, many years ago by Hermann Phleps. Beautifully published and translated into English by Lee Valley Tools. Every serious log-builder should have this book. The drawings are gorgeous. Buildings are centuries old; hundreds of notching styles are shown. Text is dry and academic (concerns the history of the craft in Europe and Scandinavia), but the drawings and captions are two thirds of the book and are very useful. Buy it from Lee Valley Tools, though any bookstore can order it.

Architecture in Wood
Edited by Hans Jurgen Hansen, $40 (in 1971) The Viking Press. Out of print, but available used (see sources at end of this book list). Large format book, but not a coffeetable book—it has chapters written by experts: Norwegian log building, wooden homes and barns in France, Germany, US; churches and cathedrals, Russia, and so on. Many drawings, B&W photos, and some color photos.

A Reverence for Wood
By Eric Sloane, 1965, American Museum of Natural History, Ballantine Books, NY. Sloane loves old wooden buildings—especially barns. There is some interesting stuff in here about draw-pegging, timber building, shake roofs. It is not a text book, but you will learn some things. He has also written a book on early American tools.

Building the Chateau Montebello

By Allan and Doris Muir, great story of the construction of one of the largest (the largest??) log building in the world—a hotel in Canada. Many GREAT old photos. $20.

The Wooden Architecture of Russia

By Alexander Opolovnikov. 1989, Harry Abrams, Inc. Fantastic picture book of spectacular Russian log buildings. Churches with 23 domes, bridges, barns, forts, homes. Wow. Out of print—so search Amazon.com (Amazon sells used books, too). Large format, loads of color photos. If you're crazy about log construction this is a *must have*. Expect to have a hard time finding a copy, and then cherish it when you get it! Try Biblio or Amazon for a copy.

Durant, The Fortunes and Woodland Camps of a Family in the Adirondacks

By Craig Gilborn, 1981, North Country Books. This is not the tale of a guy building his log home. This is the story of a dynasty and the extraordinary log lodges he had built for him in the New York lake country between 1875 and 1900. The Durants owned more than 700,000 acres. They had their own steamboats to get from one camp to the next. Fair black and white photos—The Adirondack Furniture book photos are much better. We could learn much by studying the design of these buildings and use what we learn today. They have a scale that is nearly perfect. They fit their settings in a way that few homes today do. The roof of Kamp Kill Kare is the epitome of shelter for a log home. Learn from these designs.

Tomsk—Texture in Wood

By Werner Blaser. Birkhauser Verlag, 1994. Fantastic picture book of spectacular Russian log homes—these black and white photos are artwork. The city of Tomsk is a living museum of log building treasures—carvings all over.

Norwegian Wood, A Tradition of Building

By Jerri Holan, 1990, Rizzoli International Publications. The best black and white photos of log homes anywhere. The text is of more interest to architects or historians than to builders. But it is the best book about Norwegian log building traditions that is available in English. Definitely worth owning.

The Real Log Cabin

By Chilson D. Aldrich, new edition with Harry Drabik, Nodin Press, 525 North Third St, Mpls, MN 55401, 1928 and 1994. A classic, long out of print, now back. Not up-to-date construction, but a faithful telling of how log buildings were made in the heyday of the 1920's and 1930's. The new author has added a lot of commentary material, and it runs side by side with Aldrich's writing.

3. Magazines and Newsletters

Log Building News

Newsletter published by the International Log Builders Association (it was the CLBA for 20+ years). Emphasis on technical stuff. The best, state-of-the-art information on scribe fit log building published anywhere in the world—written by builders for builders. Back issues #1-58 available for $6 each. Comes free with annual membership in ILBA, 800-532-2900, or 250-547-8776, PO Box 775, Lumby, British Columbia, Canada V0E 2G0, e-mail: info@logassociation.org

The Journal of Light Construction

A bimonthly journal of light construction. A damn fine magazine. Each issue covers a special topic (computers, roofing, chimneys, plumbing, etc.) in depth, plus has a dozen regular columns on design, construction law, new products, the business of building. I read it cover to cover. It holds no punches—and is not afraid of hurting advertiser's feelings. PO Box 278, Montpelier, Vermont 05602.

Fine Homebuilding Magazine

The magazine of fine homebuilding. How-to on design, tools, construction. Written by builders for builders. They love timber frames, but are a bit cool towards log homes—a New England regional bias against log homes that they market throughout North America. The Taunton Press, Newtown, CT 06470.

Joiners Quarterly

A small-scale specialty magazine that deals with timber, log, straw, clay, and boat joinery. Technical articles. Had my articles on full-scribe dovetail log homes. Snowville Road, West Brownfield, Maine 04010.   www.foxmaple.com

Log Home Living

A magazine with great photos and ads (many of them manufactured log homes). Home Buyer Publications, 4451 Brookfield Corporate Drive, Suite 101, Chantilly, VA 22022

Log Home Design Ideas

Magazine subscription 800-310-7047, PO Box 500, Missouri City, TX 77459 www.loghomemag.com or www.lhdi.com

4. Engineering, Preservation, and Maintenance

Understanding Wood

By Bruce Hoadley, Fine Homebuilding Book, 1980, $20. A beautiful book—well written, well-edited, great drawings and photos (a trademark of Fine Homebuilding). Contains the scientific info we need on shrinkage, preservation, strength, etc. written so we can understand it. The only one of its kind.

Log Span Tables

By Allan Mackie, Norman Read, and Tom Hahney. Published by the ILBA. Revised with much info added by Tom Hahney. Good little pocket-sized book of span tables and other engineering data. A must have for every serious log builder—it has the engineering stuff for round log beams that is found no where else. Revised edition 2001.

Wood Handbook: Wood As an Engineering Material

USDA, Forest Products Lab, Ag Handbook # 72. This should be in every log builder's home. The Forest Products Lab (in Madison, WI) puts out lots of information on wood. Much of it is in here: drying, shrinkage, decay, cutting, gluing, beam strength, etc. When I don't know where to find something about wood, this is where I look first. The 1999 edition is available free on the web (and you can download each chapter as a PDF file). Web site: www.fpl.fs.fed.us

Wood Structural Design Data

National Forest Products Association, 1619 Massachusetts Ave NW, Washington, DC 20036. A great resource for the properties of wood beams, spans, loading, strength of plank (2x6 T&G) floors and roof decks, etc, etc. I use this book when I am checking on floor joist sizes, beams, etc. Very useful.

The Log Home Owner's Manual:
A Guide to Protecting and Restoring Exterior Wood

By Jim Renfroe, Advance Marketing, 1994. Distributed by Wood Care Systems, 1075 Bellevue Way NE, Suite 181, Bellevue, WA 98004. A good, practical handbook on how to care for a log home. Plenty of basics and useful tips: pressure washing, chemicals, preservatives, insects, stains, oil finishes, and much much more. How to avoid problems and how to fix problems—written by a professional in the field. $20

Design Values for Wood Construction

National Forest Products Association, 1619 Massachusetts Ave NW, Washington, DC 20036. Every tree has different structural properties, and this lists all them all. A must for engineers.

The Log Home Maintenance Guide

By Gary Schroeder, The Countryman Press 2003. A very useful book on how to maintain and repair log homes. Plenty of color photos and drawings; straightforward and practical descriptions of insect, bird, and water damage; how to repair and replace chinking. Really excellent information for understanding finish problems and surface repairs like sanding, waterblasting, sandblasting, bleaching, and more. Every log home owner should read this book every few years. If they did, they just might avoid having to hire a log repair specialist. ($25).

5. Timber Framing

The Timber-Frame Home, Design, Construction, Finishing

By Tedd Benson, 1998, A Fine Homebuilding Book, Taunton Press. A follow-up to his earlier book. This one is more about concept and design of homes and how this affects the design of bents; the earlier book more nuts and bolts on how to lay out and cut joinery. Plenty of good pictures, and drawings. Good information on applying a skin to the frame, routing plumbing and heating, and electric. A very good general overview for contractors, or home owners, too. $35

The Timber Framing Book

By Stewart Elliott and Eugene Wallas, Housesmiths Press, 1977, PO Box 157, Kittery Point, Maine 03905 $12. This was one of the first books on the subject written since the 1800's. It is still a pretty good book. Good photos and drawings. But buy Benson's if you're going to own just one book.

Building the Timber Frame House

By Tedd Benson, 211 pages, $20. This is the book that re-started timber framing. A good book, not a great one (it lacks organization). But it is the book I use for truss design, floor calculations, etc. Tells you how to do the numbers on loads in a way you can understand and do yourself. Great details on how to cut the joints—the ins and outs of mortise and tenon joinery.

Timber Frame Construction

By Jack Sobon, Garden Way Publishing 1984. After Benson's book, this is the other book on timber framing. More emphasis on hand tools—Sobon's interest. A good book . Sobon also wrote *Build a Classic Timber Frame* that some prefer to *Timber Frame Construction*. ($20).

A Timber Framer's Workshop

By Steve Chappell, Fox Maple Press 1995, 120pages. Steve has taught some of the best timber framers around—this is his spiral bound workshop manual for beginners. Well organized, very good drawings (especially of the exploded joints), some engineering. Probably best if used along with a hands-on class, but good by itself, too. Steve understands it, does it, and can teach it and write about it.

Timber Framing

Like *Log Building News*, this has the best, most timely, technical info. A classy newsletter published by the Timber Framer's Guild, sent free to TFGNA members, or can be ordered by subscription for about $20 per year. www.tfguild.org

6. Carpentry, Rustic Furniture and More

Adirondack Furniture, and the Rustic Tradition

By Craig Gilborn. 1987, Harry N. Abrams, Inc. Sometimes called "twig furniture," this is fantastic, rustic, truly American. Many historic pieces beautifully photographed. Too bad there is not more on how they were constructed, or at least a few photos from the underside. Still, if you want a dream book of log, bark-on, or twig beds, tables, mirrors, chests, chairs, and the like, there is no better book. Much of this furniture is in log homes, so the photos show the interiors of significant historical log homes, an added benefit.

Rustic Traditions

By Ralph Kylloe, Gibbs-Smith Publisher, 1993, $25. Rustic furniture, log furniture, bark and twig. Good color photos will you give you ideas about the history and the possibilities of making rustic furniture. Also by Kylloe: *Rustic Furniture Makers*, but I like *Traditions* better.

Modern Carpentry

By Willis Wagner, The Goodheart-Wilcox Company, 1983 $45. The best textbook on carpentry I have read—covers all you should need to know about doors and windows, wall-framing, roof framing, kitchens, foundations, stairs, chimneys. It has it all, and so is a great place to start when you know only a little. Highly recommended. 600 pages.

Roof Framing

By Marshall Gross, Craftsman Book Company, 1986. Roof framing is probably the most complex subject in house-building. I have seen many books on the topic, but this is the first one that I understood. It explains all the math you need to know, how to use a rafter square and hand-held calculator, and how to cut rafters from dimensional lumber. Step by step you go from the most simple roofs to the most complex. If you want to cut your own roof (instead of hiring someone to put up pre-manufactured trusses made from 2X material), then this book can teach you how. But plan ahead and start learning before your log walls are done!

7. Being an Owner-Builder, House Design, Tools, the Business of Building and the Kitchen Sink

The Alternative Building Sourcebook

By Steve Chappell, Fox Maple Press 1998, 140 pages. Steve has collected information on building with wood, straw, clay in one place. This is sort of a Whole Earth Catalogue of natural/recyclable/low toxic materials, tools, techniques, and schools. Very little on logs. Fox Maple 207-935-3720.

The Antique Tool Collector's Guide to Value

By Ronald Barlow, 1985, Windmill Publishing Co., El Cajon, CA 92020. If you are interesting in owning old tools (and using them), this book will help you get a feel for what they're are worth at auction and in flea markets. Lots of interesting reprints from old tool catalogues.

The Real Goods Independent Builder

Designing and Building a House Your Own Way, by Sam Clark, Chelsea Green, 1996, 515 pages, $30. An excellent handbook on how to go about designing and building your own home: a manual for owner-builders. Starts with the basics and builds on these. Includes info on making cost estimates, negotiating a contract, but plenty of practical stuff on how to build. As much here as an encyclopedia on construction, but it is very easy to read and understand. I highly recommend this book for everyone who wants to build their own home.

A Pattern Language

By Christopher Alexander and others, Oxford University Press 1977, $45+, 1200 pages. Here's the idea: if you look at the buildings made by people in many cultures, throughout history, you will find that some things are used often (archways, private bedrooms, greenhouses, windows, columns, doors). By observing these patterns and taking note of which work well (make life secure, comfortable, or enjoyable) and which don't work (make people nervous, uncomfortable, or are ugly), then you can help people design their own homes. So this book is a dictionary of 253 patterns. It includes instructions on how to use it to design your own home. It will sure help you avoid some easy mistakes; and I have no doubt that your home will be easier and more enjoyable to live in if you follow the Patterns that appeal to you.

The Passive Solar Energy Book

By Edward Mazria, Rodale Press, 1979, $15. 430 pages of everything you need to know on the topic. The recognized authority. As a reference book it is unbeatable.

The Natural House

By Dan Chiras, Chelsea Green Publishing, 2000. One chapter on straw bale, one on log, one on adobe, cob, cordwood, rammed earth, and so on. A good introduction to each method, with plenty of pointers on where to learn more. An encyclopedia of natural building materials, that is not very detailed on any, but the best first place to look.

The Builder's Guide to Running a Successful Construction Company

By David Gerstel, A Fine Homebuilding Book, 1991. Loads of good, practical information on how to organize and operate your small construction business: bidding, contracts, bookkeeping, managing subs, getting good jobs. The first book you should read if you're going into the log home business for yourself. For a FIRST rate building contract, though, the ILBA has one for sale 800-532-2900.

The Sauna

By Rob Roy, Chelsea Green Publishing, 1996 $20. How to build freestanding saunas of cordwood, earth or timber, sauna stoves, and how to take a sauna. Good stuff. (By the way, the author tells us, it's pronounced "sow na".)

Finnish Fireplaces

By Albert Barden and Heikki Hyytiainen, Building Book Ltd, 1988. The Finnish fireplace (which is really a masonry stove, or furnace) is beautiful, efficient, and proven—they are GREAT. I have a soapstone heater in my home. Good photos, good drawings.

Where to Buy Books

Your local bookstore — They can order it for you, if it is in print.

Schroeder Log Home Supply 800-359-6614, 218-326-4434 Good one-stop shopping for many log books, and more. Good catalogue available. Web: www.loghelp.com

The Log Home Store 503-324-0922, 800-827-1688 Log home supplies, books, and etc. Book list available at: www.aloghomestore.com

Magard Ventures Maurice Gardy has it all: log home supplies, books, and tools. phone 250-962-9057, fax 250-962-9157 Prince George, BC, Canada. No website.

Log Home Guide Information Center 800-345-5647, or 218-326-4626 Order on-line at www.lhgic.com Grand Rapids, MN.

Biblio On the web at www.biblio.com they say they have more than 30 million out-of-print, rare, and used books for sale. I found copies of Thiede's "The Log Home Book" for $4, and Obomsawin's "Best Log Home Plans" for $6.

Amazon www.amazon.com Has every book, almost all of them at a discount—only available on the Internet, so you must have a computer to buy from Amazon. Also has an extensive list of, and access to, used and out-of-print books.

Summerbeam Books 315-462-3444 or 877-272-1987 Specializes in books about building (especially timber frame), and has 1400 titles. Catalogue available. Charlotte Cooper www.SummerbeamBooks.com email: char@summerbeambooks.com

Log Building
Schools and Workshops

Radomske School of Log Building
Del & Helen Radomske
1231 Philpott Road
Kelowna, British Columbia, Canada, V1P 1J7
250-765-5166 phone
email: info@okslb.ca   web: www.OkanaganSchoolOfLogBuilding.com
The best long workshops that I know of—6 to 12+ weeks. Write or call for a free brochure.

Robèrt Savignac
217 Mabel Lake Road
Lumby, British Columbia, Canada, V0E 2G5
phones: 250-547-8750
email: logbob@telus.net
Robèrt teaches in Canada, where he lives, and also in the USA and overseas. He has taught many workshops for Native Americans, First Nations, and tribal peoples. He often travels to teach, unlike many log building instructors. I have authorized Robèrt to teach my patented Accelerated Log Building methods— Robèrt and I are the only two people who are allowed to teach Accelerated Log Building.

Lloyd Beckedorf
Moose Mountain Log Homes
PO Box 26, Bragg Creek, Alberta T0L 0K0 Canada
403-932-3992 (phone), 403-932-9299 (fax)
email: info@moosemountain.com   web site: www.moosemountain.com
Lloyd teaches once a year in his building yard, often a 2-week course. Recommended.

International Log Builders Association
PO Box 775, Lumby, British Columbia, Canada V0E 2G0
Contact: 800-532-2900 (US and Canada) or 250-547-8776;
email: info@logassociation.org   web site: www.logassociation.org
The ILBA offers hands-on workshops at its annual conferences (in the spring) and rendezvous (in the autumn). The ILBA has tremendous log builder's get-togethers twice each year at various locations in the US and Canada. Also day-long seminars for log home customers. Eye-opening, and worth every penny.

Architectural Design for Log Buildings

Tom Hahney
Designing Change
7928 Lynwood Dr.
Ferndale, WA 98248 USA
phone/fax: 360-354-5840
tmhstuff@earthlink.net
Drafting and design, structural analysis

Mira Jean Steinbrecher
Jean Steinbrecher Architects
P.O. Box 788
Langley, WA 98260 USA
phone: 360-221-0494
jsa@whidbey.com   www.JeanSteinbrecher.com

Ed Levin
5 Crowley Terrace
Hanover, NH 03741 USA
phone: 603-643-2002 fax: 603-643-5651
elevin@valley.net
Timber frame design and engineering

Murray Arnott
PO Box 425
Guelph, Ontario N1H 6K5 Canada
phone and fax: 519-829-1758 or 866-603-3889
mail@designma.com   www.designma.com
Architectural design of log buildings

Cyril Courtois
RCM CAD Design
8285 Chelmsford Place
Chilliwack, BC V2R 3X2 Canada
phone: 604-702-1188   www.LogHomeDesign.ca
Log home design and drafting

Adrian Kelly
Kaila Drafting & Design
Box 216
Enderby, BC V0E 1V0 Canada
phone: 604-838-2172 fax: 604-838-6408
Kaila@junction.net   www.kaila.junction.net
Design services, AutoCad

Robert S. Miller
RSM Drafting
2730 Auto Road SE
Salmon Arm, BC V1E 2H4 Canada
phone and fax: 604-832-0165
Drafting and design

Ted Murray
Ted Murray , Architect
209-2211 West 4th Ave.
Vancouver, BC   V6K 4S2 Canada
phone: 604-734-4050 fax: 604-739-2514
Architectural design

Evan Wilson
Wilson Building Technologies, Ltd.
120 Crieff Pl.
Prince George, BC V2M 6W3 Canada
phone: 250-562-2504 fax: 250-562-2674
*Design, planning, tech. consult,
project management and inspections*

Leo Zagwyn
Deleoza Log Home Design
Box 472
Harrison Hot Springs, BC V0M 1K0 Canada
phone: 604-796-9779 fax: 604-796-9689

Patricia & Jeff Wiseman/Keller
Wiseman-Keller Design Corp.
PO Box 1021
Woodland Park, CO 80866-1021 USA
phone: 719-687-9009 fax: 719-687-1434
*Log building design specializing in hand-crafted
and timber frame . Also, log accents for frame
construction*

Earl Hilchey
Earl Hilchey, Architect
4805 McKnight Ave NE
Albuquerque, NM 87110 USA
phone: 505-255-3120 fax: 505-255-1710
Architectural and design services

David Salmela
852 Grandview Ave
Duluth, MN 55812 USA
phone: 218-724-7517
Award winning architect

Marjie Rozumalski
9433 County Rd. J., Minoqua, WI 54548 USA
Planning and Design Services

Robbin Obomsawin
Beaver Creek Log Homes
35 Territory Rd.
Oneida, NY 13421 USA
phone: 315-245-4112
www.beavercreekloghomes.com
Many good stock plans for handcrafted log homes

Structural Engineering for Log Buildings

Bruno Franck, PE, PhD
Carroll, Franck & Associates
1357 Highland Parkway
St. Paul, MN 55116 USA
phone: 651-690-9162 fax: 651-690-9156
carrfran@gold.tc.umn.edu
Structural design, engineering, wood structures

Curt J. Egerer, PE
Egerer & Associates
5455 Midland Rd.
Freeland, MI 48623 USA
phone: 517-695-6191 fax: 517-695-6672
Structural design, damage assessment, litigation support

Richard Rock
Rock Engineering
3525 Surrey Ct. SE.
Port Orchard, WA 98366 USA
phone: 360-871-8660 fax: 360-871-8661
Structural analysis, reports & design
also, *Plan preparation and assistance*

Robert Hodnett
Hodnett Engineers, Inc.
111 Green St.
Huntsville, AL 35801 USA
phone: 205-533-2771 fax: 205-533-2772
Structural engineer and consulting

Mark J. Mueller, PE
MJ Mueller Co. Inc.
PO Box 2747
Vail, CO 81658 USA
phone: 970-476-2627 fax: 970-476-2637
Structural engineering

Mike Thele, PE
Structural Engineering Services, Inc.
0296 Seven Oaks Road
Carbondale, CO 81623 USA
phone: 970-963-3181

Barry Houseal
BLH, Inc
10 Del Azul
Irvine, CA 92614 USA
phone: 714-756-9350 fax: 714-756-9250
109 W. Colorado Ave.
Telluride, CO 81435 USA
phone: 970-728-6922 fax: 970-728-1021

Tom Beaudette
Beaudette Consulting Engineers
131 West Main Street
Missoula, MT 59802 USA
bce@bceweb.com www.bceweb.com
phone: 406-721-7315 fax: 406-542-8955

Ed Levin
5 Crowley Terrace
Hanover, NH 03741 USA
phone: 603-643-2002 fax: 603-643-5651
elevin@valley.net
Engineering, 3-D graphics, software, timber frame

Tom Hahney
7928 Lynwood Dr.
Ferndale, WA 98248 USA
phone/fax: 360-354-5840
tmhstuff@earthlink.net
Drafting and design, structural analysis

Jennifer Anthony, PE
Fearless Engineers PLLC
201 South Fourth Street West #2
Missoula, MT 59801 USA
phone/fax: 406-721-7833
fearlessengineers@msn.com
www.fearlessengineers.com

Other Resources

Log Inspection and Log Grading

Timber Products Inspection, Inc.
1641 Sigman Road
Conyers, GA 30012 USA
phone: 770-922-8000; fax: 770-922-1290
www.tpinspection.com

Log Homes Council
National Association of Home Builders
1201 15th Street, NW
Washington, DC 20005 USA
phone: 800-368-5242, ext. 8577; fax: 202-266-8141
www.loghomes.org

Gaskets

Emseal Joinst Systems Inc.
23 Bridle Lane, Suite 3
Westborough, MA 08581 USA   www.emseal.com
phones: 508-836-0280; 800-526-8365

Sof Rod and Sof Seal
John Boys at Nicola LogWorks
phone: 877-564-4667   email: gasket@logworks.ca

Natural Log Homes
Rubber-fin tape seal
email: info@naturalloghomes.com

Denarco Inc.
301 Industrial Drive
Constantine, MI 49042 USA
phone: 616-435-8404; fax: 616-435-8404

Resource Conservation Technology Inc.
2633 North Calvert Street
Baltimore, MD 21218 USA
phone: 410-366-1146

Chinking Manufacturers

Perma Chink Systems, Inc
www.permachink.com
17635 NE 67th Court
Redmond, WA 98052 USA
phone: 800-548-1231 or 425-885-6050; fax: 425-869-0107

Sashco—Log Jam
www.sashco.com
10300 East 107th Place
Brighton, CO 80601
phones: 800-767-5656; 303-286-7271; fax: 303-286-0400

Log Builders Associations

International Log Builders Association
www.logassociation.org
PO Box 775, Lumby, British Columbia, Canada V0E 2G0
phone: 250-547-8776
Email: info@logassociation.org

Great Lakes Logcrafters Association
www.gllca.com
Kay Sellman
24355 Esquire Blvd.
Forest Lake, MN 55025 USA
phone: 612-464-6506

Log Builders Association of New Zealand
www.logbuildingnz.org.nz
Ray Bremer, Secretary
Email: ray@handcraftedloghomes.co.nz

Association of Latvian Craft
www.lak.lv
Rozu-iela 21-27
Riga, Latvia LV-1056

Korean Log Builders Association
www.logbuilder.co.kr
Jay Wan Yu
Email: jaylog@hanafos.com

Association Bois Sacre TCB (France)
www.boisbrut.free.fr/indexbuildenglish.html
Thierry Houdart
La Nouaille, Lamaziere-Basse F19160 France

Web Sites

Robert W. Chambers
    www.LogBuilding.org
    www.BuildYourOwnLogHome.com
    www.NaturalLogHomes.com
    www.LogConstructionManual.com
Free downloads, sales of "Cut-It-Yourself" handcrafted pre-scribed log home kits, free videos, log selection spreadsheets, and more.

Oikos
www.oikos.com
A great way to search for building products, especially those that are environmentally sensitive, or save energy.

Expert Witness —Mediation or Litigation

Robert W. Chambers
    robert@logbuilding.org

Brian G. Lloyd
    bg_lloyd@hotmail.com
    +1-250-549-3545   British Columbia, Canada

2000 Log Building Standards
for Residential, Handcrafted, Interlocking, Scribe-fit Construction

The International Log Builders Association, founded in 1974 as the Canadian Log Builders Association, is a world-wide organization devoted to furthering the craft of log construction. The ILBA is a registered non-profit Society in Canada and the United States. The ILBA writes and distributes educational material on log construction to individuals, institutions, and the industry. The organization is dedicated to the advancement of Log Builders and to promoting the highest standards of the trade.

It is the responsibility of every builder to understand and to conform to the best practices of the trade. These are minimum Standards for residential, hand-crafted, interlocking, scribe fit log construction. Log Building Standards are revised by the ILBA Building Standards Committee. Changes to this edition were made in January, 2000.

The ILBA has endeavored to prepare this publication based on the best information available to the Association. While it is believed to be accurate, this information should not be used or relied upon for any specific application without competent professional examination and verification of its accuracy, suitability, and applicability. The publication of the material herein is not intended as a representation or warranty on the part of the International Log Builders Association, its affiliates, or any person named herein that this information is suitable for any general or particular use or is free from infringement on any patent or patents. Anyone making use of this information assumes all liability arising from such use.

The Log Building Standards are founded on performance principles that allow the use of new materials and new construction systems. Anyone may propose amendments to these Standards. These Standards are not intended to prevent the use of any material or method of construction not specifically prescribed by these standards, provided the proposed action is satisfactory and complies with the intent of the provisions of these standards and that the material, method or work offered is, for the purposes intended, at least the equivalent of that prescribed in these Standards in suitability, strength, effectiveness, fire resistance, durability, safety and sanitation.

For further information, or additional copies of these Standards, please contact:

International Log Builders Association
P.O Box 775
Lumby, British Columbia, Canada V0E 2G0
250-547-8776 phone
800-532-2900 toll-free phone
250-547-2900 fax
info@logassociation.org e-mail

2000 Log Building Standards

Standards

Preface

1. In these Standards the word "shall" means mandatory, and the word "may" means discretionary.

2. The 2000 Log Building Standards are comprised of both the Standards and the Commentary.

Section 1 FOUNDATIONS

Shall conform to applicable building codes and accepted engineering practice.

Section 2 LOG WALLS

2.A. LOG SPECIFICATIONS

2.A.1. The minimum diameter of wall logs shall be 20 centimeters (8 inches).

2.A.2. Green or dry logs may be used for construction.

2.A.3. Logs shall have all bark removed.

Table 2.A

	RIGHT HAND	LEFT HAND
straight	less than 1:20	less than 1:30
moderate	1:20 to 1:10	1:30 to 1:20
severe	greater than 1:10	greater than 1:20

2.A.4. Spiral Grain

The following restrictions apply to the use of green logs. (Refer to Table 2.A for definitions of spiral grain categories):

a. Left-hand severe spiral grain logs shall be used as wall logs only as cut-in-half sill logs. However, left-hand severe spiral logs may be used as a whole-sill log if all four of the conditions listed below are also met.

b. Left-hand moderate spiral grain logs shall be used only in the lowest one-third (1/3) of the vertical height of a wall. However, moderate left-hand spiral logs may be used in the lowest one-half (1/2) of the vertical height of a wall if all four of the conditions listed below are also met.

c. Right-hand severe spiral grain logs shall be used only in the lower one-quarter (1/4) of the vertical height of a wall. However, right-hand severe spiral logs may be used in the lowest one-third (1/3) of the vertical height of a wall if all four of the conditions listed below are also met.

Conditions:

1) the log has two or more corner notches, and

2) the log is not spliced, and

Commentary

Section 1 FOUNDATIONS

Like all buildings, the foundation of a log building must be of sufficient design to support safely the loads imposed as determined from the character of the soil. In addition to the loads imposed by gravity, the foundation is important in connecting the building to the ground as it resists wind or seismic forces and accelerations. Therefore the connection between the building and the foundation must also be capable of resisting the sliding, uplift and overturning associated with local wind and seismic conditions.

Section 2 LOG WALLS

2.A. LOG SPECIFICATIONS

2.A.1. *Logs smaller than 20 centimeters (8 inches) in diameter are unsuited to residential construction.*

2.A.2. *For the purposes of this Standard, "dry" means moisture content equal to or less than 19%, and "green" means moisture content greater than 19%. Dry and green logs have different requirements for preventing sapstain, and have different shrinkage and structural properties that must be appropriately accounted for in design and construction.*

2.A.3. *Leaving the bark on logs promotes insect attack and makes scribe-fitting difficult. Eventually, the bark will fall off by itself, though by that time the wood has usually been degraded by fungus or insects, or both.*

2.A.4. *Spiral grain is the condition in which the alignment of wood fibers is at an oblique angle to the long axis of the log. Spiral grain is expressed as the slope of the direction of fiber alignment to the length of the log-this slope is shown in Figure 2.A.*

To determine fiber alignment, examine the log for surface checks caused by drying-surface checks are parallel to fiber alignment. Another option is to use a sharply pointed timber-scribe instrument designed for detecting spiral grain.

To determine whether a log has left-hand or right-hand spiral grain, place your right hand on the log, fingers pointing down the length of the log. You can stand at either end of the log. If the grain spirals around the trunk like a barber pole in the direction your thumb is pointing, then the tree has left-hand spiral grain. If the grain spirals in the direction your little finger is pointing, then the tree has right-hand spiral grain.

Scientific studies have shown that left-hand spiral grain logs undergo more severe distortions during drying than right-hand spiral grain logs, and this is one reason why greater restrictions are placed on the use of left-hand spiral logs (Table 2.A). Also, left-hand spiral grain logs are considerably weaker in bending and deflect more than straight-grain or right-hand spiral grain logs, though this is more critical in using logs as structural elements (joists, rafters, and timber members for example), than as wall logs.

3) no more than two-thirds (2/3) of the log's diameter is cut or removed at any opening, and

4) if any portion of this log extends beyond a notch in a wall, then the length of this extension is not more than 4'-0" (122 cm), measured from the center of the closest notch to the end-cut of this log.

d. Right-hand moderate spiral grain logs may be used as a wall log at any location in the building, except shall not be used in the top round of wall logs.

e. Straight grain logs may be used in any location.

f. The top round of logs shall be straight grain only, see also Section 2.I.4.

Figure 2.A

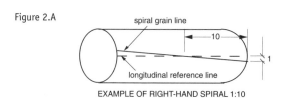

EXAMPLE OF RIGHT-HAND SPIRAL 1:10

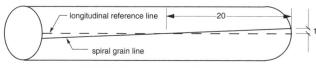

EXAMPLE OF LEFT-HAND SPIRAL 1:20

2.B. LOG WALLS
Shall be constructed of logs laid in horizontal courses, scribe-fit one to another, with interlocking notches at the corners.

2.C. SILL LOGS
2.C.1. Shall be not less than 20 centimeters (8 inches) in diameter.

2.C.2. Shall be flattened on their bottom side for their entire length to a width of not less than 10.2 centimeters (4 inches).

2.C.3. Shall not be in direct contact with masonry.

2.C.4. Shall be set on a vapor, weather, and air barrier.

2.C.5. Shall have a drip cut or flashing that directs water away from the underside of the sill log.

2.C.6. Shall be anchored to resist applicable wind and seismic loads.

2.C.7. Shall be a minimum of 30.5 centimeters (12 inches) above grade.

2.D. LONG GROOVES
2.D.1. Logs in walls shall have a continuous scribe-fit long groove along the length of each log. A long groove is required wherever a log wall separates unheated from heated space, or heated space from the exterior of the building.

Section 2.A.4 *describes the ways builders can help hold spiral logs in place in the walls. Logs that are more likely to twist are used lower in the wall, where there is more weight on them. For spiral types (a), (b), and (c) the standards can be relaxed somewhat if you use four additional methods to restrict twisting. A log that has at least two full notches is more likely to stay put than a log with one notch (Condition # 1). A full notch is more stabilizing than a spliced notch (Condition # 2). A spiral log with a window sill cut-out will not behave like it is one piece if more than two-thirds of a log's diameter is removed (Condition # 3). And, when a wall log extends beyond a notch more than 48 inches (122 centimeters) to a door or window opening, this portion of the log is more likely to twist (Condition # 4).*

Table 2.A refers to green logs, slope of grain may change as logs dry.

2.B. LOG WALLS
These Standards do not apply to walls constructed of vertical logs, or logs that are not fully scribe-fit to one another (e.g. chinked), or to Piece en Piece, or to manufactured log home kits. For more on notches see Section 4.

2.C SILL LOGS
2.C. *Sill Logs are the bottom logs of the building, the first logs above the foundation in each wall.*

2.C.1. *See also the log specifications in Section 2.A.*

2.C.2. *A continuous sawn flat provides bearing area and stability for sill logs.*

2.C.3. *Untreated wood should not be in direct contact with masonry because of the likelihood of decay.*

2.C.4. *Caulks, sealants and gaskets can provide vapor, air and water barriers.*

2.C.5. *To avoid decay, it is important that rainwater be directed away from under the sill logs.*

2.C.6. *The amount and kind of anchoring depends upon local conditions and codes. In areas of extreme wind and seismic load conditions, continuous through-bolting the full height of the log wall to the foundation can be an effective technique.*

2.C.7. *Sill logs can be prone to decay if they are too close to grade and rainwater and soil splashes on them.*

2.D. LONG GROOVES
Also known as "lateral," "lateral groove," "cope," "Swedish cope," and "long notch." The long groove is a notch cut into a log to fit two logs together along their length and between intersecting corner notches.

2.D.1. *The long groove must be continuous between notches, or between openings, such as doors. Other styles of log construction do not have a long groove, or have a groove that is not continuous—the gaps between logs are then filled with a chinking material. Scribe-fit logwork, in contrast, has a continuous long groove, and no chinking is required because there are no gaps to fill. The interior edges of the long groove are often sealed with a gasket material, and the interior cope is commonly insulated.*

Standards

2.D.2. Long grooves shall be self-draining or shall have gaskets, and in all cases shall restrict water, air, and insect infiltration.

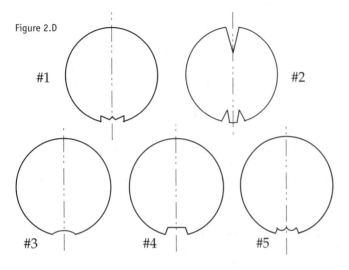

Figure 2.D

#1 #2 #3 #4 #5

2.D.3. The minimum width of the long groove shall be 6.3 centimeters (2.5 inches) and this minimum width shall extend for no more than 30.5 centimeters (12 inches) in continuous length. At all times, however, the long groove shall conceal and protect through-bolts, pins, dowels, kerfs, electrical holes, and the like, and shall be wide enough to restrict weather and insect infiltration.

2.D.4. The maximum width of the long groove shall be three-eighths (3/8) of the log diameter at each point along the log. In cases of extremely irregular log contours the width may be increased to one-half (1/2) of the log diameter, but this increased allowance shall extend for no more than 46 centimeters (18 inches) in continuous length.

2.D.5. The long groove may have the following cross-sectional profiles: rectangular, shallow cove, "W" shaped, or double-scribed.

2.D.6. The depth to which the groove is cut shall be less than one-quarter (1/4) the diameter of the log (see also Section 2.J.2).

2.E. LOG EXTENSIONS

2.E.1. The maximum length of log extensions shall be based on weather protection criteria described in Section 7.D.

2.E.2. The minimum length of log extensions shall be 23 centimeters (9 inches) measured from the edge of the notch to the end of the log overhang. This standard applies to both interior and exterior log extensions. Dovetail corner notches are exempt from this requirement.

2.E.3. Exterior log extensions shall not have a tight fit to the log extensions below. See Figure 3.B.3

Commentary

2.D.2. *Some profiles are not self-draining, that is they could trap water, and so promote decay. Such long grooves shall have gaskets to restrict water from getting into the groove. Being visibly tight is not sufficient to restrict air or water infiltration.*

2.D.3. *Narrow long grooves are difficult to seal from the weather. The groove must always be wide enough to restrict weather infiltration into kerfs, electrical holes, and the like.*

2.D.4. *Wide long grooves remove so much wood that the log is unduly weakened and may check only on the bottom of the log, which is not desired. (See also Section 2.J.)*

2.D.5. *There are many shapes, or cross-sectional profiles, for long grooves. Figure 2.D illustrates some of these. Desired traits are: sharp and strong edges along the scribe line; a reasonable minimum amount of wood removed from the groove so that the groove touches the log below only along its scribed edges with no internal "hang-ups;" and a reasonable assurance that the log will check on its top (that is, in the kerf) as it dries. (See Section 2.J for more on kerfs.)*

2.D.6. *Deep long grooves are not necessary, and can weaken a log. Note that at least one-half of the diameter of the log must remain intact after both the kerf and long groove are cut (Section 2.J.2).*

2.E. LOG EXTENSIONS
Also known as "flyways" or "log overhangs," are the short part of the log that extends past a notched corner.

2.E.1. *Overly long log extensions can be prone to decay unless adequately protected by roof overhangs, or by other means.*

2.E.2. *Overly short log extensions can be prone to having wood split off, severely weakening the notch and the corner. Interior log extensions are those that project inside a building, and exterior log extensions extend towards the outside of a building. The stability of a dovetail corner does not depend upon log extensions, and is not susceptible to having wood split off, and so is exempt from any minimum length requirement.*

2.E.3. *The end-grain of exterior log extensions can take on moisture seasonally, shrinking or swelling more than the rest of the log. If the long grooves of extensions fit tightly, then during periods of high moisture the tight fit of the long grooves along the rest of the log could be compromised. This has, in fact, been observed—tight long grooves in the log extensions and gaps in the grooves everywhere else.*

Since log extensions are not kerfed (Section 2.J.7), it is probable that log extensions will check on their bottoms—from their long grooves towards the center of the log. When logs check in this location, internal hang-ups are common. To avoid this, the grooves of exterior log extensions should have enough wood removed to avoid hang-ups after checking and slumping. See Figure 3.B.3.

Standards

2.E.4. Where a log extension acts as a support for a structural member this extension and the structurally supporting logs shall be exempt from the requirement in 2.E.3 (see also Section 7.J).

2.F. DISTANCE BETWEEN CORNERS

2.F.1. When using logs with a diameter less than 30.5 centimeters (12 inches) the distance between intersecting log walls with corner notches shall be no more than 7.3 meters (24 feet). When using logs with a minimum diameter of 30.5 centimeters (12 inches) the distance between corner notches shall be no more than 9.75 meters (32 feet). Log walls with spans in excess of these distances shall have reinforcement such as wood keys, dowels, smooth-shaft steel, through-bolts, lag screws, steel bar, or log stub-walls. All such reinforcement shall allow for settling (see Section 6).

2.F.2. Log walls with openings cut for doors, windows and passageways may require additional bracing. The loads on a log wall, and the openings cut into a log wall, will affect its structural performance and may require structural analysis.

2.G. JOINING LOGS LENGTHWISE

2.G.1. Spliced logs shall be secured to each other with bolts or other fasteners, and to adjoining courses of logs above and below with steel pins, wooden dowels, lag bolts or through-bolts in a manner that preserves the structural integrity of the wall.

2.G.2. When more than half of the logs in a corner are spliced, then engineering analysis shall be required.

2.G.3. The notch and long-groove shall at all times completely hide a splice and its fasteners, and help protect splices against weather and insect infiltration.

2.H. HEADER LOGS

2.H.1. A header log shall have no more than half of its vertical height removed at the location of openings, unless it is covered by at least one more log. In all cases, the header log shall be adequate for structural requirements.

Commentary

2.E.4. *Where roof overhangs, outriggers, or balconies are supported by log extensions, it may be necessary to have two or even three log extensions fit tightly so as to gain the structural strength needed to support the cantilevered load put on these logs.*

2.F. DISTANCE BETWEEN CORNERS

2.F.1. *Log walls gain lateral stability from corner notches at stub walls and intersecting log walls, and this is the reason for limiting the distance between notched corners. Larger logs are laterally more stable than small logs and so are allowed a longer maximum distance between notches.*

2.F.2. *Openings cut into a log wall, especially numerous, tall, or wide openings reduce the lateral stability of the wall. Some stability is gained by door and window framing (see Section 5), but in most cases other steps must be taken to stabilize the wall, especially when the wall is supporting the load of floors or roofs.*

2.G. JOINING LOGS LENGTHWISE

2.G.1. *Some walls are straight and too long to be spanned with single logs, and so logs are joined end-to-end. Better design may be to step a long wall in or out to add corner notches and allow the use of wall-length logs, thereby eliminating end-to-end splices. End-to-end butt splicing of wall logs is an acceptable practice, however, so long as steps are taken to maintain the strength and stability of the walls and corners, and the spliced joint is completely covered from view.*

2.G.3. *The completed wall must appear to be made of only continuous, full-length logs. No exposed splices or joints are allowed. All joints and splices must be completely covered by corner notches or stub-wall notches.*

2.H. HEADER LOGS

Are logs at the head, or top, of window and door openings cut into log walls.

2.H.1. *A header log has a level, sawn cut facing the opening. A header cut should not remove more than half the vertical diameter of the header log at this point unless the strength of the wall is sufficient to support the roof and floor loads placed upon it.*

Figure 2.H.1

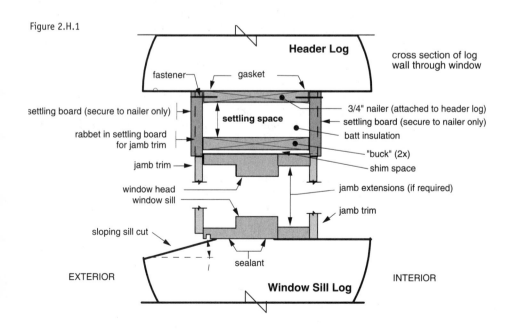

Standards

2.H.2. Openings in header logs shall be cut so as to completely cover door and window head jambs and exterior trim in order to restrict water infiltration.

2.I. PLATE LOGS

2.I.1. Wall plate logs shall be notched, drifted, pegged, lag-bolted or through-bolted to the log below to prevent movement caused by drying stress and roof thrust. Wall plate logs shall be attached with lag or through-bolts to one or more rounds of logs below the plate log so as to resist the uplift forces associated with local wind and seismic conditions.

2.I.2. Where conventional framing meets a plate log this intersection shall have an expandable gasket to accommodate anticipated shrinkage of the log plate and to restrict weather and insect infiltration.

2.I.3. The ceiling vapor retarder, where required by local code, shall be permanently sealed to the plate log with caulk or sealant.

2.I.4. Plate logs shall be straight grained wood (see Section 2.A.4.f).

2.J. KERFING

2.J.1. When building with green logs, a longitudinal kerf shall be cut on the top of each wall log.

2.J.2. The depth of the kerf shall be at least one-quarter (1/4) of the diameter of the log, and shall be no deeper than one-half (1/2) the diameter. In no case shall more than one-half (1/2) the diameter of the log be removed by the kerf and long groove combined.

2.J.3. Kerfs shall at all times be protected from weather by being fully covered by the long groove of the log above, or by a notch.

Commentary

2.H.2. *Figure 2.H.1 illustrates one way to install settling boards and avoid water infiltration.*

2.I. PLATE LOGS

Are the top logs on each wall. The roof framing rests on the plate logs.

2.I.1. *Wall plate logs are prone to twisting and shifting and need extra steps to keep them in place. Square notches and lock notches can provide restraint, as can any number of methods using bolts, threaded rod, and pegs. The number, type, size, and spacing of mechanical fasteners used for this purpose must be determined by accepted engineering practice. Continuous gable-end plate logs are very effective at resisting roof thrust, and so are recommended when it is necessary to counteract these forces. When continuous gable-end plate logs are not used, or are not used in a manner that will resist roof thrust, then this force must be restrained or eliminated by other methods.*

Roof uplift caused by wind, for example, can be counteracted by locking together the top rounds of each wall. Smooth pins such as dowels, smooth shaft steel, and wooden pegs are not sufficient for preventing uplift, and this is why lag bolts and through-bolts are specifically mentioned.

2.I.2-3. *A study of Minnesota log homes found the intersection of roof framing and the plate log to be the source of considerable air infiltration. Special steps are required to make this area weather-tight. Permanently sealing the vapor barrier to the plate log is an accepted method of reducing air infiltration and retarding the migration of water vapor. Stapling the vapor retarder to the plate log is, by itself, not sufficient.*

2.J. KERFING

2.J.1. *The kerf is usually, though not always, a cut made with a chainsaw. Logs are known to check, or crack, in those places where wood has been removed closest to the pith, or center, of the log. Kerfing is therefore an effective way to control the location of checks as green logs dry.*

Because dry logs already have seasoning checks, kerfing will not change the location of checks, and therefore kerfing usually is not required for dry logs.

2.J.2. *The kerf must be deep enough to promote checking. Note that even those long groove profiles that do not require kerfing (like the double-cut) are nevertheless required to be the depth of at least one-quarter of the diameter of the log at every point along the top of the log. (See also Section 2.D.5.)*

After a log has both the kerf and the long groove cut, there must still be at least one-half of the diameter of the log remaining un-cut. Removing more than half the diameter of the log for kerf and groove combined would weaken the log, and so should be avoided.

The amount of wood removed by the kerf (or special long groove profile) must be between 1/4 and 1/2 of the log diameter (Section 2.D.6). When the kerf is 1/4 of the diameter of the log deep, then the groove must be no more than 1/4 of the log diameter deep (1/4 plus 1/4 equals 1/2). When the kerf is 1/3 of the log diameter deep, then the groove must be no more than 1/6 of the log diameter deep (1/6 plus 1/3 equals 1/2).

2.J.3. Because kerfs are not self-draining, that is, they can catch rainwater and hold it, kerfs must always be protected by being fully covered by the groove of the log above (also see Section 2.D.3). In practical terms, this means that kerfs are never visible in a completed wall.

2.J.4. The kerf shall be continuous, or shall start 15 centimeters (6 inches) from the edge of all notches, and shall be continuous between the notches, except that kerfs need not extend into openings in log walls, or at the ends of log extensions, where they would be seen.

2.J.5. No kerf shall be required when the long-groove profile encourages checking on the top of wall logs as in Figure 2.D #2, as long as the groove and kerf along the top of the log is at least 1/4 of the diameter of the log.

2.J.6. No kerf shall be required on the top of the half-log sill logs.

2.J.7. No kerf shall be cut in exterior log extensions.

2.K. LOG WALL-FRAME WALL INTERSECTIONS

2.K.1. Log walls shall be cut as little as necessary when joined to non-log partition walls.

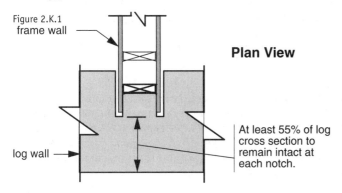

Figure 2.K.1
frame wall

Plan View

log wall

At least 55% of log cross section to remain intact at each notch.

2.K.2. Where wood is removed at the intersection of a log wall and frame wall, the log wall shall have 55% or more of its cross-sectional area remain intact and uncut. See Figure 2.K.2.

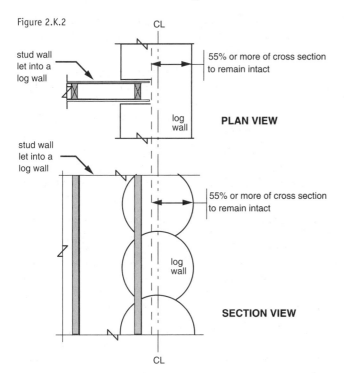

Figure 2.K.2

stud wall let into a log wall

55% or more of cross section to remain intact

log wall

PLAN VIEW

stud wall let into a log wall

55% or more of cross section to remain intact

log wall

SECTION VIEW

CL

2.J.4. The kerf should run the full length of the top of every log, either stopping before reaching a notch, or continuing through a notch. In the case of openings or passageways cut in log walls that are not covered by jambs or doors, the kerf would be unsightly—and in these areas the kerf need not extend all the way to the opening.

2.J.5. Some long-groove profiles encourage checking without kerfing. For example, the long-groove known as double-cut or double-scribed (see Section 2.D.5), removes a "V" shaped section from the top of every log. Long-groove profiles that promote checking on top of wall logs do not require a kerf, but they must comply with Section 2.J.2.

2.J.6. Half-logs do not usually check, and so do not require a kerf.

2.J.7. No kerf should be cut on any log extensions outside the building because this upward-facing cut could catch and hold moisture from rain and promote decay. The long grooves of exterior log extensions shall not be tight-fitting (Section 2.E.3), and so do not protect the kerf from water, and this is why log extensions should not be kerfed.

2.K. LOG WALL-FRAME WALL INTERSECTIONS

It is common for some interior, non-bearing partition walls to be conventionally framed with studs. This section describes how stud walls and other non-log walls should be attached to logs walls.

2.K.1. It is common for a plumb groove, dado, or rabbet to be cut in the log wall and the first stud of the frame wall to be attached to the log wall in this groove. One problem is that to have the frame wall completely seal against the log wall, the groove must be cut as deep as the narrowest long groove, and this is often close to the mid-point of the log wall. One way to avoid removing too much wood from the log wall, and unduly weakening it, is shown in Figure 2.K.1.

2.K.2. Enough wood must be left in the log wall that it is not weakened by the dado. The dado must leave 55% or more of the cross-sectional area at this intersection uncut, Figure 2.K.1.

Standards

2.K.3. Where frame partition walls are notched into opposite sides of a log wall there shall be a minimum of 122 centimeters (4 feet) between the end of one notch and the beginning of the next notch on the opposite side of the log wall, or, if closer than 122 centimeters (4 feet), a minimum of one-third (1/3) of the wall cross-sectional area shall remain intact and uncut.

Figure 2.K.3

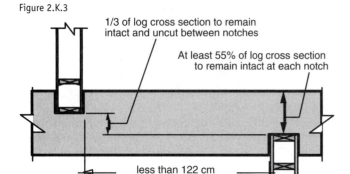

Plan View

1/3 of log cross section to remain intact and uncut between notches

At least 55% of log cross section to remain intact at each notch

less than 122 cm (4 feet)

2.K.4. In no case shall cuts go past the centerline or midpoint of the log wall.

2.K.5. Log wall-frame wall intersections must allow for unrestricted settling of the log wall (see also Section 6).

2.L. HEIGHT OF LOG WALLS
Log walls taller than two stories, or 6.1 meters (20 feet) in height, shall require engineering analysis.

2.M. BEARING WALLS
Bearing walls shall be designed and constructed to structurally accommodate horizontal and vertical forces which are anticipated to act upon the building.

2.N. PRESERVATION OF LOG WALLS
Where necessary, steps should be taken to restrict the growth of mildew and fungus on logs while the building is under construction.

Section 3 NOTCHES
3.A. SELF-DRAINING & WEATHER-RESTRICTING NOTCHES
All forms of interlocking notches and joinery shall be self-draining and shall restrict weather and insect infiltration.

3.B. NOTCHING STANDARDS
3.B.1. Notches shall have a concave profile across the notch not less than 1.5 centimeters (5/8 of an inch) and not more than 3.5 centimeters (1 and 3/8 inches).

Commentary

2.K.3. *Where two frame walls are closer than 122 centimeters (4 feet) to each other, and on opposite sides of a log wall, the cross section of the log wall, after both dadoes are cut, must have at least one-third of the wall area remain un-cut, Figure 2.K.3 Note, also, that Section 2.K.1 still applies—each single cut shall leave 55% or more of the cross sectional area at each intersection un-cut and intact. See Figure 2.K.3.*

2.K.4. *Cutting past the center of a log wall weakens it, and should be avoided.*

2.K.5. *The first stud attached to the log wall must be fastened in such a way as to allow the log wall to shrink and settle. One common method is for lag screws to be attached to the logs through vertical slots cut in the stud, not just round holes. The lag screw and washer should be attached near the top of the slot, and allowed to slide down the slot as the log wall behind shrinks in height.*

The frame wall must also allow a second floor, or the first floor ceiling, to lose elevation as the log walls shrink in height. (See Section 6 for more on settling.)

2.L. TALL LOG WALLS *should be evaluated for stability.*

2.M. BEARING WALLS
Bearing walls can be exterior or interior log walls. Roof and floor loads are the most common loads to design for, but uplift and lateral loads from winds and seismic activity may have to be considered as well.

2.N. PRESERVATION OF LOG WALLS
Green logs, in particular, are prone to attack by mold, mildew, and fungus during construction. Dry wood will not decay, and so good roof protection is very effective in prolonging the life of log walls. During construction, and until roof protection is complete, it may be advisable to use sapstain and mold preventative chemicals or processes. Additionally, the use of a sealant on all exposed end grain during log storage, construction and after all work is completed will slow the loss of moisture and reduce checking.

Section 3 NOTCHES

3.A. SELF-DRAINING & WEATHER-RESTRICTING NOTCHES
Means that notch surfaces slope in a way that restrict water from getting into areas where it can be held, promoting decay. Interlocking means that notches will tend to be stable when exposed to stresses and loads that the corner can reasonably be anticipated to experience. Shrink-fit and compression-fit notches are designed to remain tight fitting as the wall logs shrink in size as they dry. (Note that a round notch which is designed to function as a compression-fit notch also meets this criteria.)

3.B. NOTCHING STANDARDS
3.B.1. *When a straightedge is held across a notch so that it is approximately perpendicular to the long axis of the log and so that the straightedge touches the scribed edges of the notch, then the straightedge should not touch the inside of the notch at any place. In fact, the gap between the straightedge and the inside of the notch should be between 1.5 and 3.5 cm (5/8" and 1-3/8").*

continued...

<table>
<tr><td>

Standards

3.B.2. Notches shall be clean in appearance and have no ragged edges.

3.B.3. To maintain tight notches with green logs the following apply:

a. Space shall be left at the top of the notch to allow for compression.

b. Sapwood from the sides of the log should be removed to create a saddle scarf. These saddle scarfs shall be smoothly finished.

Figure 3.B.3

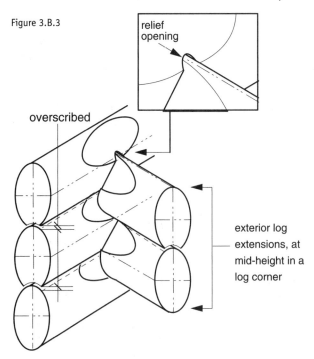

3.B.4. The amount of log to remain uncut at a notch shall not be less than one-third (1/3) the original diameter of the log, or not less than one-third (1/3) of the original cross-sectional area.

3.B.5. All forms of dovetail notches are exempt from the requirements of Section 3.B.

3.C. BLIND-NOTCHES
A blind-notch shall resist the separation of the two log members it joins, or shall have mechanical fasteners that resist separation.

Section 4 JOISTS AND BEAMS
4.A. Joists and beams, if dimensional material, shall conform to applicable building codes.

4.B. Joists and beams, if log or timber, shall conform to the following standards:

4.B.1. Shall have straight grain, or shall be right-hand spiral grain, with spiral no more than 1:12. (See Section 2.A.4 for more on spiral grain.)

</td><td>

Commentary

3.B.1…continued. *This means that the notch, when in place over the log below, should touch the log below only on its scribed edges, and should touch at no other place. (If it touches on some inside place it causes a "hang up.") The concave area created by scooping out the notch in this way not only prevents internal hang-ups, but also can be used to place materials that will prevent air infiltration through the notch (gaskets and insulation, for example)—an important consideration in all climates.*

3.B.2. *The scribed edge of notches should be sharp, strong, and cleanly cut. The edges should not crush or permanently deform under the load they support. Ragged wood fibers may indicate weak notch edges or a notch that was cut past the scribe line.*

3.B.3. *There are techniques that help keep notches tight as green logs season and dry. One technique is to remove wood at the top of a notch to allow the notch to compress onto the log below as it dries. The extra wood removed from the top of a notch creates a gap that should be nearly invisible when the corner is assembled, that is, the gap should be covered by the notch of the next log. Figure 3.B.3.*

Cutting saddles, or saddle scarfs, is another technique that helps. Saddle scarfs should not be simply chainsawed off, but should be finished to a smoother surface. See Figure 3.B.3.

3.B.4. *After a notch has been cut there shall be no less than one third of the log's original cross-sectional area or diameter at the notch remaining uncut.*

Removing more than two-thirds of the log area or diameter by notching weakens a log, sometimes even to the point where the log extensions may break off. Good log selection avoids the problem of notches that remove more than two-thirds the diameter of the log at the notch.

3.B.5. *Dovetail notches are unlike most other notches, and are not required to follow the standards of Section 3.B.*

3.C. BLIND NOTCHES
A blind notch is a log joint in which one log does not cross over or beyond the other log. Because one log does not continue past or over, it can be prone to separating from the log it is joined to. To resist separation the following methods are recommended:

1. A dovetail or half-dovetail on the blind notch to interlock with the intersecting log.

2. Hidden dowels that accommodate settling.

3. Hidden metal straps, fasteners or bolts to join the intersecting log walls together.

Section 4 JOISTS AND BEAMS
4.A. *Dimensional joists and beams (including rafters, purlins, ridges, and the like) shall conform to local applicable building codes for dimensions, load, and span.*

4.B. *Log joists and beams, including sawn timber members, shall be sized to adequately support the loads they carry.*

4.B.1. *Studies have shown that left-hand spiral grain logs and timbers are significantly weaker than straight and right-hand grain members, but it is not yet known precisely how much weaker. Therefore, left-hand grain is not allowed for these members unless it can be shown that it is structurally adequate. Straight-grain, and right-hand spiral grain up to a slope of 1:12, is allowed.*

</td></tr>
</table>

The figure is labeled with: "relief opening", "overscribed", and "exterior log extensions, at mid-height in a log corner".

| **Standards** | **Commentary** |

Standards

4.B.2. Shall be designed to resist all loads according to applicable building codes and accepted engineering practice.

4.C. Where log or timber beams are notched at an end, on the bottom face, the depth of the notch shall not exceed one-fourth (1/4) of the beam depth at the location of the notch, or less if calculations so indicate.

Figure 4.C

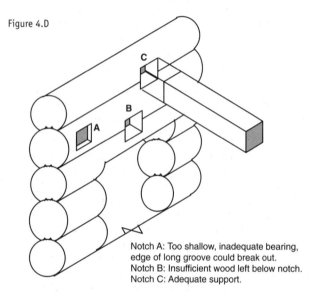

4.D. Where log or timber joists are supported by a log wall, the wall logs shall be notched to receive the joists in such a way as to prevent failure in the supporting log wall.

Figure 4.D

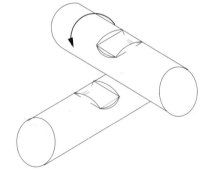

Notch A: Too shallow, inadequate bearing, edge of long groove could break out.
Notch B: Insufficient wood left below notch.
Notch C: Adequate support.

4.E. The distance, after settling is complete, from the bottom of ceiling joists and beams to the finished floor shall conform to applicable building codes.

4.F. Where a beam or joist passes through a wall to support additional floor areas or other loads, the beam or joist shall be notched in such a way that the structural integrity of both the beam and the supporting wall are maintained.

Figure 4.F

Commentary

4.B.2. *At all times, log and timber beams and joists must be designed and installed to adequately resist the loads they will experience. Joists and beams with excessive deflection can cause uncomfortable, and in some cases, unsafe, springiness in floors and roofs. Long spans are prone to excessive deflection, and in some cases a deflection limit of 1/360 of the span may not be sufficient. It is prudent to consult with an engineer familiar with wood structures for assistance in the design of complex load carrying systems.*

4.C. *Where joists and beams are notched at their ends (for example, to be supported by a log wall), no more than one-quarter (1/4) of the height of the beam shall be removed from the bottom of the beam. Less than one-quarter (1/4) shall be removed if engineering calculations require. See Figure 4.C.*

4.D. *It is also important to not remove so much wood from a log wall that is supporting a beam or joist such that the log wall itself is unreasonably or unsafely weakened. One example would be a joist above a door or window opening, see Figure 4.D.*

4.E. *Joists and beams (whether log, timber, or dimensional material) that are supported by log walls will get closer to the floor as the logs dry and shrink and the log wall gets shorter in elevation. Many local building codes specify the minimum height from the floor to joists and beams above. The height of joists and beams off the floor must conform to local building codes, if any, after settling is complete. (See Section 6.A for more on calculating settling allowances.)*

4.F. *One common log building design has floor joists that cantilever through an exterior log wall to support a balcony or roof load outside the building. It is not uncommon for the stresses which this type of beam must withstand to be at a maximum where the beam passes through the log wall. It is therefore important that all such cantilevered beams not be substantially weakened due to notching at this location. A square notch is one way to help protect the strength of the beam, Figure 4.F. Square notching does remove more wood from the log wall than other notches, and so it is important to ensure that the wall is not weakened past its ability to support the loads placed upon it.*

Log Construction Manual

Standards

4.G. Where an interior beam extends through a wall to the exterior it shall be protected from the weather so that its structural integrity is maintained. The intersection of the beam and wall shall be constructed to restrict weather and insect infiltration. See also Sections 7.F and 7.G.

4.H. Log joists and beams shall be flattened on top to a minimum of 2.5 centimeters (1 inch) where they support flooring or framing.

Section 5 WINDOW AND DOOR OPENINGS

5.A. Settling space shall be provided for all doors and windows placed in walls constructed of horizontal logs.

5.B. The settling space for windows and doors shall be covered by a cladding or trim to restrict weather and insect infiltration. In order to not restrict settling and to avoid damage to windows or doors this covering shall not be attached to both the log wall and to the window or door frame until all settling is completed. A vapor barrier shall be installed within this space, on the heated side of the insulation.

5.C. Trim at jambs shall not restrict settling.

5.D. Both sides of each opening shall be keyed vertically to withstand lateral loads, and in such a way as to allow unrestricted settling.

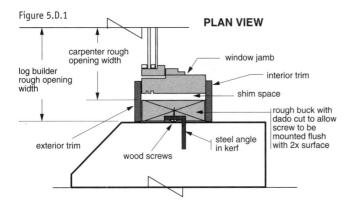

Figure 5.D.1 — PLAN VIEW

carpenter rough opening width
log builder rough opening width
window jamb
interior trim
shim space
rough buck with dado cut to allow screw to be mounted flush with 2x surface
exterior trim
wood screws
steel angle in kerf

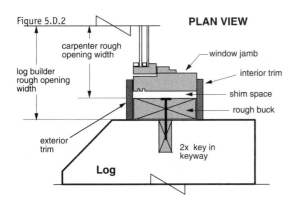

Figure 5.D.2 — PLAN VIEW

carpenter rough opening width
log builder rough opening width
window jamb
interior trim
shim space
rough buck
exterior trim
2x key in keyway
Log

5.E. All exterior sills shall be beveled to allow water to drain to the outside face of the log wall.

Commentary

4.G. *Cantilevered log beams that extend outside the building (even if they are only notched through the wall and have relatively short log extensions) need protection from decay. Metal flashings, waterproof membranes and wide roof overhangs are recommended. The top of any deck supported by logs or other structural members must slope so that water will drain in a manner that protects the house from damage. This type of detailing is important because of the susceptibility of unprotected log-ends to decay, and the great difficulty and expense in repairing or replacing such logs once degradation occurs.*

Section 5 WINDOW AND DOOR OPENINGS

5.A. *Openings cut in log walls become shorter over time as the logs dry to an in-service condition. The settling space must not have any materials in it that does not allow for the space to become vertically shorter over time. (See also Section 6 for more about shrinkage and settling.)*

5.B. *Settling spaces are typically covered by settling boards, which are pieces of trim that are wide enough to span the settling space. The settling boards can be attached to the log or to the window or door framing, but not to both. Attaching the settling board to both would not allow for the settling space to get smaller over time, and would either cause the logs to hang up, or the windows or doors to deform.*

5.C. *The sides of doors and window trim must allow for logs to settle unhindered. This means that the jamb trim on the sides of doors and windows cannot be attached to the log wall. Side trim can be attached to the window or door and to bucks, see Section 5. D.*

5.D. *Openings in log walls for door and windows need special framing to install the jambs of doors or windows; and this framing is usually called a "buck." The bucks must allow for logs to shrink and settle-typically this means that the height of the bucks is less than the height of the log opening, and the difference in these heights is equal to, or greater than, the settling allowance. (See Section 6.A for help calculating settling allowances.) The bucks are usually attached to keys of wood or angle iron that are let into the log ends of openings. Keys are required because they hold the bucks in place and because they laterally stabilize the log wall at openings: they restrict logs from moving horizontally while still allowing logs to move vertically. See Figures 5.D.1 and 5.D.2 .*

5.E. *Where a log acts as an exposed exterior window or door sill, it must shed water and slope so that it drains away from the window or door.*

Standards

5.F. The position of openings in walls constructed of horizontal logs shall conform to the following:

5.F.1. The distance from the side of window and door openings to the centerline of an intersecting log wall shall be a minimum of 25.4 centimeters (10 inches) plus one half the average log diameter.

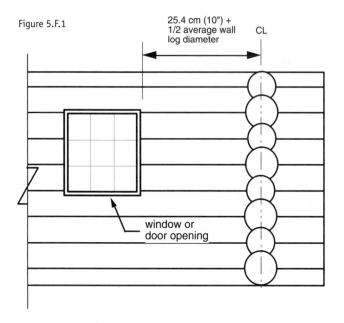

Figure 5.F.1

25.4 cm (10") + 1/2 average wall log diameter

CL

window or door opening

5.F.2. Wall sections between openings shall be a minimum of 92 centimeters (36 inches) long, or shall be provided with support in addition to the required keyways (see Section 5.D).

Section 6 SETTLING

6.A. Settling Allowance

6.A.1. The minimum allowance for settling when using green logs is 6% (3/4 inch per foot of log wall height).

6.A.2. The settling allowance for dry logs may be up to 6%, but may be less than this, depending upon the moisture content of the logs.

Commentary

5.F. *Window and Door Location*

5.F.1. *It is undesirable to have door and window openings cut too close to intersecting log wall and stub wall notches. The notched log is weakened and may split off if it is too short. (This situation is comparable to log extensions that are required to be a certain minimum length, see Section 2.E.2.) Therefore, window and door openings shall be cut no closer to the centerline of an intersecting log wall or log stub wall than 25.4 centimeters (10 inches) plus half the average log diameter, see Figure 5.F.1.*

5.F.2. *Sections of log shorter than 92 centimeters (36 inches) are prone to split, and are also unstable (since they do not contain a log corner), especially if they support loads such as those of a second floor or roof. Therefore, it is best if the sections of log wall between doors, between windows, and between a door and a window, be longer than 92 centimeters (36 inches). Sections of log wall can be shorter than this minimum if there is sufficient additional support used, but the keys required by Section 5.D do not qualify as additional support, unless they are part of a column-and-screwjack settling system.*

Section 6 SETTLING

6.A. *Settling is the term that describes the loss of log wall height over time. The principal causes of settling are:*

1) shrinkage of log diameter as logs dry to an in-service condition (also known as equilibrium moisture content, or EMC), and 2) compression of wood fibers under the load of the building. A third component is slumping, which occurs if logs check only in the long groove. Slumping is nearly eliminated by kerfing, which is one reason why kerfing is required, see Section 2.J.

6.A.1. *Green logs (defined in Section 2.A.2 as logs with greater than 19% moisture content) must be allowed to settle 6% (6 centimeters per meter, or 3/4 inch per foot) of wall height. Note that logs cannot be expected to shrink to equilibrium moisture content or completely settle by air-drying alone, but must be expected to complete settling only after a period of up to 5 years as part of a heated building. The time needed to reach equilibrium moisture content depends on a number of variables, including wood species, log diameter, initial moisture content, interior temperature and humidity, and climate.*

In general, logs do not shrink much in length, and so only the loss of diameter must be considered for settling. With extremely long logs (more than 15 meters, or 50-feet), however, it is advisable to investigate the loss of length as they dry.

6.A.2. *Dry logs (defined in Section 2.A.2 as logs with moisture content equal to, or less than, 19%) may settle nearly as much as green logs. In part, this is because of the nature of the definitions of dry and green—19% MC is a "dry" log and 20% MC is a "green" log, but these two logs will obviously differ very little in the amount they actually shrink in diameter as they approach EMC.*

It must be assumed that log walls made of dry logs will settle. Further, it should be assumed that logs stored outside, not covered by a roof, are not at EMC, and will shrink. The amount of shrinkage depends upon the difference between the actual moisture content of the logs (as determined by a moisture meter, for example) and the final in-service EMC.

continued...

Standards

6.B. Adequate provisions shall be made for settling at all openings, load bearing posts, chimneys, fireplaces, interior frame partition walls, electrical entrance boxes and conduits, plumbing vents and drains, second story water and gas pipes, staircases, downspouts, heating and air conditioning ducts, and all other non-settling portions of the building.

6.C. The log contractor shall provide information to the general contractor to help guide subcontractors in the use of techniques applicable to their trade to deal with the unique characteristics of log construction, and specifically how each trade should accommodate for settling.

6.D. All caulking and weather-sealing must account for the change in diameter and shape of the logs as they dry.

Section 7 ROOFS AND ROOF SUPPORT SYSTEMS

7.A. If constructed of dimensional material, shall conform to applicable building codes.

7.B. If constructed of log or timber, roof systems shall conform to the following standards:

7.B.1. Shall be constructed only of straight-grain, or moderately right-hand spiral grain material (see Section 2.A.4 for definitions of spiral grain).

7.B.2. Shall be designed to resist loads according to applicable building codes and accepted engineering practice.

7.B.3. Where beams are notched at an end, on their bottom face, the depth of the notch shall not exceed one-fourth (1/4) the beam depth at the location of the notch, or less if calculations so indicate.

7.C. The distance from the bottom of roof beams to the finished floor must conform to applicable building codes after settling is complete.

7.D. Roof overhang shall help protect log walls from the weather associated with the site of the building. Figure 7.D illustrates how to calculate the minimum roof overhang.

Commentary

6.A.2…continued. *Settling allowance for dry logs may be reduced from the required 6%, and the amount of the reduction allowed is proportional to the actual moisture content of the logs. Note, however, that even if the initial moisture content of the logs is equal to EMC, and the logs are not expected to shrink, the logs will still compress somewhat, and there must be a settling allowance for this compression.*

6.B. *Everything that is attached to a log wall must accommodate settling. Also, settling problems must be investigated even between two non-log items. For example, there is settling that must be accommodated between a second floor framed of 2x10's and a plumbing vent stack. Neither is log, but the floor framing is attached to and supported by log walls, and will settle. The plumbing vent stack is anchored to non-settling members in the basement or crawl space and does not settle.*

Another example is the settling between a roof framed of 2x12's and a chimney. Again, neither is made of logs, but because the roof rafters are supported by log walls, this means that the rafters will get closer to the ground as the log walls settle. Therefore, roof framing must not be attached to a chimney unless special steps are taken to accommodate settling.

The list in Section 6.B. is far from exhaustive. Every non-log, non-settling, part of a building must be examined to see if there needs to be an accommodation for settling.

6.C. *The log builder knows the special techniques involved in completing a log house and should share this knowledge with the general contractor so that the subcontractors are properly educated about settling and other potential problems.*

6.D. *Where caulks, sealants, gaskets, and the like are used in contact with logs, these joints must be designed to accommodate shrinkage of the logs without having the joint fail. Trim boards that are scribe-fit to logs shall allow for settling.*

Section 7 ROOFS AND ROOF SUPPORT SYSTEMS

7.B. *Log roof systems include, but are not limited to, log posts and purlins, ridgepoles, log trusses, and log common rafters. In Section 7, "log" also means "timber."*

7.B.1. *Severely spiral-grained logs are significantly weaker in bending strength and shall be avoided. Left-hand spiral grain logs are significantly weaker than right-hand spiral grain of equal angle. (See Section 2.A.4 for more on spiral grain.)*

7.B.2. *All log roof members shall be designed to sufficiently resist all expected loads.*

7.B.3. *Notches cut into, and any wood removed from a log beam will weaken the beam. One example of this is at the ends of a log beam, no more than one-quarter (1/4) of the depth of the beam, and less if calculations so indicate, shall be removed for a notch (Figure 4.C). It is best to consult an engineer who is familiar with wood structures for help designing log roof systems, and especially for complex roof systems.*

7.C. *Consider the original height of the beam, the involved settling height and the settling allowance (6% for green logs) to calculate the height of roof beams after settling is complete.*

7.D. *Roofs for log homes shall protect log beams and log walls from degradation caused by the weather. One good way to accomplish this is to use wide roof overhangs. The effectiveness of roof overhangs also depends upon the height of the wall and the height of the roof drip-edge. Figure 7.D shows how the amount of roof overhang shall be calculated.*

Standards

7.E. The roof shall protect all roof structural members from the weather associated with the site of the building.

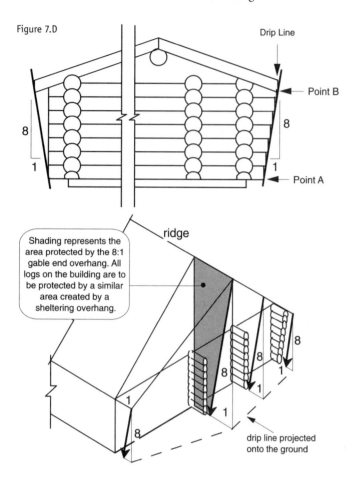

Figure 7.D

Drip Line

Point B

8

8

1

1

Point A

ridge

Shading represents the area protected by the 8:1 gable end overhang. All logs on the building are to be protected by a similar area created by a sheltering overhang.

8

8

8

1

1

1

8

1

drip line projected onto the ground

7.F. Log roof beams shall be flattened on top to a minimum width of 3.8 centimeters (1-1/2 inches) where they support lumber or finish materials.

7.G. Where log structural members pass through exterior frame walls they shall be notched slightly to receive interior and exterior wall coverings. Expandable gaskets shall be installed to restrict weather and insect infiltration. Roof members shall be designed to meet structural requirements even after such notching.

7.H. Flashing and an expandable gasket shall be used where conventionally framed gable end walls meet a plate log.

7.I. Roof structures shall be designed and constructed to resist the uplift loads associated with local wind and seismic events.

7.J. Where roof structures are supported on outriggers, which are in turn supported on log extensions, the extension log carrying the outrigger shall be supported by additional log extensions (a minimum of two extensions below the extension carrying the outrigger) in such a way as to support all loads from the outrigger in a manner other than by cantilever action, unless the log extension carrying the outrigger is designed and constructed as a structural cantilever. (See also Section 2.E.4.)

Commentary

Notes for Figure 7.D:

The criteria set forth in Figure 7.D is a minimum. This approach to calculating roof overhang is independent of roof pitch and wall height, and relies on a ratio (8:1) to define the relationship between the roof overhang and the logs to be protected. If, for example, the distance that the end of a sill log projects beyond the notch (Point A) is known, then the drip line defined by the roof overhang can be calculated by projecting a line from Point A up and out from the building at the 8 to 1 ratio as illustrated, until this line intersects the bottom of the roof plane (bottom of the rafters), then measure out horizontally here (Point B) to find the minimum roof overhang distance.

Or, if the roof overhang is known, then the maximum projection of log ends beyond the notch can be calculated by reversing the process and beginning at Point B. A reference line is then constructed down and inward toward the building at the 8 to 1 ratio until it intersects the plane of the bottom logs (usually the first floor), then measure out horizontally to Point A to find the maximum allowed length of log extensions. Also check that the log extensions are not shorter than required in Section 2.E.2. Note that the allowed length of log extensions increases as you go higher on the building. That is, log extensions may corbel out at the 8:1 ratio, if desired, though they are not required to do so. At all points around a building, this 8:1 reference line should be used, and no log or log end should project beyond this reference line.

7.E. *Log roof beams that extend to the outside of a building need protection from the weather. Purlins, ridgepoles, and posts must not extend outside the drip line of the roof unless special steps are taken, for example wrapping the log-end with a durable metal flashing. Preservative chemicals by themselves are insufficient.*

7.F. *It is impractical to attach framing lumber or finish materials to the irregular, waney round of a log. Therefore, round log roof beams shall be flattened to a width of 3.8 cm (1-1/2 inches) or more where they support other materials.*

7.G. *It is common to extend log roof beams, like purlins and ridgepoles, outside over posts to support roof overhangs. This can be a difficult spot to seal from weather infiltration as the log roof beams shrink in diameter. Gaskets help, as do shallow notches to house the sheathing and inside finished wall materials. Make sure that the roof beams are still sufficiently strong even after notching and removing wood.*

7.H. *The plate log of gable end log walls is flattened on top, often to receive conventional stud framing. It is important that the flat sawn on the plate log does not hold or wick water. A metal flashing is an effective way to direct water away from this intersection.*

7.J. *Log outriggers are roof plates outside of, and parallel to, log eave walls. Do not use just one log extension (log flyway) to support the outrigger unless it can be shown that one extension is sufficiently stiff and strong. In any case, no matter how the outrigger is supported, its means of support must be sufficient. (See Section 2.E for more on log extensions.)*

Standards

Section 8 ELECTRICAL

Shall comply with applicable codes, with accommodations where necessary for pre-wiring and wall settling allowance. (See also Section 6.B.)

Figure 8

cross-section of log wall

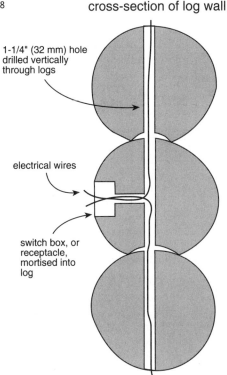

1-1/4" (32 mm) hole drilled vertically through logs

electrical wires

switch box, or receptacle, mortised into log

Section 9 PLUMBING

9.A. To comply with applicable codes, with settling considerations. See also Section 6.

9.B. A plumbing pipe shall travel through a log wall only perpendicular to the long axis of the logs, and shall be level or nearly level.

Commentary

Section 8 ELECTRICAL

Common practice is to pre-drill vertical holes in the log wall, from long groove to long groove, so that the holes are completely hidden from view and no electrical wiring is exposed inside or out. Do not use rigid conduit inside a log wall. Do not attach conduit to a log wall without allowing for settling.

Outlets and switch boxes are usually mortised into a log so that the cover plate is even with the surface of the log so they are flush with a portion of the log surface that has been flattened for this purpose, see Figure 8.

Section 9 PLUMBING

9.A. *Investigate carefully the need for settling allowances in all plumbing for log homes. It is usually preferable to run plumbing in frame walls vertically without horizontal offsets, though offsets are possible, if settling considerations are carefully made. Supply pipes to a second floor can allow for settling by incorporating a loop that opens as the second floor loses elevation. Waste and vent pipes can have a slip joint. See Figures 9.A, 9.B, 9.C, and 9.D.*

9.B. *It is usually not advisable to run plumbing waste, vent or supply pipes through or within log walls. If they must, however, pipes can run perpendicular and level through a log wall. A pipe that runs vertically up through a log wall, or a pipe that runs horizontally within a log wall (for example, lying in a long groove) can never again be serviced without cutting the log wall apart—a drastic event that is difficult to repair.*

Because supply lines are known to age, fill with scale and sometimes to leak, and because the venting of sewer gases is a matter of health and safety, it is best to not locate plumbing in log walls.

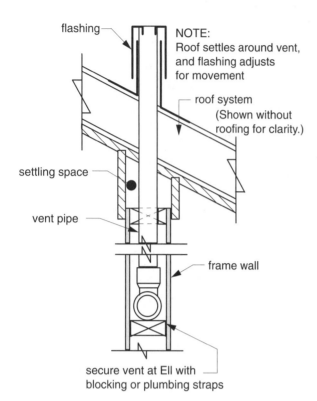

flashing

NOTE:
Roof settles around vent, and flashing adjusts for movement

roof system
(Shown without roofing for clarity.)

settling space

vent pipe

frame wall

secure vent at Ell with blocking or plumbing straps

FIGURE 9.A

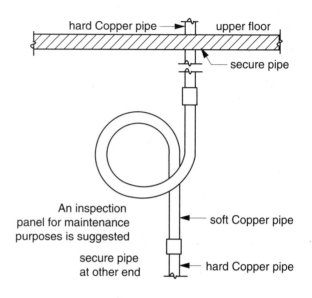

hard Copper pipe — upper floor

secure pipe

An inspection panel for maintenance purposes is suggested

soft Copper pipe

secure pipe at other end

hard Copper pipe

FIGURE 9.C

Using a combination of hard and soft Copper pipe to allow for settling in water supply lines.

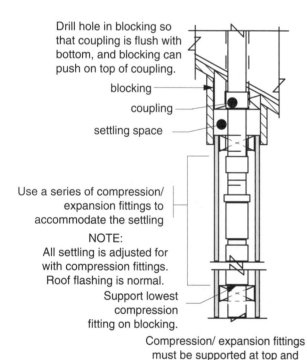

Drill hole in blocking so that coupling is flush with bottom, and blocking can push on top of coupling.

blocking

coupling

settling space

Use a series of compression/ expansion fittings to accommodate the settling

NOTE:
All settling is adjusted for with compression fittings. Roof flashing is normal.

Support lowest compression fitting on blocking.

Compression/ expansion fittings must be supported at top and bottom in some manner similar to the blocking shown.

FIGURE 9.B

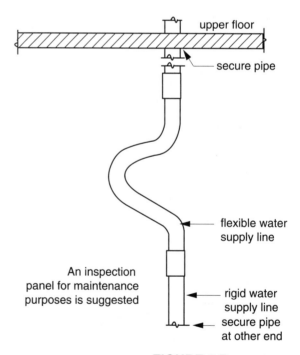

upper floor

secure pipe

flexible water supply line

An inspection panel for maintenance purposes is suggested

rigid water supply line

secure pipe at other end

FIGURE 9.D

Using a combination of flexible water supply line and rigid pipe to allow for the needed settling

Standards

Section 10 FIREPLACES AND CHIMNEYS

10.A. Shall conform to applicable codes.

10.B. No combustible materials, including log walls, shall be closer than 2 inches (5.1 centimeters) to a masonry chimney.

10.C. Flashing to conform to applicable codes, and to accommodate settling, see Figure 10. See also Section 6.

10.D. No portion of the building shall come into contact with a masonry column unless the assembly is specifically designed to accommodate structural and settling considerations.

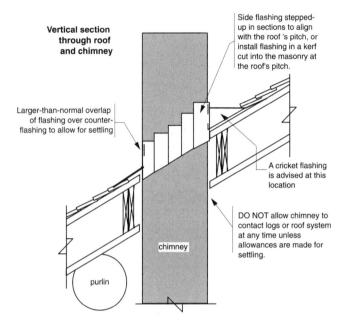

Vertical section through roof and chimney

Side flashing stepped-up in sections to align with the roof 's pitch, or install flashing in a kerf cut into the masonry at the roof's pitch.

Larger-than-normal overlap of flashing over counter-flashing to allow for settling

A cricket flashing is advised at this location

chimney

DO NOT allow chimney to contact logs or roof system at any time unless allowances are made for settling.

purlin

Figure 10

Commentary

Section 10 FIREPLACES AND CHIMNEYS

10.C. *The flashings used where a chimney goes through the roof must accommodate settling and protect against water and weather penetration at all times, including after the building has fully settled. The roof, when supported by log walls, will lose elevation while the chimney will remain the same height. This effect requires that chimneys be flashed and counterflashed (see Figure 10).*

Further, the flashing must be tall enough, and must have sufficient overlap when the logs are green, so that even after all settling is complete the counterflashing still overlaps the flashing at least 5.1 centimeters (2 inches), or more if local building codes require or the situation dictates.

Note: Because such tall areas of flashing can be exposed (12 inches—30.5 cm—is not uncommon), it is recommended that flashing material be thicker than normal to protect the flashing from degradation. Remember that the flashing and counterflashing cannot be attached to each other in any way (solder, rivets, or etc.) because they must freely slide vertically past each other to allow settling.

10.D. *This refers especially to a common practice in stick-frame buildings-supporting roof or floor beams on the masonry column of the chimney. This must not be done in a log home unless special measures are taken to allow for settling.*

It is desirable to position masonry columns during the design process so that they avoid areas in floors and roofs that require structural members. For example, position the chimney so that it avoids the ridgepole.

Affiliated Organizations:

Great Lakes Logcrafters Association
PO Box 86
Grand Rapids, Minnesota 55744 USA

Log Builders Association of New Zealand
Justin Long
PO Box 8
Masterton, New Zealand
tree-hut@clear.net.nz

Association of Latvian Craft
Rozu-iela 21-27
Riga, Latvia LV-1056

Korean Log Builders Association
Jay Wan Yu
Woorim Building, 90-10 Banpo-Dong
Su Cho-ku, Seoul, Korea

Inquiries, comments, suggestions, and additional recommendations are welcome.

© ILBA, 2000

Technical questions about the Standards may be addressed to:

Tom Hahney
7928 Lynwood Drive
Ferndale, Washington 98248 USA
360-354-5840
tomhahney@cs.com

Robert W. Chambers
421 North Main Street
River Falls, Wisconsin 54022 USA
robert@logbuilding.org

Index

V

valley rafters, 213-220, C-32 (*photo*)

    backing angles, 216, 219

    cutting, 220

    Hawkindale formulas, 220

    layout, 217

    settling, 150

W

wane, 206, 210

web sites, 247

widest gap, 88, 89, 90, 100-101, 103-104

windows, 5, 142, 259-260, C-13, C-14, C-24 (*photos*)

workshops, 19, 244

Z

zoning, 14

Dear Reader:

Thank you for purchasing my book!

Whether you are preparing to build your own log home or you would like to learn about the many techniques of handcrafted log construction, I hope my book will help you.

I am considering writing a newsletter to provide you with important tips related to log building, advise you of the latest techniques in log construction, and notify you of any new books.

I am also interested in knowing more about you and your log construction project!

Please take a moment to complete the information on one of these postage-paid postcards and send it back to me.

I'll be in touch with you now and again.

Thanks!

Robert W. Chambers
robert@LogConstructionManual.com

PO Box 283
River Falls, Wisconsin 54022
USA

Register your *Log Construction Manual* today!

Name _____

Street Address _____ Apt. No. _____

City _____ State _____ Zip _____

Country _____

Email _____

Describe your plans:
☐ To build my own log home
☐ To buy from a log home company
☐ To buy an existing log home
☐ Other: _____

Reason you purchased my book:
☐ To learn to build my own log home
☐ To train my log building staff
☐ To become familiar with log construction
☐ Other: _____

Timeframe:
☐ Less than 1 year
☐ 1 - 2 years
☐ 2 - 5 years
☐ Greater than 5 years

Privacy Notice: Your privacy is important to me. All information received will be held strictly confidential and will not be sold, rented or distributed to any other business.

Register your *Log Construction Manual* today!

Name _____

Street Address _____ Apt. No. _____

City _____ State _____ Zip _____

Country _____

Email _____

Describe your plans:
☐ To build my own log home
☐ To buy from a log home company
☐ To buy an existing log home
☐ Other: _____

Reason you purchased my book:
☐ To learn to build my own log home
☐ To train my log building staff
☐ To become familiar with log construction
☐ Other: _____

Timeframe:
☐ Less than 1 year
☐ 1 - 2 years
☐ 2 - 5 years
☐ Greater than 5 years

Privacy Notice: Your privacy is important to me. All information received will be held strictly confidential and will not be sold, rented or distributed to any other business.

Register your *Log Construction Manual* today!

Name _____

Street Address _____ Apt. No. _____

City _____ State _____ Zip _____

Country _____

Email _____

Describe your plans:
☐ To build my own log home
☐ To buy from a log home company
☐ To buy an existing log home
☐ Other: _____

Reason you purchased my book:
☐ To learn to build my own log home
☐ To train my log building staff
☐ To become familiar with log construction
☐ Other: _____

Timeframe:
☐ Less than 1 year
☐ 1 - 2 years
☐ 2 - 5 years
☐ Greater than 5 years

Privacy Notice: Your privacy is important to me. All information received will be held strictly confidential and will not be sold, rented or distributed to any other business.

BUSINESS REPLY MAIL
FIRST-CLASS MAIL PERMIT NO. 14 RIVER FALLS WI

POSTAGE WILL BE PAID BY ADDRESSEE

ROBERT W. CHAMBERS
LOG CONSTRUCTION MANUAL
PO BOX 283
RIVER FALLS WI 54022-9801

BUSINESS REPLY MAIL
FIRST-CLASS MAIL PERMIT NO. 14 RIVER FALLS WI

POSTAGE WILL BE PAID BY ADDRESSEE

ROBERT W. CHAMBERS
LOG CONSTRUCTION MANUAL
PO BOX 283
RIVER FALLS WI 54022-9801

BUSINESS REPLY MAIL
FIRST-CLASS MAIL PERMIT NO. 14 RIVER FALLS WI

POSTAGE WILL BE PAID BY ADDRESSEE

ROBERT W. CHAMBERS
LOG CONSTRUCTION MANUAL
PO BOX 283
RIVER FALLS WI 54022-9801